W9-BIN-339

INTERMEDIATE ALGEBRA FOR COLLEGE STUDENTS

HARPER'S MATHEMATICS SERIES

CHARLES A. HUTCHINSON, *Editor*

INTERMEDIATE ALGEBRA FOR COLLEGE STUDENTS

Revised Edition

THURMAN S. PETERSON, Ph.D.

PROFESSOR OF MATHEMATICS
PORTLAND STATE COLLEGE

Harper & Brothers, Publishers
New York

INTERMEDIATE ALGEBRA FOR COLLEGE STUDENTS, REVISED EDITION

D-H

This book was first published under the title of

Elements of Algebra

Library of Congress catalog card number: 53-9414

CONTENTS

PREFACE TO THE REVISED EDITION

In the preparation of the revised edition the author has endeavored to maintain those features which have given to the textbook such a large measure of success. In addition to a new format, many illustrative examples have been revised and added to improve the clarity for self-study, and to make the text thoroughly modern in all respects. The exercises and problems are almost completely new, and additional problems of special interest to superior students are given at the ends of most of the chapters.

Purpose. This book is designed to serve as a text for college students who have had not more than one year of secondary-school algebra. Being intended for students of somewhat varied preparation, the material is arranged so that the initial chapters may be reviewed hurriedly or studied in considerable detail according to individual demands.

Fundamentally, the objectives of the book are twofold: in the first place, to serve as a terminal course in algebra preparing for nonscientific studies; and secondly, to serve as a foundation course in algebra preparing for more advanced college mathematics. In view of this dual purpose, it is assumed that individual instructors will emphasize those sections most applicable to their studies. In some instances the author has purposely neglected to stress certain principles which are more effectively studied in *college algebra* and which have little value for the student who does not intend to continue work in mathematics.

Special Features. The principles of algebra are clearly stated in elementary terms. Detailed points of common error and technique of computation are called to the student's attention in the many illustrations and notes.

The exercises are unusually extensive in order to provide adequate material for classwork as well as for outside preparation. The adaptability to varying classes is facilitated by the segregation of certain material and by the rigorous grading of all sets of exercises. The additional exercises at the end of each chapter may be used for periodic reviews or for testing material. Answers to the odd-numbered problems are given in a separate section at the end of the text, and the answers to the even-numbered problems are available in a separate pamphlet.

In order to emphasize the practical aspects of algebra, stated problems are comprehensively classified and analyzed.

Acknowledgments. The author wishes to take this opportunity to express his appreciation to his many friends and colleagues who have so graciously criticized and assisted in the revision of this text. In particular the author would like to thank the many users of the original edition who have contributed many worth-while suggestions for the revision.

T. S. Peterson

University of Oregon
July, 1953

INTERMEDIATE ALGEBRA FOR COLLEGE STUDENTS

CHAPTER I

INTRODUCTION

1. General Numbers. In algebra as in arithmetic the quantities with which we deal are numbers. But whereas in arithmetic the numbers used are represented by figures (1, 2, 3, 4, etc.), each of which has a definite value, in algebra they are represented by other symbols (usually letters of the alphabet) which may have any numerical values assigned to them or whose numerical values are to be found. Such quantities are called *general numbers.*

2. Processes of Algebra. We assume that the four fundamental processes, indicated by the symbols $+$, $-$, $\times$, $\div$, will have the same meaning when applied to general numbers that they had in arithmetic.

3. Algebraic Expressions. Any combination of numbers (numerical or general) and the four processes is called an *algebraic expression.* Parts of such expressions which are separated from each other by the signs $+$ or $-$ are called **terms.**

Illustration. $2a + 3b - 5c$ is an algebraic expression consisting of three terms.

NOTE. When no sign precedes a term the sign $+$ is understood.

4. Simple and Multiple Expressions. If an algebraic expression consists of one term, it is called a *simple expression.* If it contains two or more terms, it is called a *multiple expression.*

Such algebraic expressions are also distinguished otherwise. Simple expressions are called **monomials.** An expression of two terms, such as $a + b$, is called a **binomial,** and an expression of three terms, such as $a - b + c$, is called a **trinomial.** In general, a multiple expression is called a **multinomial.**

5. Products. When two or more quantities are multiplied together the result is called the *product* of the quantities.

Since, in using letters to represent numbers, there is confusion between the multiplication sign and the letters x and X, we will often indicate multiplication in algebra either (1) by using a dot halfway up between the two letters, or (2) by writing the letters one after the other with no symbol between them. Thus, the product of a and b may be written either as $a \times b$, as $a \cdot b$, or as ab.

6. Factors. Each of the quantities multiplied together to form a product is called a **factor** of the product. Thus 3, a, and b are the factors of the product $3ab$. 3 is called the *numerical factor*, and a and b are called *literal factors.*

7. Coefficients. The numerical factor of an expression is called the **coefficient** of the remaining factors. Thus, in the expression $3ab$, 3 is the coefficient. In a more general sense, $3a$ may be considered as the *literal coefficient* of b.

NOTE. The quantity a has the same meaning as $1a$. Hence, if the coefficient of a quantity is not explicitly written, it is understood to be 1.

8. Powers and Exponents. A **power** of a quantity is the product obtained by multiplying the quantity by itself any number of times. It is expressed by writing to the right and above the quantity an index which indicates the number of factors to be taken. Thus,

$a \times a$ is called the second power of a, and is written a^2;
$a \times a \times a$ is called the third power of a, and is written a^3;
and so on.

The small number which expresses the power of any quantity is called an **exponent.** Thus, 2, 4, 6 are the exponents of a^2, a^4, a^6 respectively.

a^2 is usually read "a square";
a^3 is usually read "a cube";
a^4 is usually read "a to the fourth"; and so on.

NOTE. The quantity a has the same meaning as a^1. Hence, if the power of a quantity is not explicitly written, it is understood to be 1.

Example 1. What is the difference in meaning between $2a$ and a^2?

Solution. By $2a$, we mean the product of 2 and a. By a^2, we mean the product of a and a.

Thus, if $a = 5$,

$$2a = 2 \cdot a = 2 \cdot 5 = 10,$$

whereas,

$$a^2 = a \cdot a = 5 \cdot 5 = 25.$$

Example 2. If $p = 1$, find the value of $4p^5$.

Solution.

$$4p^5 = 4 \cdot p \cdot p \cdot p \cdot p \cdot p = 4 \cdot 1 \cdot 1 \cdot 1 \cdot 1 \cdot 1 = 4.$$

NOTE. Observe that every power of 1 is 1.

Example 3. If $x = 2$, $y = 5$, $z = 7$, find the value of $xy + yz - xz$.

Solution.

$$xy = 2 \cdot 5 = 10;$$
$$yz = 5 \cdot 7 = 35;$$
$$xz = 2 \cdot 7 = 14.$$

Hence the value of the expression equals $10 + 35 - 14 = 31$.

NOTE. Each term is evaluated before adding and subtracting.

Example 4. If $a = 5$, $b = 3$, $c = 1$, evaluate $4(a - b)(b + c)$.

Solution.

$$(a - b) = (5 - 3) = 2;$$
$$(b + c) = (3 + 1) = 4.$$

Hence, $4(a - b)(b + c) = 4 \cdot 2 \cdot 4 = 32$.

NOTE. The parenthesis ( ) is used to indicate a single quantity.

EXERCISE 1 (1–8 Oral)

What is the greatest number of factors contained in each of the following products? Name each factor.

1. (*a*) 6, (*b*) 30, (*c*) $2x$, (*d*) $15h$, (*e*) $14ab$.
2. (*a*) 12, (*b*) 16, (*c*) $8y$, (*d*) $18cd$, (*e*) $24h^3$.

Read each of the following aloud. Give the coefficient of each term and the exponent of each letter:

3. (*a*) $6x^5$, (*b*) a^2b^3, (*c*) $2mn^4$, (*d*) $8x^4yz^7$, (*e*) $52q^{12}$.
4. (*a*) x^5y^2, (*b*) $9abc$, (*c*) $3p^5q$, (*d*) $\frac{1}{2}mv^2$, (*e*) uv^2w^3.

Express each of the following products in exponent form:

5. (*a*) $2aab$, (*b*) $xxxxx$, (*c*) $3yyy$, (*d*) $2 \cdot 2 \cdot 2mmnnn$,
6. (*a*) $6aaaa$, (*b*) $10bcb$, (*c*) $7xyyz$, (*d*) $4sssstttt$.

7. Evaluate each of the following expressions when $a = 2$, $b = 3$:

(*a*) b^2, (*b*) a^3, (*c*) $5ab$, (*d*) $2a^2b$, (*e*) $a + b$.

8. Evaluate each of the following expressions, when $x = 5$, $y = 1$:

(*a*) $7x$, (*b*) $2xy$, (*c*) y^5, (*d*) $x^2 + y^2$, (*e*) $2x - 7y^2$.

Evaluate the following expressions when $a = 1$, $b = 2$, $c = 3$, $d = 6$:

9. b^3c^2
10. $4b^2 - 5$
11. $bd - ac$
12. $2ac^2 + b^4$
13. $abcd$
14. $6bc - d^2$
15. $6a^{12}$
16. $(d - b)^2$
17. $d(a + b + c)$
18. $(a + b)(d - 1)$
19. $a^2 + b^2 + c^2$
20. $abc - d$
21. $c^2 - b^3$
22. $2a + 3b - c$
23. $\frac{1}{2}(2d - b)$
24. $\dfrac{cd}{b}$
25. $\dfrac{d - b}{c - a}$
26. $\dfrac{d - a}{5}$
27. $\dfrac{c + d}{a + b}$
28. $d\left(\dfrac{1}{b} - \dfrac{1}{c}\right)$
29. $\dfrac{c - b}{a} + \dfrac{d - c}{b - a}$
30. $bc^2 + (bc)^2$
31. $3(b - a)(d - c)$
32. $(c - b - a)^9$
33. $(d - c - b)^9$
34. $a^2 + 2ab + b^2$
35. $a + b(c + d)$
36. $b^2c^2 - a^2d^2$
37. $2b + 3(a + 4)$
38. $(a + b)(c + d)$

9. Formulas. A rule or principle expressed in algebraic terms is called a *formula*. In our study of arithmetic, we learned many rules for the computation of unknown quantities. A few of these rules are as follows:

1. The area of a rectangle equals the product of its length and its width:

$$A = lw.$$

2. The area of a triangle equals one-half the product of a base and its corresponding altitude:

$$A = \tfrac{1}{2}bh.$$

3. The area of a circle equals the product of π ($\pi = \frac{22}{7}$ approximately) and the square of the radius:

$$A = \pi r^2.$$

4. The perimeter (or boundary) of a rectangle equals the sum of twice the width and twice the length:

$$P = 2l + 2w.$$

5. The circumference of a circle equals the product of π and the diameter:

$$C = \pi d.$$

6. The area of a trapezoid equals the product of one-half the altitude and the sum of the bases:

$$A = \tfrac{1}{2}h(b + b').$$

7. The distance traveled equals the product of the rate of travel and the time:

$$D = RT.$$

8. Simple interest on money invested equals the product of the principal invested, the rate of interest per year, and the time in years:

$$I = PRT.$$

9. The total value of a set of identical objects equals the product of the number of objects and the price per object:

$$V = np.$$

10. Evaluation of Formulas. By substituting definite numbers for the letters of a formula and then applying the algebraic operation indicated, we can find the value of the unknown quantity sought.

Example 1. Find the area of a triangle having a base 6 feet long, if the corresponding altitude is 4 feet.

Solution. Substituting the numbers in the formula for the area of a triangle, we have

$$A = \tfrac{1}{2} \cdot 6 \cdot 4 = 12 \text{ square feet.}$$

Example 2. Find the area of a circle whose radius is 7 feet.

Solution. Substituting the numbers in the formula for the area of a circle, we have

$$A = \pi 7^2 = \tfrac{22}{7} \cdot 7 \cdot 7 = 154 \text{ square feet.}$$

Example 3. Find the area of a trapezoid whose bases are 8 feet and 12 feet, and whose altitude is 3 feet.

Solution. Substituting the numbers in the formula for the area of a trapezoid, we have

$$A = \tfrac{1}{2} \cdot 3 \cdot (8 + 12) = \tfrac{1}{2} \cdot 3 \cdot 20 = 30 \text{ square feet.}$$

EXERCISE 2

1. Find the area of a rectangle if its length is 436 and its width is 217.
2. Find the area of a rectangle if its length is 4.37 and its width is 3.6.
3. Find the perimeter of the rectangle in Problem 1.
4. Find the perimeter of the rectangle in Problem 2.
5. The radius of a circle is 63; find its area.
6. The radius of a circle is $3\frac{1}{2}$; find its area.
7. An automobile travels 40 miles per hour; how far does it go in (*a*) 2 hours, (*b*) $\frac{1}{2}$ hour, (*c*) 10 minutes?
8. An airplane travels 189 miles per hour; how far does it go in 5 hours and 20 minutes?
9. A trapezoid has bases of 22 inches and 34 inches, and an altitude of 1 foot. What is its area in square inches?
10. A trapezoid has bases of 2.52 feet and 3.9 feet, and an altitude of 1.6 feet. What is its area?
11. What is the value of 5 horses and 3 cows, if horses sell for \$85 and cows sell for \$163?
12. If eggs sell for \$6.21 a gross, how much would $7\frac{1}{3}$ gross cost?
13. What is the simple interest on \$980 at 5% per annum for 10 years?
14. What is the simple interest on \$6742 at 4% per annum for 7 years?
15. What is the circumference of a circle whose diameter is 3.29 inches?
16. What is the circumference of a circle whose radius is 259 feet?
17. The volume of a circular cylinder of radius r and height h is given by the formula $V = \pi r^2 h$. Find the volume when $r = 14$ inches and $h = 16$ inches.
18. Find the volume of a cylinder when $r = 2.6$ feet and $h = 3.5$ feet.
19. The surface of a sphere of radius r is given by the formula $S = 4\pi r^2$. Find the surface of a sphere of radius 2.45 feet.
20. Find the surface of a sphere of radius 3.85 inches.

21. The volume of a cone of radius r and height h is given by the formula $V = \frac{1}{3}\pi r^2 h$. Find the volume of a cone of radius 3.3 feet and height 4.2 feet.

22. Find the volume of a cone of radius $3\frac{1}{2}$ inches and height $4\frac{1}{2}$ inches.

23. The area of an ellipse whose semiaxes are a and b is given by the formula $A = \pi ab$. Find the area of an ellipse whose semiaxes are 2.1 and 6.9 respectively.

24. Find the area of an ellipse whose semiaxes are $3\frac{1}{2}$ and $4\frac{1}{3}$.

11. Operations of Algebra Applied to General Numbers

Addition. Just as

$$2 \text{ feet} + 3 \text{ feet} = 5 \text{ feet},$$

and

$$\$2 + \$3 = \$5,$$

so

$$2a + 3a = 5a.$$

We think of the literal part of the general number as indicating the nature of the quantities being added, and the numerical coefficients are actually added.

NOTE. It is essential that the literal parts be the same in order to add in this manner. Terms having the same literal parts are called **similar** or **like terms.**

Subtraction. Just as

$$5 \text{ feet} - 3 \text{ feet} = 2 \text{ feet},$$

and

$$\$5 - \$3 = \$2,$$

so

$$5a - 3a = 2a.$$

As in addition, the literal part of the general number indicates the type of quantities being subtracted and the numerical coefficients are actually subtracted.

Multiplication. Just as

$$3 \times 2 \text{ feet} = 6 \text{ feet},$$

and

$$3 \times \$2 = \$6,$$

so

$$3 \times 2a = 6a.$$

We need only recall the meaning of multiplication to verify these statements. For example, $3 \times \$2$ means $\$2 + \$2 + \$2$, which, by addition, is \$6. So, too, $3 \times 2a$ means $2a + 2a + 2a$, which adds to $6a$.

It is to be noted in the above multiplication that only one of the two factors contains letters. The case in which both factors contain letters will be discussed later.

Division. Just as

$$6 \text{ feet} \div 3 = 2 \text{ feet},$$

and

$$\$6 \div 3 = \$2,$$

so

$$6a \div 3 = 2a.$$

The quotient obtained in division is by definition a quantity which when multiplied by the divisor yields the dividend. That this condition is satisfied in the above examples is evident by comparing with the examples under multiplication.

NOTE. Division in algebra is denoted more often by writing the dividend and divisor as a fraction, than by using the symbol ÷. For example, the above relationships, indicating division, could be written as follows:

$$\frac{6 \text{ feet}}{3} = 2 \text{ feet}, \quad \frac{\$6}{3} = \$2, \quad \frac{6a}{3} = 2a.$$

ORAL EXERCISE

Add at sight:

1. $3a$ $2a$	$7x^2$ $5x^2$	$9mn$ $6mn$	$5y^3$ $4y^3$	$7abc$ $6abc$	$8xy^2$ $8xy^2$	$5p$ p	$8c^4$ $6c^4$
2. $13xy$ xy	$17a^2$ $9a^2$	13¢ 8¢	$23ab$ $7ab$	$39z^3$ $3z^3$	$14x^2y^2$ $8x^2y^2$	$25m$ $6m$	$48st^2$ $5st^2$
3. $23c$ $11c$	$41y^2$ $17y^2$	$25xyz$ $15xyz$	$42n$ $13n$	$32p^2$ $14p^2$	$29bc$ $21bc$	$32b^2$ $27b^2$	$42xy$ $38xy$
4. $39n$ $28n$	$47bd$ $19bd$	$36r$ $25r$	$74x$ $27x$	$67y^2$ $37y^2$	$46t$ $19t$	$57ab$ $35ab$	$66x^3$ $18x^3$

Subtract at sight:

5. $9x$ $5x$	$8ab$ $6ab$	$5y^2$ $4y^2$	$7c^3$ c^3	$6m$ $5m$	$2abc$ abc	$7bc$ $5bc$	$4x^2$ $3x^2$

6.	$19xy$	$11a$	$18p^2$	$23ax$	$13by$	$17n^2$	$25az$	$16q$
	$7xy$	$4a$	$9p^2$	$3ax$	$6by$	$8n^2$	$9az$	$8q$
7.	$37mn$	$45ac$	$39x^3$	$43cx$	$38r$	$41z$	$65bd$	$59xy^2$
	$13mn$	$11ac$	$17x^3$	$42cx$	$27r$	$31z$	$21bd$	$33xy^2$
8.	$85bz$	$67d$	$37y^2$	$41abc$	$57xy$	$81z^3$	$45x^2y$	$83ay$
	$17bz$	$39d$	$19y^2$	$29abc$	$48xy$	$67z^3$	$17x^2y$	$56ay$

Multiply at sight:

9.	$6a$	$8x$	$9t$	$8a^2$	$4xy$	$3b$	$4ac$	$7p$
	3	2	3	4	4	3	2	7
10.	$12y^3$	$16ac$	$15b^2$	$18xy$	$14x^2$	$15ab$	$12xz$	$10c^2$
	3	4	3	3	2	5	4	5
11.	$2.4x$	$2.5ab$	$3.6c$	$1.8bd$	$1.6z$	$2.1xy$	$3.3ab^2$	$2.8q$
	3	5	6	9	8	7	3	4
12.	$7.2b^2$	$6.4ac$	$3.5mn$	$3.6r$	$4.5a$	$3.9x^3$	$4.2pq$	$2.6y^2$
	8	4	7	9	5	3	6	2

Divide at sight:

13.	$\frac{8b}{2}$	$\frac{10c}{5}$	$\frac{9r}{3}$	$\frac{6s}{2}$	$\frac{12x^2}{4}$	$\frac{9xy}{9}$	$\frac{8m}{4}$	$\frac{6yz}{6}$
14.	$\frac{18ab}{9}$	$\frac{36z^3}{4}$	$\frac{56m}{7}$	$\frac{81mn^2}{9}$	$\frac{42p^2}{6}$	$\frac{45r}{5}$	$\frac{64bc}{8}$	$\frac{49st}{7}$
15.	$\frac{2.4xy}{6}$	$\frac{3.9y}{3}$	$\frac{6.8m}{4}$	$\frac{5.0r}{2}$	$\frac{8.1abc}{9}$	$\frac{6.6n}{6}$	$\frac{4.2c^2}{7}$	$\frac{5.2pq}{4}$
16.	$\frac{7.2bc}{9}$	$\frac{2.7a}{3}$	$\frac{9.5x}{5}$	$\frac{3.5mn}{7}$	$\frac{9.6a^2}{8}$	$\frac{6.5xy}{5}$	$\frac{6.8z}{4}$	$\frac{8.7an}{3}$

12. Equations. In the study of arithmetic we solved many problems of "missing numbers." Having four fundamental operations at our disposal these problems always fell into one of four basic groups, illustrated as follows:

I. What number increased by 3 equals 6?
II. What number multiplied by 3 equals 6?

III. What number decreased by 3 equals 6?

IV. What number divided by 3 equals 6?

To answer these queries, we would probably picture the questions in our minds in the following form:

I. ? + 3 = 6, II. 3 × ? = 6, III. ? − 3 = 6, IV. $\frac{?}{3}$ = 6.

In algebra, we shall answer such questions by actually carrying through our mental processes. That is, we let x (or any letter) represent the unknown quantity which we seek. Then we actually write down what we pictured in our minds, namely,

$$\text{I. } x + 3 = 6, \quad \text{II. } 3x = 6, \quad \text{III. } x - 3 = 6, \quad \text{IV. } \frac{x}{3} = 6.$$

A statement of equality between two expressions such as each of these represents is called an **equation.** The terms to the left of the sign of equality are said to form the *left member* or *left side* of the equation, and the terms to the right are said to form the *right member* or *right side* of the equation. A number which when substituted for the unknown in an equation reduces both members to the same number is said to *satisfy the equation.* A value of the unknown which satisfies an equation is called a **root** or **solution** of the equation.

Thus, the roots of the four equations given above are respectively:

$$\text{I. } x = 3, \quad \text{II. } x = 2, \quad \text{III. } x = 9, \quad \text{IV. } x = 18.$$

13. Fundamental Principle Concerning Equations. Consider the numerical equality 6 = 6. No matter what number we either add to or subtract from both members of this equality, the resulting numbers will be equal.

Illustrations. (1)

$6 = 6,$	given equality;
$6 + 3 = 6 + 3,$	adding 3 to both members;
$9 = 9,$	the results are equal.

(2)

$6 = 6,$	given equality;
$6 - 3 = 6 - 3,$	subtracting 3 from both members;
$3 = 3,$	the results are equal.

Also, no matter by what number we multiply or divide both members of this equality, the resulting numbers will be equal. (Division by zero is excluded.)

Illustrations. (1)

$$6 = 6, \quad \text{given equality;}$$
$$6 \times 3 = 6 \times 3, \quad \text{multiplying both members by 3;}$$
$$18 = 18, \quad \text{the results are equal.}$$

(2)

$$6 = 6, \quad \text{given equality;}$$
$$6 \div 3 = 6 \div 3, \quad \text{dividing both members by 3;}$$
$$2 = 2, \quad \text{the results are equal.}$$

This apparently obvious property of equalities is of fundamental importance.

> ***If equal numbers are increased, decreased, multiplied, or divided by the same number (division by zero excluded), the resulting numbers will be equal.***

14. Solution of Equations. Consider the four equations in Article 12. Each of these may be solved for the unknown x by applying the above principle; thus,

I.

$$x + 3 = 6 \quad \text{given equation;}$$
$$x + 3 - 3 = 6 - 3, \quad \text{subtracting 3 from both sides;}$$
$$x = 3, \quad \text{simplifying.}$$

II.

$$3x = 6, \quad \text{given equation;}$$
$$\frac{3x}{3} = \frac{6}{3}, \quad \text{dividing both sides by 3;}$$
$$x = 2, \quad \text{simplifying.}$$

III.

$$x - 3 = 6, \quad \text{given equation;}$$
$$x - 3 + 3 = 6 + 3, \quad \text{adding 3 to both sides;}$$
$$x = 9, \quad \text{simplifying.}$$

IV.

$$\frac{x}{3} = 6, \quad \text{given equation;}$$
$$3 \cdot \frac{x}{3} = 3 \cdot 6, \quad \text{multiplying both sides by 3;}$$
$$x = 18, \quad \text{simplifying.}$$

In general, if a and b represent any numbers whatsoever, these four basic types of equations and their solutions are as follows:

Type	Equation	Operation	Result
Additive	$x + a = b$	Subtracting a	$x = b - a$
Multiplicative	$ax = b$	Dividing by a	$x = \frac{b}{a}$
Subtractive	$x - a = b$	Adding a	$x = b + a$
Divisive	$\frac{x}{a} = b$	Multiplying by a	$x = ab$

Having obtained what we believe to be a solution in the above manner, it is important to see if it actually does satisfy the given equation. This is called the process of *checking* the solution.

Example 1. Solve the equation $x + 8 = 15$ and check your result.
Solution.

$x + 8 = 15$, given equation;
$x = 15 - 8$, subtracting 8 from both sides;
$x = 7$, simplifying.

Check.

$7 + 8 = 15$, substituting 7 for x in the given equation;
$15 = 15$, simplifying.

Example 2. Solve the equation $5x = 35$ and check your result.
Solution.

$5x = 35$, given equation;
$x = \frac{35}{5}$, dividing both sides by 5;
$x = 7$, simplifying.

Check.

$5 \cdot 7 = 35$, substituting 7 for x in the given equation;
$35 = 35$, simplifying.

Example 3. Solve the equation $x - 6.3 = 9.5$ and check your result.
Solution.

$x - 6.3 = 9.5$, given equation;
$x = 9.5 + 6.3$, adding 6.3 to both sides;
$x = 15.8$, simplifying.

Check.

$15.8 - 6.3 = 9.5,$ substituting 15.8 for x in the given equation;
$9.5 = 9.5,$ simplifying.

Example 4. Solve the equation $\frac{x}{4} = 3\frac{1}{2}$ and check your result.

Solution.

$\frac{x}{4} = 3\frac{1}{2},$ given equation;

$x = 4 \cdot 3\frac{1}{2},$ multiplying both sides by 4;
$x = 14,$ simplifying.

Check.

$\frac{14}{4} = 3\frac{1}{2},$ substituting 14 for x in the given equation;
$3\frac{1}{2} = 3\frac{1}{2},$ simplifying.

ORAL EXERCISE

At sight, give the root:

1. $x + 2 = 8$ **2.** $x + 7 = 19$ **3.** $x + 4 = 13$
4. $x + 5 = 14$ **5.** $x + 9 = 17$ **6.** $x + 1 = 11$
7. $x + 2 = 3\frac{1}{2}$ **8.** $x + 3 = 5\frac{2}{3}$ **9.** $x + 6 = 9\frac{1}{3}$
10. $x + 4 = 8.5$ **11.** $3x = 15$ **12.** $2x = 18$
13. $4x = 16$ **14.** $5x = 30$ **15.** $7x = 42$
16. $6x = 24$ **17.** $2x = 4\frac{2}{3}$ **18.** $3x = 6\frac{3}{4}$
19. $5x = 11.5$ **20.** $4x = 12.4$ **21.** $x - 3 = 7$
22. $x - 6 = 13$ **23.** $x - 5 = 6$ **24.** $x - 1 = 15$
25. $x - 7 = 9$ **26.** $x - 4 = 7$ **27.** $x - 2 = 3\frac{1}{2}$
28. $x - 3 = 2\frac{1}{3}$ **29.** $x - 4 = 1.3$ **30.** $x - 7 = 0.23$
31. $\frac{x}{2} = 8$ **32.** $\frac{x}{5} = 7$ **33.** $\frac{x}{9} = 5$
34. $\frac{x}{6} = 6$ **35.** $\frac{x}{3} = 9$ **36.** $\frac{x}{4} = 20$
37. $\frac{x}{2} = 2\frac{1}{2}$ **38.** $\frac{x}{3} = 3\frac{1}{3}$ **39.** $\frac{x}{5} = 1\frac{1}{5}$
40. $\frac{x}{4} = 2\frac{3}{4}$ **41.** $\frac{x}{2} = 3.7$ **42.** $\frac{x}{3} = 1.06$

EXERCISE 3

Solve the following equations and check your result:

1. $x + 7 = 16$ **2.** $5x = 85$ **3.** $x - 6 = 17$
4. $\frac{x}{7} = 7$ **5.** $7x = 63$ **6.** $x - 12 = 19$

7. $x + 10 = 23$
8. $\frac{x}{6} = 12$
9. $x - 11 = 34$
10. $\frac{x}{10} = 7$
11. $9x = 54$
12. $x + 3 = 13$
13. $12x = 132$
14. $x + 12 = 29$
15. $x - 57 = 69$
16. $\frac{x}{11} = 17$
17. $x + 37 = 61$
18. $15x = 255$
19. $x - 43 = 38$
20. $\frac{x}{13} = 9$
21. $x + 5 = 7\frac{1}{2}$
22. $8x = 70$
23. $x - 6 = 7\frac{1}{3}$
24. $\frac{x}{3} = 4\frac{1}{2}$
25. $4x = 34$
26. $x - 2\frac{1}{2} = 8\frac{1}{2}$
27. $x + \frac{1}{2} = 6\frac{1}{2}$
28. $\frac{x}{8} = 6\frac{1}{3}$
29. $x + 2\frac{1}{2} = 4\frac{1}{4}$
30. $3x = 50$
31. $x - 1\frac{1}{3} = 2\frac{1}{2}$
32. $\frac{x}{7} = 4\frac{2}{3}$
33. $x - 3\frac{3}{4} = 4\frac{1}{2}$
34. $x + \frac{1}{3} = \frac{1}{2}$
35. $\frac{x}{5} = 2\frac{3}{4}$
36. $2x = 4\frac{1}{2}$
37. $\frac{x}{6} = \frac{1}{2}$
38. $x - \frac{1}{2} = \frac{1}{3}$
39. $3x = 16\frac{1}{2}$
40. $x + 3\frac{1}{3} = 7\frac{1}{2}$
41. $x + 6 = 12.3$
42. $5x = 6.9$
43. $x - 3 = 2.6$
44. $\frac{x}{5} = 1.3$
45. $9x = 1.26$
46. $x - 2.7 = 4.8$
47. $\frac{x}{3} = 3.6$
48. $x + 2.5 = 7.5$
49. $x - 6.9 = 7.1$
50. $\frac{x}{9} = 0.02$
51. $x + 0.7 = 3.2$
52. $2x = 3.3$
53. $\frac{x}{2} = 3.07$
54. $x + 1.21 = 3.7$
55. $3x = 3.12$
56. $x - 1.21 = 3.64$
57. $x + 3.8 = 7.95$
58. $4x = 8.1$
59. $x - 2.5 = 7.12$
60. $\frac{x}{6} = 4.56$

15. Equations Requiring More Than One Operation. Solving some equations requires the use of more than one of the four basic methods just described.

Example 1. Solve the equation $2x + 3 = 11$ and check your result.
Solution.

$2x + 3 = 11,$	given equation;
$2x = 8,$	subtracting 3 from both sides;
$x = 4,$	dividing both sides by 2.

Check.

$2 \cdot 4 + 3 = 11,$ substituting 4 for x in the given equation;
$8 + 3 = 11,$ simplifying terms;
$11 = 11,$ combining terms.

Example 2. Solve the equation $5x - 7 = 5 + x$ and check your result.
Solution.

$5x - 7 = 5 + x,$ given equation;
$5x = 12 + x,$ adding 7 to both sides;
$4x = 12,$ subtracting x from both sides;
$x = 3,$ dividing both sides by 4.

Check.

$5 \cdot 3 - 7 = 5 + 3,$ substituting 3 for x in the given equation;
$15 - 7 = 5 + 3,$ simplifying terms;
$8 = 8,$ combining terms.

Example 3. Solve the equation $\frac{3x}{4} - 7 = 8$ and check your result.
Solution.

$\frac{3x}{4} - 7 = 8,$ given equation;

$\frac{3x}{4} = 15,$ adding 7 to both sides;

$3x = 60,$ multiplying both sides by 4;
$x = 20,$ dividing both sides by 3.

Check.

$\frac{3 \cdot 20}{4} - 7 = 8,$ substituting 20 for x in the given equation;
$15 - 7 = 8,$ simplifying terms;
$8 = 8,$ combining terms.

EXERCISE 4

Solve and check:

1. $3x + 5 = 23$ **2.** $2x - 3 = 11$ **3.** $\frac{x}{4} - 1 = 6$

4. $\frac{x}{3} + 5 = 9$ **5.** $\frac{3x}{5} = 6$ **6.** $\frac{2}{3}x = 16$

7. $5x = 15 + 2x$ **8.** $6x = 27 - 3x$ **9.** $7 + 5x = 42$

10. $4 = 7 - x$
11. $9 + \frac{x}{8} = 13$
12. $6 = 7 - \frac{x}{5}$
13. $\frac{5x}{6} = 10$
14. $\frac{3}{4}x = 15$
15. $7x - 12 = 4x$
16. $17x - 13x = 60$
17. $12x + 17 = 53$
18. $9 = 21 - 2x$
19. $3 = 5 - \frac{x}{2}$
20. $7 + \frac{x}{5} = 10$
21. $\frac{7x}{3} = 42$
22. $\frac{3x}{5} = 15$
23. $13x - 36 = 7x$
24. $5x = 30 - 5x$
25. $3x - 9 = 7 - x$
26. $2x + 1 = 17 - 2x$
27. $\frac{x}{2} - 3 = 5 - \frac{x}{2}$
28. $\frac{3x}{2} + 5 = \frac{x}{2} + 9$
29. $2\frac{1}{2}x - 7 = 2x - 4$
30. $\frac{3x}{4} + 5 = \frac{x}{2} + 9$
31. $\frac{2x}{3} + 5 = 23$
32. $\frac{4x}{5} - 3 = 21$
33. $\frac{x}{2} + \frac{x}{3} = 20$
34. $5x - 8 = 13 - 2x$
35. $x + 5 = 17 - x$
36. $\frac{2x}{3} - 1 = 5 - \frac{x}{3}$
37. $\frac{3x}{4} + 3 = \frac{x}{4} + 4$
38. $7\frac{1}{2}x + 3 = 2\frac{1}{2}x + 13$
39. $5\frac{1}{3}x - 16 = 12 - 1\frac{2}{3}x$
40. $\frac{3}{4}x + 13 = 25$
41. $\frac{5}{6}x - 2 = 18$
42. $\frac{x}{2} - \frac{x}{4} = 6$
43. $10x - 12 = 7x + 12$
44. $14x + 9 = 34 - 11x$
45. $\frac{x}{2} + 1 = \frac{x}{3} + 2$
46. $\frac{3x}{4} - 8 = 13 - \frac{3x}{4}$
47. $4\frac{2}{3}x + 7 = 2\frac{2}{3}x + 15$
48. $10\frac{1}{2}x - 40 = 35 - 4\frac{1}{2}x$
49. $\frac{5x}{3} + 7 = 22$
50. $\frac{7x}{2} - 5 = 23$
51. $\frac{x}{2} + \frac{x}{4} = 33$
52. $3x = 1 - x$
53. $x = 23 - x$
54. $\frac{3x}{4} + 6 = 7 - \frac{5x}{4}$
55. $5x + 7 = x + 9$
56. $\frac{2x}{3} + 7 = 10$
57. $\frac{x}{2} + \frac{x}{5} = 8\frac{2}{5}$
58. $2x + 17 = 23 - 2x$
59. $\frac{3}{4}x - 2 = 2$
60. $\frac{x}{3} = 1\frac{2}{3} + \frac{x}{4}$

16. Algebraic Representations. In solving algebraic problems which are stated in words the chief difficulty that one has, in the beginning, is expressing the conditions of the problem by means of symbols. A statement proposed in general numbers will frequently be puzzling, whereas a similar arithmetical statement presents no difficulty whatsoever.

Thus, the answer to the statement, *Find a number which is* x *more than* a, may not be self-evident, whereas the answer to a similar arithmetical statement, *Find a number which is* 3 *more than* 5, is found with no difficulty at all. Just as the number which is 3 more than 5 is $5 + 3 = 8$, so the number which is x more than a is $a + x$.

In general, when in doubt as to the exact meaning of any statement involving letters, choose suitable numbers for the letters and investigate the meaning of the statement with respect to these numbers.

Example 1. By how much does x exceed 12?

Solution. Take a numerical instance: "By how much does 15 exceed 12?" Clearly the answer is 3, which is obtained by subtracting 12 from 15.

In the same manner then, x exceeds 12 by $x - 12$.

Example 2. If x is an integer, what is the next larger integer?

NOTE. The whole numbers 1, 2, 3, 4, etc., are called **integers.**

Solution. Numerical instance: "If 17 is an integer, what is the next larger integer?" The answer is evidently 18, which is obtained by adding 1 to 17.

Likewise, the next larger integer following x is $x + 1$.

In the following examples, the choice of numerical instances is left to the student.

Example 3. What number is 10% more than h?

Solution. Here we must increase h by 10% of h. Since the numerical value of 10% is 0.1, the increase is to be 0.1 of h or $0.1h$. Hence, the number is $h + 0.1h$, or $1.1h$. Always simplify representations in this manner when it is possible to do so.

NOTE. In algebra, percentages should always be reduced to their equivalent numerical values. Thus $7\% = 0.07$, or $\frac{7}{100}$; $r\% = 0.01r$, or $\frac{r}{100}$.

Example 4. Jack is twice as old as Bill. If Bill is y years old, how old was Jack three years ago?

Solution. Since Jack is now twice as old as Bill, his present age is $2y$ years. Three years ago he was 3 years younger, or $2y - 3$ years old.

Example 5. How many cents are there in q quarters and d dimes?

Solution. Since each quarter contains 25 cents, q quarters will contain q times as many cents, or $25q$ cents. Also, since each dime contains 10 cents, d dimes will contain d times as many cents, or $10d$ cents. Taken together, there will be a total of $25q + 10d$ cents.

EXERCISE 5

Express in algebraic symbols:

1. The sum of x and $x + 3$.
2. The product of a and b.
3. What number exceeds a by b?
4. A man traveling r miles per hour for 3 hours travels how far?
5. What number is x more than $2x + 1$?
6. If $x + 1$ is an integer, what is the next larger consecutive integer?
7. What was John's age four years ago, if he will be y years old in five years?
8. What number is 25% larger than x?
9. If $2n$ is an even integer, what is the next larger even integer?
10. What number is 3 less than twice x?
11. If b is the larger of two numbers a and b, what is their difference?
12. The sum of two numbers is s and the larger is l; what is the smaller?
13. What number is 3 more than twice x?
14. $5r\%$ is equal to what number?
15. If x pens cost \$20, what is the price of one pen?
16. The difference between two numbers is d and the larger is l. What is the smaller?
17. What number is 5% less than x?
18. The sum of two numbers decreased by their product. (Let a and b be the two numbers.)
19. What is the average of a and b?

20. The number of days in w weeks and d days.

21. What is the length of a rectangle whose perimeter is P, if the width is 6?

22. What is the reciprocal of x? (See a dictionary.)

23. A stick l feet long is broken into two parts, one of which is twice as long as the other. How long is the shorter piece?

24. The larger of two numbers is twice the smaller. If the larger is x, what is the smaller?

25. A man leaves one-half of his money to his son and one-third to his daughter. If the daughter receives D dollars, how much does the son receive?

26. How many cents in n nickels and c cents?

27. The larger of two numbers is 4 more than the smaller. If the larger is x, what is the smaller?

28. What number exceeds y by 4 less than x?

29. Jane is 3 years older than Mary, and Mary is twice as old as Kay. If Kay is x years old, how old is Jane?

30. Tom sold 26 more papers than Dick, and Dick sold three times as many as Harry. If Harry sold p papers, how many did Tom sell?

31. A has $32 more than B, and B has five times as much money as C. If C has d dollars, how much does A have?

32. John has as much money as George; then they bet 5 cents and George loses. If, after the bet, George has x cents, how much does John have?

17. Formulas Made from Statements. Fundamentally a formula expresses a certain type of relationship which exists between two or more varying quantities. To make a formula, we need to know something of this relationship. This we may ascertain either from the nature of the quantities themselves or from an explicit statement regarding their relationship. The following examples illustrate these methods of making formulas:

Example 1. Obtain a formula for the total area (A) of a cube.

Solution. Each face of a cube is a square. If we let the letter e represent the length of each edge of the cube, then the area of each face of the cube will be represented by e^2. Since, however, there are six faces to a cube, the total area is six times the area of one face, or

$$A = 6e^2.$$

Example 2. What is the cost (C), in dollars, of 50 books at x dimes each?

Solution. Since one dime is valued at $\frac{1}{10}$ of a dollar, x dimes are valued at x times as much, or $\frac{x}{10}$ dollars. Thus, the cost of one book is $\frac{x}{10}$ dollars. The cost of 50 books is fifty times the cost of one book. Hence,

$$C = 50 \cdot \frac{x}{10} = 5x,$$

where C is understood to be in dollars.

Example 3. What is the area (A) in square feet of a rectangle whose length is x yards and whose width is y feet?

Solution. Since 1 yard contains 3 feet, x yards will contain x times as many feet, or $3x$ feet. The area in square feet is the product of the length and width, both expressed in feet. Hence,

$$A = 3xy.$$

Example 4. Make a formula for the total cost (C), in cents, of a telegram, when the first fifteen words of the telegram cost 80 cents and each additional word costs 5 cents. (Assume the telegram to contain more than 15 words.)

Solution. Let the letter n represent the number of words in the telegram. The first fifteen of these words as a group cost 80 cents and the remaining $(n - 15)$ of the words each cost 5 cents. Hence, the cost of the remaining words is $5(n - 15)$ cents and the total cost is

$$C = 80 + 5(n - 15).$$

EXERCISE 6

Obtain the formulas in each of the following:

1. The cost (C) of n pencils at p cents each.
2. The sum (S) of twice a number x and three times a number y.
3. The product (P) of three numbers a, b, and c.
4. The cost (C) of n books at $1.20 each.
5. A man's age (A) now, who was x years old 7 years ago.
6. A man's age (A) now, who will be x years old in 5 years.
7. The area (A) of a square whose sides are s.
8. The perimeter (P) of a triangle whose sides are a, b, and c.
9. The yearly interest (I) on $1000 at $r\%$. $\left(\text{Note } r\% = \frac{r}{100}\cdot\right)$

10. The yearly interest (I) on P dollars at 6%.

11. The distance (D) traveled at the rate R for 2 hours and 30 minutes.

12. The rate of speed (R), when D miles are traversed in 2 hours.

13. The value (V) in cents of 10 nickels and q quarters.

14. The value (V) in cents of q quarters, d dimes, and c cents.

15. The value (V) in dollars of d dimes and n nickels.

16. The value (V) in dollars of 25 half dollars, 15 quarters, and d dimes.

17. The cost (C) in cents of s pounds of sugar at 9 cents a pound and t pounds of tea at 70 cents a pound.

18. The cost (C) in cents of 12 pounds of sugar at p cents a pound and 5 pounds of tea at q cents a pound.

19. The amount (A) of alcohol in a solution S which is 35% alcohol.

20. The amount (A) in pounds of copper in T tons of ore which is 1% copper.

21. The distance (D) in feet traveled at the rate of 10 miles per hour for t hours.

22. The distance (D) in yards traveled at the rate R miles per hour for 30 minutes.

23. The volume (V) of a cube whose edges are e.

24. The perimeter (P) of a semicircle whose radius is r.

25. The area (A) of a ring whose outside radius is R and whose inside radius is r.

26. It costs $25 to make the printing plates for a circular and 2 cents per copy for the circulars. Find the total cost (C) for n circulars.

27. The first fifteen words of a telegram cost 90 cents and each additional word costs 7 cents. Find the total cost (C) of a telegram of n words. (Assume n more than 15.)

28. The first fifteen words of a telegram cost P cents and each additional word costs p cents. Find the total cost (C) of a telegram of 100 words.

29. The first 100 pounds of baggage costs $5 to ship and each additional 10 pounds costs 25 cents. Find the total cost (C) for $10n$ pounds of baggage. (Assume n more than 10.)

30. The first 250 pounds of a shipment costs $7.50 to ship and each additional 25 pounds costs 30 cents. Find the total cost (C) of $25n$ pounds. (Assume n more than 10.)

31. It costs $100 to prepare a manuscript for printing and 80 cents for each copy made. Find the total cost (C) for N copies.

32. The first hundred copies of a pamphlet cost 5 cents each, the second hundred cost 4 cents each, and any over 200 cost 3 cents each. Find the cost (C) of n pamphlets. (Assume n more than 200.)

33. In Problem 32, what will the formula be if n is a number greater than 100 but less than 200?

34. Electricity costs $1.00 for the first 15 kilowatt-hours, 5 cents for each of the next 20 kilowatt-hours, 3 cents for each of the next 40 kilowatt-hours, and 2 cents for each kilowatt-hour thereafter. Find the cost (C) of n kilowatt-hours. (Assume n more than 75.)

35. In Problem 34, what will the formula be if n is a number greater than 35 but less than 75?

36. In Problem 34, what will the formula be if n is a number greater than 15 but less than 35?

18. Stated Problems. A written or verbal statement expressing some condition or conditions of equality which exist among one or more unknown quantities is called a *stated problem.*

The method of procedure in solving such problems is as follows:

1. Represent some one of the unknown quantities by a symbol, such as x.
2. Give the algebraic representations of the other unknown quantities in terms of the unknown x.
3. From a condition of the problem, obtain a simple equation involving the unknown quantities.
4. Solve the equation obtained in 3 for the unknown, x. When x is known, the other unknowns may be determined by substitution in 2.

NOTE. Unknown quantities are usually represented by the last letters of the alphabet.

Example 1. The larger of two numbers is 7 more than the smaller, and their sum is 41. Find the numbers.

Solution. Let $x =$ the smaller number;

then $x + 7 =$ the larger number.

Since the sum of these two unknown numbers must equal 41, we have

$$x + x + 7 = 41.$$

Hence, $2x + 7 = 41$, combining terms;

$2x = 34$, subtracting 7 from both sides;

$x = 17$, dividing both sides by 2.

Therefore, $x + 7 = 24$, substituting 17 for x in $x + 7$.

Answer. The numbers are 17 and 24.

Check. 24 is 7 more than 17; and the sum of 17 and 24 equals 41.

Example 2. The length of a rectangle is 3 feet less than twice its width, and the perimeter of the rectangle is 36 feet. Find the dimensions of the rectangle.

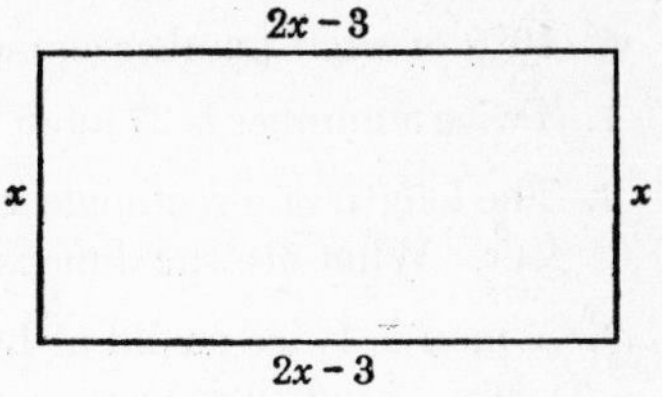

Solution. Let x = the width in feet;
then $2x - 3$ = the length in feet.

Since the perimeter is the total distance around a closed figure, we have

$$x + 2x - 3 + x + 2x - 3 = 36.$$

Hence, $6x - 6 = 36$, combining terms;

$6x = 42$, adding 6 to both sides;

$x = 7$, dividing both sides by 6.

Therefore, $2x - 3 = 11$, substituting 7 for x in $2x - 3$.

2x − 3
x x
2x − 3

FIG. 1

Answer. The dimensions of the rectangle are 7 feet by 11 feet.

Check. 3 less than twice 7 equals 11; and the perimeter, $7 + 11 + 7 + 11$, equals 36.

NOTE. The student will find that geometric problems are more easily understood when a figure is drawn.

Example 3. What number increased by one-fourth of itself equals 35?

Solution. Let x = the number;
then $\frac{1}{4}x$ = the increase.

Since the number after being increased equals 35, we have

$$x + \tfrac{1}{4}x = 35.$$

Hence, $\frac{5x}{4} = 35$, combining terms;

$5x = 140$, multiplying both sides by 4;

$x = 28$, dividing both sides by 5.

Answer. The number is 28.

Check. $\frac{1}{4}$ of 28 equals 7; and 7 added to 28 equals 35.

EXERCISE 7

1. One number is 6 more than the other, and their sum is 32. Find the numbers.
2. The perimeter of a rectangle is 132 feet, and the length is twice the width. What are its dimensions?

3. A stick 42 inches long is broken into two pieces, so that one piece is twice as long as the other. How long are the two pieces?
4. Two boys divided \$3.36 so that one boy received 12 cents less than three times as much as the other. How was the money divided?
5. The sum of two consecutive integers is 77. Find the integers.
6. 40% of what number equals 32?
7. Twice a number is 27 more than half the number. What is the number?
8. The length of a rectangle is $2\frac{1}{2}$ times the width, and the perimeter is 42 feet. What are the dimensions?
9. A man is twice as old as his son, and the difference in their ages is 23 years. Find their ages.
10. The sum of three consecutive odd integers is 69. Find the integers.
11. What number increased by 20% (of itself) equals 78?
12. A board 65 inches long is sawed into two pieces, so that one piece is 7 inches shorter than twice the length of the other piece. Find the length of the two pieces.
13. In twenty years Charles will be three times as old as he is now. How old is he now?
14. The smaller of two numbers is one-half the larger, and their sum is 27. Find the numbers.
15. Two sides of a triangle are equal in length, and the third side is 5 less than the length of the equal sides. If the perimeter of the triangle is 49, how long are its sides?
16. One number is 5 more than twice the other, and their sum is 74. Find the numbers.
17. The sum of five consecutive integers is 85. Find the integers.
18. A boy spent one-half of his money for a book and one-third of his money for a pen. The remaining \$1.25 he saved. How much money did he have originally?
19. A certain number exceeds 18 by as much as 44 exceeds the number. What is the number?
20. The width of a rectangle is 1 less than one-half its length, and the perimeter is 46. Find the dimensions.
21. To finance a bridge costing \$94,500, the state contributed twice as much as the county and the county contributed twice as much as the city. How much did each contribute?

22. The difference between two numbers is 5, and twice the smaller is 18 more than the larger. Find the numbers.

23. George is six years older than his sister, and the sum of their ages is 68 years. How old is George?

24. The sum of three consecutive even integers is 258. Find the integers.

25. $30 is to be divided among 3 boys and 2 girls, so that each of the boys receives twice as much as each of the girls. How much does each receive?

26. The larger of two numbers exceeds the smaller by 9, and their sum is 43. Find the numbers.

27. Two sides of a triangle are equal, and the third side is 5 less than the sum of the equal sides. If the perimeter of the triangle is 47, how long are its sides?

28. In twenty years Mary will be five times as old as she is now. How old is she now?

29. One-half of what number equals 12 plus one-third of that number?

30. The sum of five consecutive integers is 55. Find the integers.

31. Six less than five times a certain number is the same as 3 more than twice the number. Find the number.

32. The length of a rectangle is 5 less than three times the width, and the perimeter is 54. Find the dimensions.

33. The difference between two numbers is 14, and their sum is 72. Find the numbers.

34. Three-fourths of a number is equal to 1 more than one-fourth of the number. Find the number.

35. A newsboy sells twice as many 8¢ papers as he does 5¢ papers. If he takes in $4.83, how many papers did he sell?

36. The smaller of two numbers increased by 7 equals the larger, and twice the smaller added to the larger equals 40. Find the numbers.

37. A side of the larger of two squares is twice a side of the smaller, and the sum of the perimeters of the two squares is 72. Find the sides of the two squares.

38. A measuring tape 72 inches long is cut into three pieces, so that one piece is twice as long as the shortest piece and the longest piece is three times as long as the shortest piece. What is the length of each of the three pieces?

39. A man is five times as old as his son, and two years older than his wife. If their combined ages total 64 years, how old is each?

40. Two boys worked in a store at the same rate of pay. One boy worked 7 hours and the other worked 10 hours. Together they received $9.35. How much did each receive? (Let x = the rate of pay per hour.)

REVIEW OF CHAPTER I

A

Evaluate the following expressions when $a = 0$, $b = 1$, $c = 3$, $d = 5$:

1. $a^2 + b^2 + c^2$ **2.** $bd - c$ **3.** $c^2 - ad$
4. $2a - 3b + c$ **5.** $2b^9$ **6.** $c(a - b + d)$
7. $(bd)^2$ **8.** $abcd$ **9.** $ab + bc + cd$
10. $\frac{1}{2}d(b + c)$ **11.** $d^2 - 4d + 3$ **12.** $a^3 + b^3 + c^3$
13. $\dfrac{d - c}{c - b}$ **14.** $\dfrac{cd}{d - c}$ **15.** $\dfrac{a}{b} + \dfrac{b}{c} + \dfrac{c}{d}$
16. $(d - c)^3$ **17.** $(a + c)(b + d)$ **18.** $5(d - 3b) - 2ac$
19. $(d - c - b)^5$ **20.** $b + c(c + d)$

B

1. What is the circumference of a circle whose radius is 63 inches?

2. What is the area of a circle whose diameter is 8.4 feet?

3. What is the simple interest on $870 at $4\frac{1}{2}\%$ per annum for 5 years?

4. What is the area of a trapezoid whose bases are 10.2 inches and 7.5 inches and whose altitude is 3.4 inches?

5. An automobile traveling 45 miles per hour goes how far in 2 hours and 24 minutes?

6. Each angle of a regular polygon of n sides is given by the formula $A = \dfrac{180(n - 2)}{n}$. Find the angles of a regular pentagon.

7. The volume of a circular cylinder of radius r and height h is given by the formula $V = \pi r^2 h$. Find the volume of a circular cylinder when its radius is 2.8 inches and its height is 6.7 inches.

8. Given the formula $C = \dfrac{nE}{R + nr}$, find C when $n = 5$, $E = 3.2$, $R = 4$, and $r = 0.8$.

9. Given the formula $A = P(1 + r)^2$, find A when $P = 2500$ and $r = 0.04$.

10. Given the formula $S = \dfrac{n}{2}(a + l)$, find S when $n = 15$, $a = 2$, and $l = 26$.

C

Solve the following equations and check your result:

1. $x + 5 = 17$	**2.** $x - \frac{1}{2} = \frac{1}{3}$	**3.** $8x = 112$
4. $\frac{x}{15} = 15$	**5.** $2x - 1 = 9$	**6.** $4x = 25 - x$
7. $5 + \frac{x}{2} = 7$	**8.** $6 = 7 - 3x$	**9.** $5x = 3x + 16$
10. $x + 1 = 5 - x$	**11.** $5x + 3 = 2x + 9$	**12.** $\frac{3x}{2} + 1 = \frac{x}{2} + 3$
13. $\frac{3}{4}x - 3 = \frac{1}{2}x + 1$	**14.** $2\frac{1}{3}x + 5 = 8 - \frac{2}{3}x$	**15.** $3\frac{1}{2}x - 1\frac{1}{3} = 5\frac{2}{3}$
16. $2 + 2x = 7 - 3x$	**17.** $\frac{5x}{7} - 3 = 17$	**18.** $\frac{2}{5}x - 1 = 5 - \frac{4}{5}x$
19. $\frac{x}{4} - \frac{x}{5} = 2$	**20.** $\frac{x}{2} - \frac{1}{3} = \frac{x}{4} - \frac{1}{5}$	

D

Express in algebraic symbols.

1. What number is x more than x^2?

2. What number is $33\frac{1}{3}\%$ less than a?

3. If x books cost y dollars, what is the cost in cents of one book?

4. What is the area in square inches of a rectangle which is x inches wide and 2 feet long?

5. If $2x - 17$ is an integer, what is the next larger integer?

6. What is the value in cents of x nickels and $3x$ dimes?

7. A has 12 more marbles than B, and B has twice as many marbles as C. If C has m marbles, how many marbles has A?

8. What number exceeds a by 5 less b?

9. A boy earns c cents and then spends one-third of it for a show and one-fifth of it for some candy. How much does he have left?

10. A man is five times as old as his son. If the sum of their ages is y years, how old is each?

Obtain the formula indicated in each of the following:

11. The cost (C) in cents of N books at $2.20 each.

12. A boy's age (A) now, who was y years old 5 years ago.

13. The value (V) in dollars of h half dollars and q quarters.

14. The amount (A) in pints of alcohol contained in g gallons of a solution which is 25% alcohol.

15. The area (A) of a closed box whose dimensions are a feet by b feet by c feet.
16. The distance (D) in yards traveled, when proceeding at the rate of 4 miles per hour for t minutes.
17. The cost (C) in cents of a pounds of coffee at 72 cents a pound and b pounds of tea at 84 cents a pound.
18. It costs $12 to make plates for a circular and 3¢ a copy for the circulars. Find the total cost (C) for n circulars.
19. The first fifteen words of a telegram cost 60 cents and each additional word costs 4 cents. Find the total cost (C) for a telegram of n words. (Assume n more than 15.)
20. The first thousand copies of an article cost 6 cents each, the second thousand cost 4 cents each, and any more than 2000 cost 3 cents each. Find the cost (C) of N copies. (Assume N more than 2000.)

E

1. One number is 9 more than another and their sum is 139. Find the numbers.
2. 30% of what number equals 24?
3. The sum of three consecutive even integers is 84. Find the integers.
4. The length of a rectangle is $2\frac{1}{3}$ times the width, and the perimeter is 20 feet. What are the dimensions?
5. Twice what number is 33 more than one-half of that number?
6. The length of a rectangle is 50% longer than the width. If the perimeter of the rectangle is 105 inches, what are its dimensions?
7. The larger of two numbers exceeds twice the smaller by 3. If the sum of the numbers is 54, what are the numbers?
8. Six less than seven times a certain number is the same as 2 more than three times the number. Find the number.
9. A man is six times as old as his daughter and his daughter is two years younger than his son. If their combined ages total 58 years, how old is each?
10. Two sides of a triangle are equal and the third side is one foot shorter. If the perimeter of the triangle is 38 feet, find its dimensions.
11. A baseball team played 154 games and won 28 games more than it lost. How many games did it win?
12. Three partners divided $16,000, the second partner receiving $2000 more than the first, and the third partner receiving twice as much as the first. How much did each receive?

13. The sum of the sides of a triangle is 26 feet. The second side is 2 feet longer than the first, and the third side exceeds the first by 3 feet. Find the dimensions of the triangle.

14. A newsboy sells twice as many 5¢ papers as he does 3¢ papers. If he takes in $5.46, how many papers of each kind does he sell?

15. In an election for two candidates 28,657 votes were cast. The successful candidate had a majority of 2613 votes. How many votes did each candidate receive?

16. A bar of iron 56 inches long is to be cut into two parts, such that one part is $\frac{3}{5}$ of the other. Find the length of each part.

17. An automobile was sold for $1404, which was 35% less than what it cost. Find the cost.

18. A boy made 72 and 83 on two term tests. What must he make on the third test in order to average 80?

19. Separate 23.6 into two parts such that the larger will exceed the smaller by 8.84.

20. For every dime that a girl put into her savings bank, her father put in a quarter. If her bank contained $19.95 at the end of a year, how many dimes did the girl save during the year?

Losing Is Important

If 57 contestants are entered in a single-elimination tennis tournament, how many matches must be scheduled?

Believe It or Not

For any number of terms, it is true that

$$(1 + 2 + 3 + \cdots + n)^2 = 1^3 + 2^3 + 3^3 + \cdots + n^3.$$

Was He a Square?

Augustus de Morgan, a mathematician, who died in 1871, used to boast that he was x years old in the year x^2. In what year was he born?

CHAPTER II

SIGNED NUMBERS: ADDITION AND SUBTRACTION

19. Signed Numbers. In the study of arithmetic we deal with numerical quantities connected by the signs + and −; and in finding the value of an expression, such as $10 + 2 - 7 + 6 - 8$, we understand that the quantities to which the sign + is prefixed are *additive* and those to which the sign − is prefixed are *subtractive*, and that the first quantity to which no sign is prefixed is included among the additive terms. The same idea prevails in algebra; thus, in the expression $2a + 5b - 3c - 7d$, we understand the quantities $2a$ and $5b$ to be additive, and $3c$ and $7d$ to be subtractive.

In arithmetic the sum of the additive terms is always greater than the sum of the subtractive terms; if the reverse were the case, the result would have no arithmetical meaning. In algebra, on the other hand, not only may the sum of the subtractive terms exceed that of the additive, but a subtractive term may stand alone, and yet have a very definite meaning.

To understand more clearly just how such a quantity might occur in a practical problem, consider the following illustrations:

1. Suppose that a man were to gain \$50 and then lose \$30; his net *gain* would be \$20. But if he first gains \$30 and then loses \$50, the result of his efforts is a net *loss* of \$20.

If we agree to represent *gains* as additive (+) quantities and *losses* as subtractive (−) quantities, the corresponding algebraic statements may be written

$$\$50 - \$30 = +\$20,$$
$$\$30 - \$50 = -\$20,$$

where the subtractive quantity in the second case indicates a sum of money opposite in character to the additive quantity in the first case.

2. Suppose that a man travels eastward for a distance of 50 miles and then westward for 30 miles; his position will then be 20 miles *east* of the starting point. But if he travels 30 miles eastward and then 50 miles westward, he will be just as far from the starting point, but he will be *west* of it instead of east.

In this case, we may write

$$50 \text{ miles} - 30 \text{ miles} = +20 \text{ miles},$$
$$30 \text{ miles} - 50 \text{ miles} = -20 \text{ miles},$$

where the sign $+$ indicates the direction *east* and the sign $-$ indicates the direction *west*.

From these and other illustrations it is evident that it is of great value in the solution of algebraic problems to be able to indicate by a number not only its numerical magnitude but also the sense of direction in which it is to be taken. For this reason, we divide all general numbers, zero excepted, into two classes, *positive quantities* and *negative quantities*, and we distinguish these classes by the signs $+$ and $-$. It is to be understood that a negative quantity is considered as opposite in character to whatever we may mean by a positive quantity. Thus, if gains are considered as positive, then negative quantities will represent losses. Conversely, if losses are considered as positive, then negative quantities will represent gains.

Definition. *Absolute value* is the value of a quantity irrespective of the sign $+$ or $-$.

Illustration. The absolute values of 2, $-\frac{1}{2}$, -5.1 are respectively 2, $\frac{1}{2}$, 5.1.

ORAL EXERCISE

1. If distance north is called positive, what should distance south be called?
2. If $+\$200$ represents a credit of $200, what does $-\$100$ represent?
3. If 45° north latitude is marked $+45°$, what does $-20°$ mean?
4. If positive distances are measured above sea level, what represents 200 feet below sea level?
5. If counterclockwise rotations are called positive, what will a negative quantity represent?
6. If $+\$1.20$ represents $1.20 spent, what does $-\$1.80$ represent?
7. If pounds overweight are represented by positive quantities, what will indicate 10 pounds underweight?

8. If each of the following is considered as being positive, name its negative:

Assets	Decreases	Debits
Increases	Imports	Above sea level
Time A.D.	Income	Southwest
Rises	Above zero	Backward
Wins	Deposits	To the left
Up	Liabilities	Exports

20. Addition of Signed Numbers. To combine numbers of like sign, we may consider the numbers as though they were similar algebraic terms and add accordingly. Thus, to combine $+3$ and $+5$, we think: 3 positives and 5 positives give 8 positives, or $+8$. Similarly, to combine -3 and -5, we think: 3 negatives and 5 negatives give 8 negatives, or -8.

Since negative numbers are, by definition, opposite in character to positive numbers, they each have the effect of nullifying the other. That is to say, a number such as -7 combined with $+7$ gives zero as a result.

Illustrations. Seven dollars earned and $7 spent leaves no money. Seven yards forward and 7 yards backward leaves no change in position.

In combining a positive number and a negative number of different absolute value, the smaller of the two can only partially nullify the larger. Thus, -7 combined with $+10$ leaves an excess of $+3$; and $+7$ combined with -10 leaves an excess of -3.

In general these observations lead us to the following rule:

Rule for Addition of Signed Numbers

1. ***To add signed numbers having the same signs, find the numerical sum of their absolute values and prefix to this sum their common sign.***
2. ***To add signed numbers having different signs, add together separately the positive numbers and the negative numbers. The difference of the absolute values of these two results, preceded by the sign of the greater, will give the required sum.***

Example. Add $-6, +5, -3, -2$, and $+4$.

Solution. The sum of the negative numbers is -11 and the sum of the positive numbers is $+9$. Hence, the required sum is -2.

To indicate a sum such as the one above, the numbers are often written in the form

$$-6 + 5 - 3 - 2 + 4.$$

Such a representation is called an *algebraic sum.*

Illustration. The following indicate the simplification of three algebraic sums:

(1) $-6 - 5 - 3 - 1 = -15.$
(2) $3 + 5 + 7 + 1 = 16.$
(3) $-2 + 3 - 6 + 5 = -8 + 8 = 0.$

EXERCISE 8

Add:

1.	2.	3.	4.	5.	6.
$+27$	-37	-29	$+53$	-77	-26
-18	$-\ 9$	$+13$	$+14$	$+59$	$+26$

7.	8.	9.	10.	11.	12.
-32	$+47$	-19	$+59$	-33	-27
$+17$	$+13$	-36	-31	$+19$	-44

13.	14.	15.	16.	17.	18.
$+27$	-52	$+16$	$+21$	$+49$	-16
-18	-16	-47	$+23$	$+19$	-23
$+14$	-23	-21	$+72$	-72	$+75$
-21	-14	-19	$+47$	$+12$	-36

19.	20.	21.	22.	23.	24.
$+32$	-41	-18	-23	$+32$	-45
$+14$	$+62$	$+61$	-26	$+13$	$+39$
-27	-17	-59	-41	-55	$+27$
-29	-33	$+16$	$+67$	$+27$	-16

Evaluate the following algebraic sums:

25. $-16 - 14 + 6 - 9$
26. $+22 + 13 - 74 + 39$
27. $-23 - 11 - 15 - 36$
28. $-52 + 65 - 79 - 33$
29. $+15 + 23 - 65 + 27$
30. $-42 - 19 - 51 + 87$
31. $+23 + 17 - 21 - 30$
32. $-37 + 29 + 44 - 37$
33. $-63 + 17 + 63 + 23$
34. $-49 + 16 - 22 + 72$
35. $+51 - 25 - 25 - 11$
36. $-23 + 45 + 13 - 35$

21. Addition of Algebraic Expressions. We have already learned that to add similar terms we add their coefficients and take this sum as the coefficient of the common literal part. We now apply this new rule of addition in evaluating such sums.

Example 1. Find the sum of $2a$, $5a$, $-a$, $-8a$.

Solution. Adding the coefficients 2, 5, -1, -8, we have -2. Hence, the required sum is $-2a$.

When a sum is written in the form of an algebraic sum, the process of combining is sometimes called *collecting terms.*

Example 2. Collect terms: $3xy - 5xy + xy - 2xy$.

Solution. First combining the positive terms and the negative terms, we have

$$3xy - 5xy + xy - 2xy = 4xy - 7xy = -3xy.$$

When two or more like terms are to be added, we see that the result is expressed as a single like term. If, however, the terms are *unlike,* they cannot be collected. Thus, in finding the sum of two unlike quantities, such as a and b, all that we can do is to connect them by the sign of addition and express the sum in the form $a + b$.

Example 3. Collect like terms in the algebraic sum

$$a - 5b + c + 4a + 4b - c.$$

Solution. Rearranging and collecting like terms, we have

$$a + 4a - 5b + 4b + c - c = 5a - b.$$

To find the sum of several multinomials, it is often more convenient to observe the following procedure:

Rule for Addition of Algebraic Expressions

To add multinomials, arrange the expressions in rows so that the like terms are in the same vertical columns; then add each column separately.

Example 4. Find the sum of $2a - 3b + 5c$; $3a - 2c$; $3b + c$.
Solution. Arranging according to the above rule, we have

$$\begin{array}{l} 2a - 3b + 5c \\ 3a \qquad\ \ - 2c \\ \qquad\ \ 3b + \ c \\ \hline 5a \qquad\ \ + 4c \end{array}$$

NOTE. It is necessary that the + sign precede the term $4c$ in order to indicate that it is to be added to $5a$ for the final result.

EXERCISE 9

Simplify the following expressions by combining similar terms:

1. $\begin{array}{r} 2m \\ -5m \\ \hline \end{array}$ 2. $\begin{array}{r} -9y \\ -3y \\ \hline \end{array}$ 3. $\begin{array}{r} 2a^2 \\ 5a^2 \\ -3a^2 \\ \hline \end{array}$ 4. $\begin{array}{r} xy \\ -4xy \\ 5xy \\ \hline \end{array}$ 5. $\begin{array}{r} 7x \\ -3x \\ x \\ -5x \\ \hline \end{array}$ 6. $\begin{array}{r} -3pq^2 \\ 5pq^2 \\ -7pq^2 \\ 4pq^2 \\ \hline \end{array}$

7. $-6b^2 + 3b^2$
8. $4xy - 7xy$
9. $3x^3 - 4x^3 + 2x^3$
10. $7xy - xy - 5xy$
11. $ax - 5ax - 3ax + 7ax$
12. $3by - 5by + by - 2by$

13. $\begin{array}{r} 2x + 4y \\ -5x + \ y \\ \hline \end{array}$ 14. $\begin{array}{r} a - b \\ 5a + b \\ \hline \end{array}$ 15. $\begin{array}{r} 26a + 10b \\ -12a - \ \ 5b \\ 5a - \ \ 6b \\ \hline \end{array}$ 16. $\begin{array}{r} 14x - 6 \\ -7x + 3 \\ -9x - 1 \\ \hline \end{array}$

17. $\begin{array}{r} 6xy - 3xy^2 \\ -4xy - 7xy^2 \\ 3xy - \ \ xy^2 \\ -xy + 9xy^2 \\ \hline \end{array}$ 18. $\begin{array}{r} 4s - 6t \\ -2s - \ t \\ -11s - \ t \\ 6s + \ t \\ \hline \end{array}$

19. $6x - y + 3x + 2y$
20. $9x^2y^2 - 2x^2y - 7x^2y^2 - 3x^2y$
21. $4a - 2b + 6a + 5b - 9a - 3b$
22. $-6x^2 + y^2 - 2x^2 - y^2 + 5x^2 - 7y^2$
23. $-x + 6y - 5x - 3y + 7x - 2y - x - 5y$
24. $9a + 6ab - 7a - 4ab - 5a + 10ab + a - 2ab$

25. $\begin{array}{r} 4x^2 - 2xy - y^2 \\ 3x^2 + 2xy - y^2 \\ \hline \end{array}$ 26. $\begin{array}{r} -6a - 2b + 4 \\ 5a + 3b - 7 \\ \hline \end{array}$

27.
$$\begin{array}{r} 14x^2 - 10x - 1 \\ 5x^2 + 3x - 5 \\ -x^2 - 2x - 3 \\ \hline \end{array}$$

28.
$$\begin{array}{r} 6m - 5n - 3p \\ -4m + n + 6p \\ -7m + 4n - 4p \\ \hline \end{array}$$

29.
$$\begin{array}{rrr} 9u & + 6v & + 7w \\ -u & & + 2w \\ 4u & - 5v & \\ & 4v & - 10w \\ \hline \end{array}$$

30.
$$\begin{array}{rrr} 6x & & + z \\ -2x & - 4y & \\ -5x & - y & + 7z \\ & 9y & - 5z \\ \hline \end{array}$$

31. $7a + 9b - c - 9a + 2b + c$

32. $6x^2 - 5xy - 7y^2 + x^2 + 7xy + 8y^2$

33. $2x - 6y - 5z - 3x + 5y - z - 2x + y + 3z$

34. $-3u - 5v - w + 6u - 3v - w + u + 6v + 3w$

35. $4x^2 - 6xy + y^2 - 2x^2 + 3xy - 7y^2 - 3x^2 + 8xy - 3y^2 - x^2 + xy + 5y^2$

36. $6a + 2b - c + 5a - 7b + 5c - a - 5b + 2c - 6a + 9b - 3c$

37.
$$\begin{array}{rrr} \frac{1}{2}x^2 & + xy & - \frac{2}{3}y^2 \\ -\frac{1}{4}x^2 & - \frac{1}{5}xy & + \frac{1}{3}y^2 \\ \hline \end{array}$$

38.
$$\begin{array}{rrr} -\frac{1}{6}x & + \frac{2}{3}y & - z \\ \frac{1}{3}x & - \frac{5}{6}y & - \frac{1}{2}z \\ \hline \end{array}$$

39.
$$\begin{array}{rrr} \frac{1}{2}a & - \frac{4}{5}b & - \frac{1}{3}c \\ -\frac{1}{4}a & + \frac{1}{5}b & + \frac{1}{2}c \\ -\frac{1}{4}a & & - \frac{1}{6}c \\ \hline \end{array}$$

40.
$$\begin{array}{rrr} -\frac{2}{3}a^2 & - \frac{1}{6}b^2 & - c^2 \\ \frac{1}{6}a^2 & & - \frac{2}{5}c^2 \\ & \frac{1}{3}b^2 & - \frac{3}{5}c^2 \\ \hline \end{array}$$

41.
$$\begin{array}{rrr} \frac{1}{3}x & + \frac{1}{2}y & - z \\ x & & + \frac{1}{3}z \\ & - \frac{1}{3}y & - \frac{1}{2}z \\ \frac{1}{2}x & + y & - z \\ \hline \end{array}$$

42.
$$\begin{array}{rrr} 2x^2 & - y^2 & - 3 \\ -\frac{1}{2}x^2 & - \frac{1}{2}y^2 & + \frac{1}{3} \\ -\frac{2}{3}x^2 & & + \frac{4}{5} \\ & 2y^2 & + \frac{1}{3} \\ \hline \end{array}$$

43.
$$\begin{array}{rrr} 2.1x & - 1.3y & + 6.7 \\ -3.2x & + 1.6y & - 4.8 \\ \hline \end{array}$$

44.
$$\begin{array}{rrr} -3.2a & - 6.9b & + c \\ 1.9a & - 2.1b & - 3.6c \\ \hline \end{array}$$

45.
$$\begin{array}{rrr} 2.45a^2 & - 3.17b^2 & - 4.11c^2 \\ -0.15a^2 & + 4.92b^2 & + 6.71c^2 \\ -4.32a^2 & - 1.30b^2 & + 2.93c^2 \\ \hline \end{array}$$

46.
$$\begin{array}{rrr} 6.92x & - 3.63y & - 4.23z \\ -5.11x & + 3.47y & - 2.19z \\ 4.89x & - 2.07y & - 1.62z \\ \hline \end{array}$$

47.
$$\begin{array}{rrr} 2.95x^2 & + 6.23xy & - 9.07y^2 \\ -4.99x^2 & - 2.67xy & - 3.22y^2 \\ -1.09x^2 & - 3.74xy & + 7.77y^2 \\ \hline \end{array}$$

48.
$$\begin{array}{rrr} -6.66a & + 3.67b & - 2.48c \\ 9.01a & - 4.76b & + 5.55c \\ -4.19a & - 2.81b & - 3.07c \\ \hline \end{array}$$

22. Subtraction of Signed Numbers. The operation of subtraction is the inverse of the operation of addition. This means that the difference between two numbers is such a number that when we add it to the subtrahend (number being subtracted) we obtain the other number (minuend). Thus, 2 subtracted from 5 equals 3, *because* 3 added to 2 gives 5.

When the minuend and the subtrahend may be either positive or negative, the possible types of subtraction are four in number; they are illustrated by the following examples:

$$\begin{array}{r} +12 \\ +\ 5 \\ \hline +\ 7 \end{array} \qquad \begin{array}{r} -10 \\ -\ 7 \\ \hline -\ 3 \end{array} \qquad \begin{array}{r} +15 \\ -\ 8 \\ \hline +23 \end{array} \qquad \begin{array}{r} -13 \\ +\ 4 \\ \hline -17 \end{array}$$

In each of these examples, observe that the condition of subtraction is satisfied. That is, in each case, the difference added to the subtrahend equals the minuend.

Now observe, in each of these examples, that we may compute the difference *directly* by changing the sign of the subtrahend and adding to the minuend. Thus, we have

Rule for Subtraction of Signed Numbers

Change the sign of the number to be subtracted and add it to the other number.

Examples.

(1) $(+23) - (+17) = 23 - 17 = 6$,
(2) $(-47) - (-32) = -47 + 32 = -15$,
(3) $(+27) - (-12) - (+46) = 27 + 12 - 46 = -7$,
(4) $(-19) - (-23) - (+4) = -19 + 23 - 4 = 0$.

NOTE. Observe that an algebraic sum, such as $13 - 7$, may be considered as an addition $(+13) + (-7)$, or as a subtraction $(+13) - (+7)$.

EXERCISE 10

Subtract:

1. $\begin{array}{r} -23 \\ +17 \\ \hline \end{array}$ **2.** $\begin{array}{r} -52 \\ -47 \\ \hline \end{array}$ **3.** $\begin{array}{r} +57 \\ -19 \\ \hline \end{array}$ **4.** $\begin{array}{r} +16 \\ +35 \\ \hline \end{array}$ **5.** $\begin{array}{r} -17 \\ -17 \\ \hline \end{array}$ **6.** $\begin{array}{r} +42 \\ -52 \\ \hline \end{array}$

7. $\begin{array}{r} +27 \\ -19 \\ \hline \end{array}$ **8.** $\begin{array}{r} -45 \\ -18 \\ \hline \end{array}$ **9.** $\begin{array}{r} -39 \\ +17 \\ \hline \end{array}$ **10.** $\begin{array}{r} -42 \\ -42 \\ \hline \end{array}$ **11.** $\begin{array}{r} +15 \\ -31 \\ \hline \end{array}$ **12.** $\begin{array}{r} -37 \\ -25 \\ \hline \end{array}$

13. $(+24) - (-15)$ **14.** $(-26) - (-35)$ **15.** $(-12) - (+6)$
16. $(+79) - (+33)$ **17.** $(-71) - (-47)$ **18.** $(+43) - (-17)$
19. $(-34) - (-34)$ **20.** $(+51) - (+29)$ **21.** $(-53) - (-61)$
22. $(+25) - (+37)$ **23.** $(+46) - (-31)$ **24.** $(-52) - (-72)$
25. $(+87) - (+42) - (-27) - (-16)$
26. $(-56) - (-16) - (-39) - (+25)$
27. $(-21) - (+15) - (-32) - (+54)$
28. $(+26) - (+72) - (-66) - (+20)$
29. $(+82) - (-66) - (+49) - (+89)$
30. $(-55) - (-29) - (-34) - (+41)$
31. $(-42) - (-16) - (+31) - (-54)$
32. $(+17) - (-25) - (+32) - (+10)$
33. $(-65) - (-27) - (-18) - (-10)$
34. $(+16) - (-49) - (+56) - (-20)$
35. $(+74) - (+47) - (-44) - (+59)$
36. $(-75) - (+15) - (-38) - (-65)$

23. Subtraction of Algebraic Expressions. We have already learned that to subtract similar terms we subtract their coefficients and take this difference as the coefficient of the common literal part.

Example 1. Subtract $-3a$ from $7a$.
Solution. Arranging in vertical order, we have the following to subtract

$$\begin{array}{r} 7a \\ \underline{-3a} \end{array}$$

Changing the sign of the subtrahend, we have the following to add

$$\begin{array}{r} 7a \\ \underline{+3a} \end{array}$$

The result of the addition is $10a$. Hence, in subtracting $-3a$ from $7a$, we have $10a$.

For the subtraction of multinomials, we subtract similar terms as in the above example.

Example 2. From $2a - 5b - 3c$ subtract $3a - 6b - 3c$.
Solution. Arranging in vertical order, we have the following to subtract:

$$\begin{array}{r} 2a - 5b - 3c \\ \underline{3a - 6b - 3c} \end{array}$$

Changing the sign of *every* term in the subtrahend, we have the following to add:

$$\begin{array}{r} 2a - 5b - 3c \\ -3a + 6b + 3c \\ \hline \end{array}$$

The result of this addition is $-a + b$.

> **Rule for Subtraction of Algebraic Expressions**
>
> ***Change the sign of every term in the expression to be subtracted and add it to the other expression.***

NOTE. It is neither necessary nor advisable actually to change the signs of the terms in the subtrahend; the changing of signs should be done *mentally*.

Example 3. Subtract $3x^2 - 7xy + 5y^2$ from $4x^2 - 5xy - 3y^2$.

Solution. Arranging in vertical order, we obtain by subtraction:

$$\begin{array}{r} 4x^2 - 5xy - 3y^2 \\ 3x^2 - 7xy + 5y^2 \\ \hline x^2 + 2xy - 8y^2 \end{array}$$

To check a problem in subtraction: when the difference is added to the subtrahend the sum should equal the minuend.

Example 4. Subtract $3x^2 - 4x$ from $1 - 2x^3$.

Solution. Terms containing different powers of the same letter, being *unlike*, must stand in different columns. Thus,

$$\begin{array}{rrrr} -2x^3 & & & +1 \\ & 3x^2 & -4x & \\ \hline -2x^3 & -3x^2 & +4x & +1 \end{array}$$

The rearrangement of terms is *not necessary*, but it is convenient because it gives the result in descending powers of x.

EXERCISE 11

Subtract the second multinomial from the first:

1. $5x - 2y;\ 4x - 7y$
2. $-6a^2 + b^2;\ 4a^2 - 3b^2$
3. $3a - 5b;\ 2a - 5b$
4. $6x^2 - 3y^2;\ -x^2 - y^2$
5. $9m - 4n;\ -7m + 4n$
6. $3p - q;\ 5p + q$
7. $5x^2y^2 - 3x^2y;\ 3x^2y^2 - 5x^2y$
8. $-6xy - 2y;\ xy - 3y$
9. $5a^2 + 2a;\ -3a^2 + a$
10. $-7y^2 + 6y;\ -2y^2 + 6y$

11. $6a - 3b + 2c$; $4a - 2b - 2c$
12. $7x^2 - 9y^2 - z^2$; $-4x^2 - 8y^2 + z^2$
13. $-3x + 5y - 7$; $-2x - 5y - 7$
14. $6p + 9q - 10$; $4p - 3q - 6$
15. $10x^2 - 6xy + 2y^2$; $-x^2 - 5xy + 7y^2$
16. $8m^2 - 6mn - 2$; $6m^2 - 2mn + 2$
17. $4x^4 - 3x^3 - 2x + 1$; $3x^4 - x^3 - x^2$
18. $7x^4 - 2x^2 + 3x - 7$; $x^4 - x^3 + 10x - 7$
19. $-3x^3 + 2x^2 - 5x - 2$; $x^2 - 6x - 5$
20. $10x^3 - 7x + 13$; $3x^3 - 2x^2 + 13$
21. $5x^4 - 3x^3 - 2x + 7$; $-4x^3 - 2x + 9$
22. $-x^3 - x^2 + 7x - 9$; $7x^3 + 6x - 8$
23. $\frac{1}{2}a + \frac{1}{3}b$; $\frac{1}{3}a - \frac{1}{2}b$
24. $\frac{1}{6}x^2 - \frac{1}{2}y^2$; $\frac{2}{3}x^2 - y^2$
25. $-\frac{2}{3}x + \frac{1}{5}$; $\frac{1}{2}x + \frac{1}{3}$
26. $\frac{3}{4}x - \frac{1}{2}y$; $\frac{1}{3}x + \frac{2}{5}y$
27. $2.5x - 6.3y - 9.1z$; $3.4x + 2.6y - 7.5z$
28. $-9.3a + 2.9b - 3.6$; $7.2a + 3.3b + 4.7$
29. $-6.21a^2 + 3.27ab - 5.95b^2$; $9.27a^2 - 6.93b^2$
30. $8.88p^2 - 9.07pq + 3.21q^2$; $4.28p^2 - 16.00pq$

MISCELLANEOUS EXERCISES

1. To the sum of $a - 3b + c$ and $2b - a + 6c$ add the sum of $2a - 3c + b$ and $b - 4c$.
2. From $5x^2 - 3x - 6$ take the sum of $x^3 + x + 1$ and $x - 3x^2 + 2$.
3. Subtract $2x^2 - 5x + 3$ from x^3, and add the result to $2x^2 + 3x - 1$.
4. What expression must be subtracted from $4a - 3b + 2c$ so as to leave $2a - 5b - c$?
5. Find the sum of $3x - 4y + z$ and $2y - x$, and subtract the result from $z - 5y$.
6. Take $a^2 - b^2$ from $3ab - 4b^2$, and add the remainder to the sum of $2ab - a^2 - b^2$ and $3a^2 + 5b^2$.
7. To what expression must $5m^3 - 4m^2 + 2m + 1$ be added so as to give $9m^3 - 5m + 6m^2$?
8. From what expression must $2xy - 3xz + 4yz$ be subtracted so as to leave a remainder $6xz - yz$?
9. Add together $2a^2 - 3a + 7$ and $a^3 + 3a - 2$, and diminish the result by $2a^2 + 5$.
10. Subtract $5x^2 - 3x + 1$ from unity (one), and add $2x^2 - 3x$ to the result.
11. Subtract $3a - 2b + c$ from the sum of $a + 3b + 4c$ and $4a - b - 2c$.

12. To what expression must $7x^3 - 5x^2 + 3x$ be added so as to produce zero?
13. Increase $2ab - 3ac + 4bc$ by $2ac - ab$, then subtract $ab + ac + bc$ from this result.
14. From the sum of $x - 3y + z$ and $z - 4x + 2y$, subtract the sum of $y - 3z$ and $x - 2y$.
15. Subtract $6 - 3x + 2x^2 - x^3$ from $3x^2 - 4x - 5$, then subtract the difference from zero, and add this last result to $x^2 - 1$.
16. What expression must be added to $a - 3b - c$ to produce an expression equal to the sum of $2a - 3c$ and $c - 3b - a$?
17. Subtract $x^2 - xy$ from zero, and add the difference to the sum of $3x^2 - 5y^2$ and $xy - x^2 - y^2$.
18. What expression added to $4a - 3c + b$ gives the same result as that obtained in subtracting $b + 2c - 3a$ from $2a - 3b + c$?

24. Parentheses. Parentheses ( ), brackets [ ], and braces { } are used to indicate that the terms which they contain are grouped together to form a single quantity.

In an algebraic sum of any number of terms, we may insert parentheses wherever we wish without changing the value of the sum, provided that a plus sign (+) occurs or is understood before each pair of parentheses.

Illustration.

$$3 + 5 - 2 = (3 + 5) - 2,$$

or

$$= 3 + (5 - 2).$$

The same is true with algebraic expressions.

Illustration.

$$5x + 6y - 3x + 2y + 9 - y$$
$$= 5x + (6y - 3x) + (2y + 9) - y,$$

or

$$= (5x + 6y - 3x) + (2y + 9 - y).$$

However, we have learned that the subtraction of a quantity is equivalent to the addition of its negative. Hence, if we insert parentheses and write a minus sign (−) before it, we must change the sign of each term within the parentheses.

Illustration.

$$5x + 6y - 3x + 2y + 9 - y$$
$$= 5x + 6y - (3x - 2y - 9) - y,$$

or

$$= 5x - (-6y + 3x) - (-2y - 9 + y).$$

25. Removal of Parentheses. From the discussion on parentheses, we have the following rule for their removal.

Rule for Removal of Parentheses

1. ***When an expression within parentheses is preceded by the sign +, the parentheses may be removed without making any change in the expression.***
2. ***When an expression within parentheses is preceded by the sign −, the parentheses may be removed if the sign of every term within the parentheses is changed.***

Example 1. Remove parentheses and collect terms

$$\begin{aligned}(5x - 3y) - (2x - 4y)& \\ &= 5x - 3y - 2x + 4y, \\ &= 3x + y.\end{aligned}$$

Example 2. Remove parentheses and collect terms

$$\begin{aligned}(a^2 + 2b^2) - (3a^2 + 2ab - b^2) + (3ab - 5b^2)& \\ &= a^2 + 2b^2 - 3a^2 - 2ab + b^2 + 3ab - 5b^2, \\ &= -2a^2 + ab - 2b^2.\end{aligned}$$

When one set of parentheses is contained within a bracket, first remove the innermost parentheses and collect terms. Then continue this process until all parentheses, brackets, etc., have been removed and similar terms combined.

Example 3. Remove parentheses and collect terms

$$\begin{aligned}a - [3a - (2a - b)]& \\ &= a - [3a - 2a + b], \\ &= a - [a + b], \\ &= a - a - b, \\ &= -b.\end{aligned}$$

EXERCISE 12

Remove parentheses and combine similar terms:

1. $(2x - y) + (3x + 5y)$ **2.** $(7x - 2) + (5x + 6)$
3. $(4a - 3b) - (5a - b)$ **4.** $(a - 5b) - (2a - 3b)$
5. $(2x - 1) - (4x + 3)$ **6.** $(3x - 7) - (2x + 5)$

7. $(x + y + z) - (3x - 2y + z)$
8. $(4u - 5v + 6w) + (2u + 4v - 3w)$
9. $(a - b + c) + (a + b - c)$
10. $(5x - 3y + 7) - (2x - 7y + 5)$
11. $(6x^2 - 9x + 1) - (4x^2 - 6x - 3)$
12. $(5x^2 - x + 9) - (9x^2 - 4x - 1)$
13. $(3x - 2y) - (6x - 9y) - (2x + 3y)$
14. $(4x - y) - (2x + 5y) + (9x - 5y)$
15. $(2a - b) + (6a + 5b) - (8a + 3b)$
16. $(a - b) - (3a - b) - (2a - 3b)$
17. $2p - (3p - q) + (5p - 2q)$
18. $5n + (2m - 3n) - (m - n)$
19. $(3x^2 + 6x - 9) - (4x^2 - 6x + 3) - (-2x^2 - 6x - 1)$
20. $(4x^2 - 2x + 1) + (x^2 - 3x + 1) - (5x^2 - 2x - 3)$
21. $(6a - 3b + c) - (2a - b - c) + (a - 3b + 5c)$
22. $(2a - 3b + 6c) - (5a - 9b - c) - (-a + 2b - c)$
23. $3x^2 - (x^2 - 1) + (2x^2 - 3x + 6)$
24. $(5x^2 + 6x - 1) - (3x^2 - 5) + 2x - 7$
25. $5x + [6x - (3x + 2)]$
26. $9a - [3a + 2 - (6a - 5)]$
27. $(5a + 2b) - [(2a - b) - (3a + b)]$
28. $(6x + y) - [(2x - 5y) - (5x - 9y)]$
29. $[1 - (2x - 3)] - [x - (6x - 5)]$
30. $[(a + 3b) - a] - [a - (a - 3b)]$
31. $x - \{3x - [2x - (3x + y)]\}$
32. $4a + 3 - \{6a - [(3a + 2) - (4a + 1)]\}$
33. $\{[(6a + 5b) - 3b] - (2a - 3b)\} - b$
34. $\{[(a + b) - b] - 2a\} - (3a - b)$

REVIEW OF CHAPTER II

A

Add:

1. $\begin{array}{r} +2a^2 \\ -3a^2 \\ +5a^2 \\ \hline \end{array}$

2. $\begin{array}{r} -5x \\ +4x \\ -\ x \\ \hline \end{array}$

3. $\begin{array}{r} -3b \\ -2b \\ -4b \\ \hline \end{array}$

4. $\begin{array}{r} +2xy \\ -3xy \\ +\ xy \\ \hline \end{array}$

5. $\begin{array}{r} -2(x - y) \\ +5(x - y) \\ -2(x - y) \\ \hline \end{array}$

6. $\begin{array}{r} +\ (a^2 + b^2) \\ -3(a^2 + b^2) \\ -2(a^2 + b^2) \\ \hline \end{array}$

Subtract:

7. $\begin{array}{r} 8xy^2 \\ -2xy^2 \\ \hline \end{array}$

8. $\begin{array}{r} -4a^3 \\ +\ a^3 \\ \hline \end{array}$

9. $\begin{array}{r} -\ bc \\ -4bc \\ \hline \end{array}$

10. $\begin{array}{r} 3m^2 \\ -\ m^2 \\ \hline \end{array}$

11. $\begin{array}{r} -4(x+y) \\ -5(x+y) \\ \hline \end{array}$

12. $\begin{array}{r} 3(2a+b) \\ -4(2a+b) \\ \hline \end{array}$

Collect terms:

13. $7x^2 - 8x - 6 - 2x^2 + 3x + 4$
14. $4a^2 - 9ab - 7b^2 - 9a^2 + 7b^2$
15. $3a - 2b + 4c - 2a + 2b - 5c$
16. $5x - 6y - 3x + 5y + x - 4y$
17. $3m - n + 5m - 4n - m + 7n$
18. $3a^2 + 7a + 6 - 5a^2 - 3a + 2 + 2a^2 - a - 9$
19. $7x^2 + 2xy - y^2 - 7xy - y^2 + 2x^2 + 5y^2$
20. $3a - 2b + 6c - a - b + 2c - a - 5b - 5c$
21. $x^2y - xy^2 - 3x^2y + 7x^2y^2 - 3xy^2 + 2x^2y + 5xy^2$
22. $2a^2 - 6ab - 3b^2 + a^2 - ab + b^2 + 6ab - 7b^2$

B

Add:

1. $\begin{array}{rrr} 2a^2 & -\ 6ab & +\ 3b^2 \\ -a^2 & & +\ b^2 \\ & 3ab & -\ 4b^2 \\ \hline \end{array}$

2. $\begin{array}{rrr} 6x & -\ 5y & -\ z \\ 3x & +\ 2y & -\ 3z \\ -x & -\ y & +\ 5z \\ \hline \end{array}$

3. $2a - 3b + c,\ c - a - b,\ 4b + 2a - 5c$
4. $9x^2 - y^2 + 5z^2,\ 2z^2 - 3y^2 + x^2,\ 2y^2 - 3x^2 - 7z^2$
5. $6m - 3n - 5,\ 7 - n - m,\ 21 + 2m + 5n$
6. $3x^2 - y^2,\ 2xy - x^2 - 2y^2,\ y^2 - x^2 - 3xy$

Subtract:

7. $\begin{array}{rrr} 3a & -\ 2b & +\ 5c \\ a & +\ b & -\ c \\ \hline \end{array}$

8. $\begin{array}{rrr} 5x^2 & -\ 6xy & +\ 2y^2 \\ & 3xy & -\ 2y^2 \\ \hline \end{array}$

9. Take $3x - 2y + z$ from $z + 2y - 4x$.
10. Take $6p - 7 + q$ from $2q - 4p + 7$.
11. Take $3a^2 + 2ab + b^2$ from $b^2 - 5a^2 + 3ab$.
12. Take $4 - a + 2b$ from $3a + 2b - c$.
13. Take $4x - y$ from the sum of $5x - 3y$ and $-x + 3y$.
14. Subtract the sum of $4a^2 - 6b^2$ and $ab + 5b^2$ from the sum of $3a^2 - 2ab$ and $a^2 + 3ab - 2b^2$.

15. Add $3m^2 - 2m - 5$ and $1 + 3m - 2m^2$ and subtract the result from zero.

16. Subtract $3a^2 - ab$ from $5a^2 - 2ab + b^2$ and add the result to the sum of $3ab - 2b^2$ and $b^2 - a^2 - ab$.

17. Increase $2x^2 + 3x + 1$ by the excess of $5 - 2x + x^2$ over $3 - x^2$.

18. What quantity is $3a + 6b - c$ greater than the sum of $a - b$ and $c - 4a - 6b$?

C

Remove parentheses and collect like terms:

1. $(3a - 2b) + (4a + 5b)$
2. $(x - 3y) - (4x - 2y)$
3. $(x^2 + 2xy - 3y^2) - (4x^2 - 2xy + 3y^2)$
4. $(3m - 2n - 5) - (2m - 3n - 2)$
5. $(3a + b - c) - (2a + 6b - 4c) - (a - 5b - c)$
6. $(5y^2 - 2y + 1) - (1 - 3y - y^2) + (2y - 5 + y^2)$
7. $5x + [2x - (3x - y)]$
8. $(a - 2b) - [(3a + b) - (2a - 3b)]$
9. $[(3x^2 - 2y^2) - x^2] - [y^2 - (2x^2 - 3y^2)]$
10. $[(3x + 2y) - (3x - 5y)] - [(2x + y) - (6x - 3y)]$

Remove parentheses and solve the following equations:

11. $4x - (2x + 5) = 3$
12. $3x = 8 - (x - 4)$
13. $(5 + x) - (5 - x) = 12$
14. $(5x - 4) - (2x + 5) = 0$
15. $x - (7 - 2x) = 2$
16. $5 - (x + 3) = 9 - (2x - 1)$
17. $6 - (7 - x) = 5 + (2 - x)$
18. $x - (2x + 5) = 1 - (5x + 6)$
19. $x - [3 + (x + 1)] = 5 - x$
20. $1 - [(2x + 1) - (4x + 1)] = x + 5$

Going Up?

A and *B* slowly walk up the descending escalator in a store. *A* goes up step by step in 10 minutes, taking 660 steps, and *B* does it in 15 minutes, taking 960 steps. How many steps would they have taken if the escalator had been stationary?

Arithmetic Is Easy?

Find the remainder if 2^{100} is divided by 7.

CHAPTER III

MULTIPLICATION AND DIVISION

26. Multiplication of Signed Numbers. Basically the operation of multiplying positive integers can be considered as one that indicates briefly certain types of addition. That is to say, we may think of 3×5 as meaning $5 + 5 + 5$. The multiplier (3) indicates how many times the other number (5) is to be taken as an addend (quantity being added). For a positive multiplier, we have

$$3 \times (+5) = (+5) + (+5) + (+5) = +15,$$
$$3 \times (-5) = (-5) + (-5) + (-5) = -15.$$

We know that the operation of multiplication is commutative; that is, $6 \times 7 = 7 \times 6$, $5 \times 9 = 9 \times 5$, etc. We may use this fact to observe the effect of multiplying a number with a *negative* multiplier. Thus, first applying the commutative law, we have

$$\begin{aligned} -3 \times 5 &= 5 \times (-3) \\ &= (-3) + (-3) + (-3) + (-3) + (-3) \\ &= -15. \end{aligned}$$

However, since $3 \times 5 = 15$, it is evident that multiplying with a *negative* multiplier just changes the sign of the product obtained when the multiplier is considered as positive. Hence, the product $-3 \times (-5)$ will have the opposite sign of the product $3 \times (-5)$, or

$$-3 \times (-5) = +15.$$

This leads us to the following:

Rule of Signs for Multiplication

1. If two numbers have like signs, their product is positive.

2. If two numbers have unlike signs, their product is negative.

Illustrations. (1) $(4)(-5) = -20$.
(2) $(-\frac{1}{2})(+\frac{1}{3}) = -\frac{1}{6}$.
(3) $(-1)(-1) = +1$.

The product of more than two factors can be found by successive applications of the above rule.

Example 1. Multiply $(-3)(+2)(-1)(-5)$.
Solution. Multiplying the first two factors in turn gives

$$\begin{aligned}(-3)(+2)(-1)(-5) &= (-6)(-1)(-5),\\ &= (+6)(-5),\\ &= -30.\end{aligned}$$

Example 2. Multiply $(-2)(-2)(-2)(-2)$.
Solution. Multiplying the first two factors and the last two factors, we have

$$(-2)(-2)(-2)(-2) = (+4)(+4) = 16.$$

A study of these examples indicates the following:

General Rule of Signs for Multiplication

1. ***If an even number of factors are negative, the product is positive.***
2. ***If an odd number of factors are negative, the product is negative.***

EXERCISE 13

Multiply:

1. $(-6) \times (+7)$ 2. $(-5) \times (-9)$ 3. $(+4) \times (-2)$
4. $(+6)(+5)$ 5. $(-5)(-4)$ 6. $(-2)(+3)$
7. $(-7)(+7)$ 8. $(-8)(-9)$ 9. $(-10)(+12)$
10. $(+11)(-13)$ 11. $(-12)(-12)$ 12. $(+10)(+14)$
13. $(-3) \times (-1) \times (+2) \times (-5)$ 14. $(+2) \times (-2) \times (-2) \times (+2)$
15. $(-3) \times (-2) \times (-3) \times (-1)$ 16. $(+5) \times (-3) \times (+2) \times (-5)$
17. $(+2)(-3)(+3)(-2)$ 18. $(-2)(-3)(-3)(+3)$
19. $(-4)(+2)(+3)(+2)$ 20. $(-2)(-3)(-4)(-5)$
21. $(+2)(-6)(-5)(-2)$ 22. $(-3)(-2)(-1)(+6)$
23. $(-7)(-2)(+1)(-1)$ 24. $(+3)(-9)(+8)(+2)$

25. $(+3)(-3)(-2)(-1)$
26. $(-4)(+2)(+1)(-3)$
27. $(-1)(-1)(+1)(-1)$
28. $(-5)(+5)(-2)(-2)$
29. $(+1)(+2)(-4)(+4)$
30. $(+7)(-2)(-3)(+5)$

27. Division of Signed Numbers. The quotient of two numbers is such a number that if we multiply it by the divisor, we obtain the dividend. Thus, from the discussion on multiplication, we have

Since $(+2) \times (+3) = +6$, then $(+6) \div (+2) = +3$.
Since $(-2) \times (+3) = -6$, then $(-6) \div (-2) = +3$.
Since $(+2) \times (-3) = -6$, then $(-6) \div (+2) = -3$.
Since $(-2) \times (-3) = +6$, then $(+6) \div (-2) = -3$.

Rule of Signs for Division

1. ***When the dividend and divisor have like signs, the quotient is positive.***
2. ***When they have unlike signs, the quotient is negative.***

EXERCISE 14

Divide:

1. $(+12) \div (-2)$
2. $(-6) \div (-3)$
3. $(-14) \div (+7)$
4. $(-63) \div (-9)$
5. $(+27) \div (+3)$
6. $(+36) \div (-4)$
7. $(-25) \div (+5)$
8. $(-92) \div (-46)$
9. $\dfrac{-28}{-4}$
10. $\dfrac{-42}{+6}$
11. $\dfrac{+87}{-3}$
12. $\dfrac{+121}{+11}$
13. $\dfrac{-84}{+4}$
14. $\dfrac{-56}{-8}$
15. $\dfrac{+69}{-23}$
16. $\dfrac{-95}{+19}$

Evaluate:

17. $\dfrac{(-6)(+2)}{-4}$
18. $\dfrac{(-9)(-6)}{-27}$
19. $\dfrac{(+12)(+3)}{-9}$
20. $\dfrac{(-10)(+2)}{+4}$
21. $\dfrac{-36}{(-6)(+2)}$
22. $\dfrac{+60}{(+3)(-5)}$
23. $\dfrac{-48}{(-4)(-6)}$
24. $\dfrac{-32}{(+2)(+4)}$
25. $\dfrac{(+12)(-5)}{(-2)(-3)}$

26. $\frac{(-6)(-8)}{(-4)(+3)}$ 27. $\frac{(-14)(-10)}{(+7)(+5)}$ 28. $\frac{(+9)(-8)}{(-6)(+3)}$

29. $\frac{(-10)(-6)}{(-5)(+4)}$ 30. $\frac{(+12)(-6)}{(-8)(-9)}$

28. Numerical Reductions. In more difficult arithmetic problems necessitating the use of more than one of the four fundamental operations, it is advisable to simplify one step at a time. Unless brackets, braces, or denominator signs indicate otherwise, always remember that in arithmetic the operations of multiplication and division are to be completed before addition and subtraction.

Example 1. Find the value of $(-3)(-5) - (-4)(+2)$.
Solution. Since $(-3)(-5) = +15$ and $(-4)(+2) = -8$,

$$(-3)(-5) - (-4)(+2) = (+15) - (-8) = 23.$$

Example 2. Evaluate $[(-3) + (+7)][(-5) - (-8)]$.
Solution. Since $(-3) + (+7) = +4$ and $(-5) - (-8) = +3$,

$$[(-3) + (+7)][(-5) - (-8)] = [+4][+3] = 12.$$

Example 3. Evaluate $\frac{(-4) - (+8)}{-2} - \frac{(-3)(-12)}{6 + (-2)}$.

Solution.

$$\frac{(-4) - (+8)}{-2} - \frac{(-3)(-12)}{6 + (-2)} = \frac{-12}{-2} - \frac{+36}{+4} = 6 - 9 = -3.$$

EXERCISE 15

Evaluate:

1. $(-3)(+5) + (-4)(-3)$
2. $(-2)(-4) - (-3)(-5)$
3. $-2(-3) + 5(-7)$
4. $6(-2) - (+4)(-3)$
5. $3(-2)^2 + 5(-2) - 7$
6. $2(-3)^2 - 5(-3) + 2$
7. $-(-4)^2 - 7(-4) + 1$
8. $6 - 3(-1) - 5(-1)^2$
9. $\frac{(-6)}{(+2)} - \frac{+12}{(-3)}$
10. $\frac{18}{(-3)} + \frac{(-15)}{+5}$
11. $\frac{9}{(+3)} + \frac{(-12)}{-2}$
12. $\frac{-21}{-7} - \frac{-27}{-9}$
13. $\frac{(-4)}{(+2)} \div \frac{(-24)}{(-12)}$
14. $\frac{24}{-3} \div \frac{(-20)}{5}$
15. $\frac{-35}{-7} \div \frac{2}{-14}$
16. $\frac{13}{26} \div \frac{6}{(-36)}$
17. $[(-2) - (-3)] \cdot [5 - 10]$
18. $[(-5) + (-9)] \cdot [-2 + 7]$

19. $[(+7) - (-2)][-14 - (-5)]$

20. $[(-3) - (-5)][(-7) - (-4)]$

21. $[(-2) - 10] \div [(-7) - (-4)]$

22. $[(-1) - 11] \div [7 - (-5)]$

23. $[(-5) - (25)] \div [13 + (-7)]$

24. $[15 - (-13)] \div [-3 - 4]$

25. $\dfrac{(+5) - (-5)}{(-3) + (-2)}$

26. $\dfrac{17 - (-13)}{10 - (-5)}$

27. $\dfrac{-19 - (-31)}{1 + (-13)}$

28. $\dfrac{22 - (-14)}{-15 - (-6)}$

29. $\dfrac{(-4) + 2}{(-5) - (-3)}$

30. $\dfrac{3 - (-3)}{7 - (+3)} \times \dfrac{-14 - (-4)}{-4 - (-9)}$

31. $\dfrac{(-10) - (-2)}{+11 - 9} \times \dfrac{-12 - (-7)}{+18 + (-13)}$

32. $\dfrac{-7 - 13}{-8 + 10} \div \dfrac{-16 + (-14)}{-7 - (-1)}$

33. $\dfrac{(-4)(-3)}{+6} - \dfrac{-36}{+2(-9)}$

34. $\dfrac{10(-7)}{-15 - (-1)} - \dfrac{-14 - 6}{(-2)(-2)}$

35. $\dfrac{16}{2(-4)} + \dfrac{12 - (-8)}{-10}$

36. $\dfrac{(-1)(-18)}{11 - (+2)} + \dfrac{-3 - 9}{(-3)(-4)}$

37. $\dfrac{2[(-3) - (-12)]}{-1 - 1 - 1} + \dfrac{-2(-3)}{-6}$

38. $\dfrac{-3(-2) - 4}{(3 - 2)(-5)} \div \dfrac{(-1)(-1) + 1}{-5[(-3) + (-2)]}$

39. $\dfrac{(-1)(-2)(-3)}{-1 - (-2)} \times \dfrac{6[(-1)(-5) - (-7)]}{[(-1) + (-2)][5 + (-2)]}$

40. $\dfrac{6[(-3) - (-8)]}{-5[(-12) - (-9)]} - \dfrac{(-1)(-2)(-3)}{-1 - 2 - 3}$

29. Evaluation of Algebraic Expressions. To find the value of an algebraic expression when the letter or letters assume definite numerical values, we (1) substitute the numbers in the expression and (2) compute arithmetically the resulting value.

Example 1. Evaluate $3a^3$, when $a = -2$.
Solution. Substituting -2 for a, we have

$$3a^3 = 3(-2)^3 = 3(-2)(-2)(-2) = 3(-8) = -24.$$

Example 2. Evaluate $a^2b - 2b^2$, when $a = 3$ and $b = -2$.
Solution. Substituting for a and b, we have

$$\begin{aligned} a^2b - 2b^2 &= (3)^2(-2) - 2(-2)^2, \\ &= (9)(-2) - 2(4), \\ &= -18 - 8, \\ &= -26. \end{aligned}$$

Example 3. Evaluate $(a - 2b)(3a + b)$, when $a = -2$ and $b = 5$.

Solution.

$$\begin{aligned}(a - 2b)(3a + b) &= [(-2) - 2(5)]\,[3(-2) + (5)],\\ &= [-2 - 10]\,[-6 + 5],\\ &= [-12]\,[-1],\\ &= 12.\end{aligned}$$

NOTE. Since parentheses, brackets, and braces are merely symbols of grouping we may interchange one set for the other if we choose to do so. It was convenient to do this in the above problem in order to avoid confusion with too many parentheses.

Example 4. Evaluate $x^4 + 3x^3 - 2x^2 - 7x + 3$, when $x = -2$.

Solution. Substituting, we have

$$\begin{aligned}x^4 + 3x^3 - 2x^2 - 7x + 3 &= (-2)^4 + 3(-2)^3 - 2(-2)^2 - 7(-2) + 3,\\ &= (16) + 3(-8) - 2(4) - 7(-2) + 3,\\ &= 16 - 24 - 8 + 14 + 3,\\ &= 1.\end{aligned}$$

EXERCISE 16

1. Find $4a^3$ when $a = -2$.
2. Find $5a^2$ when $a = 4$.
3. Find $2xy$ when $x = -2,\ y = -5$.
4. Find xy^2 when $x = -3,\ y = 2$.
5. Find y^3 when $y = -3$.
6. Find x^2y^2 when $x = 3,\ y = -1$.
7. Find $x^2 - y^2$ when $x = -2,\ y = 3$.
8. Find $a^3 - 8$ when $a = -2$.
9. Find $2a^2 + 5$ when $a = 2$.
10. Find $x^3 + y^3$ when $x = -3,\ y = 3$.
11. Find $a^2 + ab + b^2$ when $a = 3,\ b = -2$.
12. Find $(a - b)(a + b)$ when $a = 7,\ b = -3$.
13. Find $(a + b)^3$ when $a = 1,\ b = -2$.
14. Find ab^2c^3 when $a = 3,\ b = 2,\ c = -1$.
15. Find $(abc)^2$ when $a = -1,\ b = 3,\ c = -2$.
16. Find $3x - 5(x + 1)$ when $x = -4$.
17. Find 2^x when $x = 3$.
18. Find 3^a when $a = 2$.
19. Find $5c^2 - a^2c$ when $a = 3,\ c = -2$.
20. Find $7pq - 3q^3$ when $p = 5,\ q = -2$.
21. Find $\frac{1}{2}at^2$ when $a = 4,\ t = 3$.

22. Find $\frac{5}{9}(F - 32)$ when $F = 5$.
23. Find $a + (n - 1)d$ when $a = 5$, $n = 10$, $d = -2$.
24. Find ar^7 when $a = \frac{1}{16}$, $r = -2$.
25. Find $3x^3 - 2x^2 + 5x - 6$ when $x = -2$.
26. Find $4x^4 - 3x^3 - 2$ when $x = -1$.
27. Find $2x^3 - 3x^2 - 3x - 1$ when $x = 2$.
28. Find $3x^3 - 7x + 15$ when $x = -3$.
29. Find $4x^4 - 3x^3 + 2x^2 + 7x + 2$ when $x = -1$.
30. Find $x^4 - 5x^2 - 7x - 15$ when $x = 3$.
31. Find $\dfrac{x^3 + y^3}{x + y}$ when $x = 2$, $y = -3$.
32. Find $\dfrac{x^2 + y^2}{xy}$ when $x = 2$, $y = -2$.
33. Find $\dfrac{a^3 + 2ab}{a^3 - 2ab}$ when $a = 2$, $b = 3$.
34. Find $\dfrac{a^2 + ab + b^2}{a^2 - ab + b^2}$ when $a = 3$, $b = -2$.
35. Find $\dfrac{1}{x} + \dfrac{1}{y}$ when $x = 2$, $y = -2$.
36. Find $\dfrac{2}{x} - \dfrac{5}{y}$ when $x = -1$, $y = -10$.
37. Find $x - \dfrac{3}{x} + 1$ when $x = 6$.
38. Find $x^2 + \dfrac{7}{x^2} - 15$ when $x = 2$.
39. Find $x^2 - y$ when $x = 2.1$, $y = 3.7$.
40. Find $x^3 - y^3$ when $x = 0.1$, $y = -0.3$.
41. Find abc when $a = 1.7$, $b = 1.8$, $c = 1.9$.
42. Find $ab - c$ when $a = 2.5$, $b = 7$, $c = -15.05$.
43. Find $(a + b)(a - 3b)$ when $a = 3.5$, $b = 2.4$.
44. Find $x + \dfrac{1}{x}$ when $x = 1.25$.
45. Find $x - \dfrac{1}{2x}$ when $x = -0.8$.
46. Find $(a - 2b)^2$ when $a = -0.4$, $b = 0.03$.
47. Find $\dfrac{6}{c - 5} + \dfrac{2}{5 - c}$ when $c = -3$.
48. Find $\dfrac{18}{x^2 - 1} - \dfrac{x}{x + 1}$ when $x = -2$.

49. Find $\dfrac{x+3}{x-8} - \dfrac{2x^2-3}{x^2-7x-5}$ when $x = 7$.

50. Find $\dfrac{3x-5}{x+5} - \dfrac{x^2-1}{x^2+4x-5}$ when $x = -3$.

30. Multiplication of Monomials. Since, by definition, $a^3 = aaa$ and $a^4 = aaaa$, it follows that

$$a^3 \times a^4 = aaa \times aaaa = aaaaaaa = a^7 = a^{3+4}.$$

That is, *the exponent of a letter in a product is the sum of its exponents in the factors of the product.* This is called the ***law of exponents for multiplication.***

Example 1. Multiply $3a^2$ and $5a^3$.
Solution.

$$3a^2 \times 5a^3 = 3aa \times 5aaa = 3 \times 5aaaaa = 15a^5.$$

When the monomials to be multiplied contain powers of different letters, a similar method is used.

Example 2. Multiply $6a^2x^3$ and $7a^2bx$.
Solution.

$$6a^2x^3 \times 7a^2bx = 6aaxxx \times 7aabx = 6 \times 7aaaabxxxx = 42a^4bx^4.$$

NOTE. The exponents of one letter cannot be combined in any way with those of another. Thus the expression $42a^4bx^4$ admits of no further simplification.

Rule for Multiplication of Monomials

To multiply two monomials, (1) multiply the coefficients together and (2) prefix their product to the product of the different letters, giving each letter an exponent equal to the sum of the exponents which that letter has in the separate factors.

Applying this rule to the product of Example 2, we have

$$6a^2x^3 \times 7a^2bx = 6 \times 7a^{2+2}bx^{3+1} = 42a^4bx^4.$$

The above rule also applies if we multiply more than two monomials.

Example 3. Find the product of $-2x^2y^3$, $3y^2z$, and $-xyz$.
Solution. Following the above rule, we have

$$(-2x^2y^3)(3y^2z)(-xyz) = (-2)(3)(-1)x^{2+1}y^{3+2+1}z^{1+1} = 6x^3y^6z^2.$$

EXERCISE 17

Multiply:

1.	2.	3.	4.	5.	6.
$4x$	$-7x$	$2x$	$10x$	$-8xy$	$7x$
-2	3	$7x$	$-3x$	$-2x$	$2xy$

7.	8.	9.	10.	11.	12.
$3xy^2$	a^2b	$4a$	$-5xy^2$	$3pq$	$-6st^2$
$-xy$	$-2b$	$-6ab$	$-x^2y$	$-p$	$-2st$

13. $(5x)(-3x)$
14. $(-7y)(-2y)$
15. $(-9x)(-x^2)$
16. $(-xy)(3x^2)$
17. $(4x^2)(-3y^2)$
18. $(-xy)(-xy)$
19. $(-3abc)(2ab^2)$
20. $(a^2b)(-7ac)$
21. $(-2xy^2z)(3x^2z)$
22. $(9xy)(-3xz)$
23. $(7uvw^2)(9uv)$
24. $(-7ab)(2xy)$
25. $(6x)(-3x)(-2x)$
26. $(5a)(-a)(-3a)$
27. $(-2ab)(-2a)(-2b)$
28. $(4xy)(-3x^2)(3y)$
29. $(-3x)(2y)(-xy)$
30. $(-4xy)(-x)(-7y^2)$
31. $(7ab)(-3ac)(-2bc)$
32. $(-2a^2)(-3ab)(8ab)$
33. $(6x)(-3y)(-4z)$
34. $(-2a)(7b)(-8c)$
35. $(9abc)(-6bc)(2c)$
36. $(5xy)(2xz)(-xyz)$
37. $(2x)(-3x^2)(-5)(-x)$
38. $(-4a)(-6)(-a^2)(2a)$
39. $(3xy)(-4x)(3y)(5xy^2)$
40. $(-2x)(9yz)(-3xz)(-y)$
41. $(6ab)(-7b)(-a)(-b)$
42. $(-3x)(-5y)(-xy)(-2x)$
43. $(9x^2)(-5xy)(2y^2)(-x)$
44. $(-3a^2b)(-a)(5b^2)(-3ab)$
45. $(-3bc)(-ac)(abc)(-4b)$
46. $(-3xy)(-3xy)(-3xy)(2x)$
47. $(-9xyz)(-2xy)(-2xy)(3yz)$
48. $(8ac)(-6ab)(-6ab)(-c)$

31. Distributive Law for Multiplication. This law states that *the product of a multiple expression and a single factor is the algebraic sum of the partial products of each term of the multiple expression by that factor.* Let a, b, and c be any numbers; then algebraically the distributive law states that

$$a(b+c) = ab+ac.$$

Illustration. Either $5(6+7) = 5 \cdot 13 = 65$,
or $5 \cdot 6 + 5 \cdot 7 = 30 + 35 = 65$.

Example 1. Find the product of 3 and $2x - 5y$.

Solution. Multiplying each term of the binomial by 3, we have

$$3(2x - 5y) = 6x - 15y.$$

It is sometimes convenient to multiply in vertical order, as in the following example.

Example 2. Multiply $3x - 1$ by $-2x$.

Solution. Multiplying each term of the binomial by $-2x$ gives

$$\begin{array}{r} 3x - 1 \\ -\ 2x \\ \hline -6x^2 + 2x \end{array}.$$

The above procedure also applies in multiplying a monomial and a multinomial.

Example 3. Multiply -2 and $3x - 6y + 2z$.

Solution. Multiplying each term of the trinomial by -2, we have

$$\begin{array}{r} 3x - \ \ 6y + 2z \\ -\ 2 \\ \hline -6x + 12y - 4z \end{array}$$

Example 4. Find the product of $3ab$ and $4a^2 - 6ab + 3a - b$.

Solution. Finding each partial product in turn, we have

$$3ab(4a^2 - 6ab + 3a - b) = 12a^3b - 18a^2b^2 + 9a^2b - 3ab^2.$$

EXERCISE 18

Multiply in the manner indicated:

1. $\begin{array}{r} 3x - 2y \\ 7 \\ \hline \end{array}$
2. $\begin{array}{r} 4a - b \\ 3 \\ \hline \end{array}$
3. $\begin{array}{r} 6a + 7 \\ -2 \\ \hline \end{array}$
4. $\begin{array}{r} -5x + 6y \\ -9 \\ \hline \end{array}$
5. $\begin{array}{r} 4x^2 - 7x \\ -2x \\ \hline \end{array}$
6. $\begin{array}{r} 5x - 3 \\ -9x \\ \hline \end{array}$
7. $\begin{array}{r} 5ab + 3b^2 \\ 2ab \\ \hline \end{array}$
8. $\begin{array}{r} a^2 - 5ab \\ -7a \\ \hline \end{array}$
9. $5(3p - 7q)$
10. $8(5y - 1)$
11. $-2x(6x - 5y)$
12. $4x^2(2y - x)$
13. $5ac(2a - 7c)$
14. $-6ab(ab - c)$
15. $-5b(2a - b)$
16. $-6xy(5x^2 - y^2)$
17. $7(3x - 5y - z)$
18. $\begin{array}{r} 2a - b - c \\ -3 \\ \hline \end{array}$
19. $\begin{array}{r} 2a^2 - 3ab + b^2 \\ -6a \\ \hline \end{array}$
20. $\begin{array}{r} 2x - y - 7 \\ -7xy \\ \hline \end{array}$
21. $\begin{array}{r} 9p + 6q - r \\ 6qr \\ \hline \end{array}$
22. $\begin{array}{r} 3x^2 - 2xy^2 - y \\ -2y^2 \\ \hline \end{array}$
23. $\begin{array}{r} x^2 - x^2y - x^2y^2 \\ -xy \\ \hline \end{array}$

24. $\begin{array}{r} -3a^3 + 6a^2b - b^3 \\ 3ab^2 \\ \hline \end{array}$

25. $\begin{array}{r} 3x^2 - 6y^2 - y \\ -2xy \\ \hline \end{array}$

26. $-2x^2(3xy - 2x - y)$

27. $3ab^2(4a - 9b^2 + 2ab)$

28. $7p^2q^2(-p^2 + 3pq - 1)$

29. $-st(3s - 2t - 1)$

30. $-2x^3(9x^2 - 2x - 1)$

31. $-5y(6y^2 - 3y + 1)$

32. $\begin{array}{r} 3ab - a + 1 \\ -ab \\ \hline \end{array}$

33. $\begin{array}{r} 5x^2 - 6xy - 9y - 4 \\ -2x \\ \hline \end{array}$

34. $\begin{array}{r} -5a^3 - 6ab^2 - 6b - a \\ -ab \\ \hline \end{array}$

35. $\begin{array}{r} -9pq - 6p + 2q - 1 \\ -7pq^2 \\ \hline \end{array}$

36. $\begin{array}{r} 3xy - 6x - 7y + 9 \\ 3x \\ \hline \end{array}$

37. $\begin{array}{r} 3a - 6b + 5ab - 2 \\ 7a^2 \\ \hline \end{array}$

38. $\begin{array}{r} pq - 6p^2 - q^2 - 1 \\ -pq \\ \hline \end{array}$

39. $\begin{array}{r} 3x^3 - 6x^2 - 2x - 5 \\ -2x \\ \hline \end{array}$

40. $-8a^2(-3a^3 - a^2 + 5a - 7)$

41. $-3x(2x - 3y + xy - 7)$

42. $5ab(7a^2 - 3ab^2 - 4a - b)$

43. $9q^2(3pq - 6p^2 - 2q - 1)$

44. $-x^3(5x^4 - 3x^2 - 2x - 1)$

45. $-2a^3(4a^3 - 2a^2 - 3a + 1)$

46. $-uvw(uv - vw - uw + 1)$

47. $3x^2y(6xz - 12z^2 - 3x - z)$

48. $-5yz^2(-x + x^2 - 4xyz - 2x^2z)$

32. Multiplication of Multinomials. Let a, b, c, and d represent any numbers and consider the product $(a + b)(c + d)$. Assuming the quantity $(a + b)$ to be a fixed single quantity and $(c + d)$ to be a binomial, we have by the distributive law for multiplication

$$(a + b)(c + d) = (a + b)c + (a + b)d.$$

Now considering $(a + b)$ as a binomial, we may apply the distributive law again to each term of the above result. Thus, we obtain

$$(a + b)c + (a + b)d = ac + bc + ad + bd.$$

In this way we see that the product of two binomials equals the sum of the partial products obtained in multiplying each term of one by each term of the other.

Likewise, we may prove a similar result for the product of any two multinomials. Hence, we have the following

Rule for Multiplication of Multinomials

The product of two multinomials is the algebraic sum of all the partial products obtained in multiplying every term of one multinomial by each term of the other.

Example 1. Multiply $2x - 3$ by $3x + 2$.

Solution. Multiplying each of the first two terms by each of the second two terms, we have

$$\begin{aligned}(2x - 3)(3x + 2) &= (2x)(3x) + (2x)(2) + (-3)(3x) + (-3)(2),\\ &= 6x^2 + 4x - 9x - 6,\\ &= 6x^2 - 5x - 6.\end{aligned}$$

It is sometimes more convenient to do the multiplication in vertical order.

Example 2. Multiply $3x^2 - 5x - 2$ by $3x - 4$.

Solution. Arranging vertically, we have

$$\begin{array}{lrrrr} & 3x^2 & -\ 5x & -\ 2 & \\ & & 3x & -\ 4 & \\ \hline \text{Multiplying by } 3x, & 9x^3 & -\ 15x^2 & -\ 6x & \\ \text{Multiplying by } -4, & & -\ 12x^2 & +\ 20x & +\ 8 \\ \hline \text{Adding,} & 9x^3 & -\ 27x^2 & +\ 14x & +\ 8 \end{array}$$

Notice that after multiplying by the individual terms, we place only the similar terms in the same column. This facilitates the addition of the terms.

EXERCISE 19

Multiply in the manner indicated:

1. $\begin{array}{r} 2x - 3 \\ 3x + 5 \\ \hline \end{array}$ **2.** $\begin{array}{r} 4a - 1 \\ 2a + 7 \\ \hline \end{array}$ **3.** $\begin{array}{r} 3m + 2n \\ m - 4n \\ \hline \end{array}$

4. $\begin{array}{r} -2y + 3 \\ y - 1 \\ \hline \end{array}$ **5.** $\begin{array}{r} 5x^2 - 2y^2 \\ x^2 - y^2 \\ \hline \end{array}$ **6.** $\begin{array}{r} s^2 - 3t \\ -s^2 + t \\ \hline \end{array}$

7. $(3b + 7)(b - 5)$ **8.** $(6m - n)(2m + 3n)$

9. $(a + b)(a + c)$ **10.** $(9x - 2)(5x - 3)$

11. $(7p - q)(p + q)$ **12.** $(5a - 2b)(5a + 2b)$

13. $\begin{array}{r} 3x^2 - 2x - 1 \\ 5x - 7 \\ \hline \end{array}$ **14.** $\begin{array}{r} 4a^2 - 3ab + 5b^2 \\ 3a + 4b \\ \hline \end{array}$

15. $\begin{array}{r} 3a - 5b + 7 \\ a - b \\ \hline \end{array}$

16. $\begin{array}{r} x^2 - xy + y^2 \\ x + y \\ \hline \end{array}$

17. $\begin{array}{r} 7m^2 - mn - 2n^2 \\ 5m - n \\ \hline \end{array}$

18. $\begin{array}{r} 5x - 7y - 1 \\ 3x - 8y \\ \hline \end{array}$

19. $(5x - 1)(3x^2 - 5x + 1)$

20. $(6a + 7b)(3a^2 - 2ab + b^2)$

21. $(m - n)(m^2 + mn + n^2)$

22. $(4x^2 - 3)(4x^4 - 3x^2 - 5)$

23. $(a^2 - b^2)(3a^4 - a^2b^2 - 4b^4)$

24. $(2x - y)(4x^2 - xy - 7y^2)$

25. $\begin{array}{r} 5x^2 - x - 2 \\ 2x^2 - 9x - 1 \\ \hline \end{array}$

26. $\begin{array}{r} -3m^2 - mn + 9n^2 \\ 4m^2 - mn - 3n^2 \\ \hline \end{array}$

27. $\begin{array}{r} a^2 - ab + b^2 \\ a^2 + ab + b^2 \\ \hline \end{array}$

28. $\begin{array}{r} 5a^2 + 6a - 3 \\ 3a^2 - 2a - 1 \\ \hline \end{array}$

29. $\begin{array}{r} 6m^2 - 3mn - 7n^2 \\ m^2 - 5mn + n^2 \\ \hline \end{array}$

30. $\begin{array}{r} 7p^2 - 9pq - q^2 \\ 2p^2 - pq - q^2 \\ \hline \end{array}$

31. $\begin{array}{r} 3x^3 + 2x^2 - x + 5 \\ 3x^2 - 2x - 1 \\ \hline \end{array}$

32. $\begin{array}{r} a^3 - a^2 + 2a + 4 \\ a^2 - a + 2 \\ \hline \end{array}$

33. $\begin{array}{r} 5p^3 + 5p^2 - 3p - 7 \\ p^2 - p + 1 \\ \hline \end{array}$

34. $\begin{array}{r} y^3 - 3y^2 + 3y - 1 \\ y^2 - 2y + 1 \\ \hline \end{array}$

35. $\begin{array}{r} 6x^3 - 3x^2y - 4xy^2 - y^3 \\ 3x^2 - 2xy + 5y^2 \\ \hline \end{array}$

36. $\begin{array}{r} 3x^3 - 4x^2y - 5xy^2 - 7y^3 \\ 9x^2 - 8xy - y^2 \\ \hline \end{array}$

33. Multiplication of Binomials by Inspection. Although the result of multiplying two binomial factors can always be obtained by the methods already explained, it is important that we be able to write such products rapidly by *inspection.*

In the example

$$\begin{aligned}(2x - 3)(x + 4) &= 2x^2 + 8x - 3x - 12, \\ &= 2x^2 + 5x - 12,\end{aligned}$$

observe that the first term of the result is the product of the first terms of the binomials, and also that the last term of the result is the product of the second terms of the binomials. The middle term of the result is the algebraic sum of the *outer* product and the *inner* product.

In general, the product of any two binomials such as the ones given above is a *trinomial*, whose terms are found as follows:

1. The first term is the product of the first terms of the two binomials.
2. The second term is the algebraic sum of the product of the two outer terms and the product of the two inner terms.
3. The third term is the product of the last terms of the binomials.

Following the above procedure, in squaring a binomial we have the formulas

$$(a + b)^2 = a^2 + 2ab + b^2,$$
$$(a - b)^2 = a^2 - 2ab + b^2.$$

Illustration. $(2x - 3)^2 = (2x)^2 - 2(2x)(3) + (3)^2,$
$= 4x^2 - 12x + 9.$

EXERCISE 20

Find each of the following products:

1. $(x + 5)(x - 3)$ **2.** $(x - 2)(x - 4)$ **3.** $(x + 4)(x + 6)$
4. $(x + 10)(x - 7)$ **5.** $(x - 1)(x + 6)$ **6.** $(2x + 3)(3x + 2)$
7. $(3x - 1)(2x - 1)$ **8.** $(x + 5)(3x - 2)$ **9.** $(4x + 3)(4x - 3)$
10. $(5x - 2)(x + 1)$ **11.** $(3x + 7)(x + 5)$ **12.** $(x + 6)(2x - 1)$
13. $(4x - 1)(5x - 1)$ **14.** $(6x - 7)(5x + 3)$ **15.** $(x + 11)(2x - 17)$
16. $(x + 3y)(5x - 6y)$ **17.** $(3x + 2y)(4x + 3y)$
18. $(x + 6y)(3x - 7y)$ **19.** $(5x + 9y)(3x - 5y)$
20. $(8x - 7y)(7x - 8y)$ **21.** $(5x - 6y)(6x + 7y)$
22. $(x - 7)^2$ **23.** $(x + 5)^2$ **24.** $(x + 2)^2$ **25.** $(x - 3)^2$
26. $(2x + 1)^2$ **27.** $(3x - 1)^2$ **28.** $(2x + 3)^2$ **29.** $(3x + 2)^2$
30. $(4x - 5)^2$ **31.** $(5x - 2)^2$ **32.** $(3x + 4)^2$ **33.** $(10x - 1)^2$
34. $(2x - y)^2$ **35.** $(4x + 3y)^2$ **36.** $(2x + 7y)^2$ **37.** $(5x + 4y)^2$

38. In multiplying the trinomial $a + b + c$ by itself, show that the product is given by the formula

$$(a + b + c)^2 = a^2 + b^2 + c^2 + 2ab + 2ac + 2bc$$

State this formula in words.

Find each of the following products, using the formula given in Problem 38:

39. $(x + y - z)^2$ **40.** $(x - 2y - z)^2$ **41.** $(2x - 3y + 2z)^2$
42. $(3x + y + 2z)^2$ **43.** $(5x + 3y - z)^2$ **44.** $(4x + 5y + 6z)^2$

34. Division of Monomials. We learned in arithmetic that in division we may remove any factors common to both the dividend and divisor.

Illustrations. (1) $$\frac{27}{9} = \frac{9 \cdot 3}{9 \cdot 1} = \frac{3}{1} = 3,$$

(2) $$\frac{40}{12} = \frac{4 \cdot 10}{4 \cdot 3} = \frac{10}{3}.$$

This principle also applies to general numbers; thus, since $a^5 = aaaaa$ and $a^3 = aaa$, it follows that

$$\frac{a^5}{a^3} = \frac{aaaaa}{aaa} = \frac{aa}{1} = a^2.$$

Again, to divide $35a^3b^3$ by $7ab^2$, we have

$$\frac{35a^3b^3}{7ab^2} = \frac{5 \cdot 7 \cdot aaa \cdot bbb}{7 \cdot a \cdot bb} = \frac{5aab}{1} = 5a^2b.$$

In each case we see that *the exponent of any letter in a quotient is the difference of the exponents of that letter in the dividend and in the divisor.* This is called the ***law of exponents for division.***

Rule for Division of Monomials

To divide two monomials, (1) divide their coefficients and (2) prefix their quotient to the quotient of the different letters. The exponent of each letter in the quotient is obtained by subtracting the exponent of that letter in the divisor from its exponent in the dividend.

Example 1. Divide $42x^4y^3z^6$ by $-7x^2yz^3$.

Solution. By the rule for division of monomials, we have

$$\frac{42x^4y^3z^6}{-7x^2yz^3} = (-6)x^{4-2}y^{3-1}z^{6-3} = -6x^2y^2z^3.$$

Example 2. Divide $-28a^3b^2$ by $-4ab^2$.

Solution.

$$\frac{-28a^3b^2}{-4ab^2} = 7a^2.$$

Note: If we apply the law of exponents to divide any power of a letter by the same power of the letter, we are led to a curious conclusion.

Thus by the law $b^2 \div b^2 = b^{2-2} = b^0$;

but we know that $b^2 \div b^2 = \dfrac{b^2}{b^2} = 1$.

Hence, b^0 must equal 1 regardless of what b may be. This result may appear strange at first, but its full significance will be seen later in the study of algebra.

35. Division of Multinomials by a Monomial. From the rule for multiplying multinomials by a monomial, we have the rule

Rule for Division of a Multinomial by a Monomial

To divide a multinomial by a monomial, divide each term of the multinomial separately by the monomial, and take the algebraic sum of the partial quotients so obtained.

Example 1. Divide $6x - 9y$ by -3.
Solution. Dividing each term of $6x - 9y$ by -3, we have

$$\frac{6x - 9y}{-3} = -2x + 3y.$$

Example 2. Divide $5x^4 - 3x^3 - \frac{1}{2}x$ by $\frac{1}{2}x$.
Solution. Dividing each term in turn by $\frac{1}{2}x$, we have

$$\frac{5x^4 - 3x^3 - \frac{1}{2}x}{\frac{1}{2}x} = 10x^3 - 6x^2 - 1.$$

EXERCISE 21

Divide:

1. $15x \div (-3)$ **2.** $-18a \div 9$
3. $-14a^2 \div 7a$ **4.** $21x^3 \div 3x$
5. $25x^2y \div (-5xy)$ **6.** $-63a^3b^2 \div (-7ab^2)$

7. $\dfrac{42p^3}{-6p^2}$ **8.** $\dfrac{-81xy^3}{27y^3}$ **9.** $\dfrac{-78abc^2}{3bc}$

10. $\dfrac{-28p^3q^2}{-4p^3q^2}$ **11.** $\dfrac{64x^3yz}{16x^2z}$ **12.** $\dfrac{-87a^2b^2c^2}{3ac}$

13. $(12m^2 - 18m^4) \div 6m^2$ **14.** $(a^4 - 6a^2b^2) \div a^2$
15. $(x^3 - x^2) \div (-x^2)$ **16.** $(6x - 5y) \div (-1)$
17. $(4xyz - 8x^2z) \div (-4xz)$ **18.** $(9m^2n^3 - 21m^3n^2) \div 3m^2n^2$

19. $\dfrac{8x^4 - 20x^2}{-4x^2}$

20. $\dfrac{-14a^2b^3 + 49a^4b^4}{-7a^2b^3}$

21. $\dfrac{16m^2n - 24m^2n^3}{8m^2n}$

22. $\dfrac{15x^2yz^2 + 25xy^2z}{-5xz}$

23. $\dfrac{-12a^3b^2 + 60a^4b}{-12a^3b}$

24. $\dfrac{14p^3q^3 + 42p^2q^2}{7pq^2}$

25. $(-3x^2y + 6xy - 12y) \div (-3y)$

26. $(7a^2b^2c^3 - 7a^2bc^2 - 14ac^3) \div 7ac$

27. $(5mn^3 - 10m^2n^2 - 15m^2n^3) \div 5mn^2$

28. $(12x^2y^2 - 6x^3y^4 + 18x^2y^3) \div (-6x^2y)$

29. $(abc^2 - ab^2c + a^2bc) \div (-abc)$

30. $(2m^3 - 6m^2 + 10mn) \div (-2m)$

31. $\dfrac{21x^2y^2z^2 - 35x^3y^2z + 63xy^2z^3}{-7xyz}$

32. $\dfrac{-18a^3b + 33a^2c - 27a^3bc}{-3a^2}$

33. $\dfrac{-7x^2y^2 + 5x^2z^2 - 3y^2z^2}{-1}$

34. $\dfrac{20p^2q^3 - 16p^2q^2 - 48p^2q}{4p^2q}$

35. $\dfrac{-18m^3n^4 - 30m^4n^3 - 42m^3n^2}{-6m^2n^2}$

36. $\dfrac{15x^2y^2z^4 - 25x^3y^2z^2 - 20x^2y^3z^2}{-5x^2y^2z^2}$

37. $\dfrac{\frac{3}{4}x^2 - \frac{1}{2}x}{-\frac{1}{4}x}$

38. $\dfrac{\frac{2}{3}a^2b^2 - \frac{5}{6}ab^3}{\frac{1}{6}ab^2}$

39. $\dfrac{\frac{1}{3}m^4n^3 - \frac{2}{3}m^3n^4 - \frac{1}{2}m^2n^2}{\frac{1}{6}m^2n^2}$

40. $\dfrac{-3x^4 + x^3 - \frac{1}{2}x^2}{-\frac{1}{3}x^2}$

41. $\dfrac{a^2 - \frac{1}{2}ab - \frac{1}{3}ac}{-\frac{1}{4}a}$

42. $\dfrac{m^2 - \frac{1}{5}mn + n^2}{-\frac{1}{3}}$

43. $\dfrac{1.5x^2y - 2.7xy^2 - 2.1xy}{0.3xy}$

44. $\dfrac{abc - 5a^2c + abc^2}{-0.5ac}$

45. $\dfrac{-1.4m^2 + 2.8mn - 4.9n^2}{-0.7}$

46. $\dfrac{1.62x^5 - 3.2x^4 - 5.02x^3}{0.2x^3}$

47. $\dfrac{-4.4xz^2 - 2.8x^2z - 5.2x^2z^2}{-0.4xz}$

48. $\dfrac{9.3x^2y^2 - 1.5x^3y + 5.1x^2y^3}{-0.3x^2y}$

36. Division of Multinomials. In Example 1, p. 57, we saw that

$$(2x - 3)(3x + 2) = 6x^2 - 5x - 6.$$

By the meaning of division, it follows that

$$\frac{6x^2 - 5x - 6}{2x - 3} = 3x + 2.$$

We would like to obtain this result without resort to the first equation. The direct process by which the quotient may be obtained is very similar to the process of long division in arithmetic. Study the following example very carefully.

Example 1. Divide $6x^2 - 5x - 6$ by $2x - 3$.

Solution. Arrange the work thus:

$$\begin{array}{rl} & 3x + 2 \qquad \text{Quotient} \\ \text{Divisor } 2x - 3 & \overline{)6x^2 - 5x - 6} \text{ Dividend} \\ & \underline{6x^2 - 9x} \\ & \qquad 4x - 6 \\ & \qquad \underline{4x - 6} \end{array}$$

Explanation. (1) Divide $6x^2$ (first term of the dividend) by $2x$ (first term of the divisor) to obtain $3x$ (first term of the quotient).

(2) Multiply the whole divisor, $2x - 3$, by the $3x$ in the quotient and write the product under the similar terms of the dividend.

(3) Subtract to obtain the new dividend, $4x - 6$.

(4) Divide $4x$ (first term of the new dividend) by $2x$ (first term of the divisor) to obtain $+2$ (the second term of the quotient).

(5) Multiply the whole divisor by the new term of the quotient and subtract.

Example 2. Divide $7a + 3a^3 - 5 - 6a^2$ by $a - 3$.

Solution. To avoid confusion in long division, it is essential to arrange the dividend and divisor in descending powers of the same variable:

$$\begin{array}{rl} & 3a^2 + 3a + 16 \qquad \text{Quotient} \\ a - 3 & \overline{)3a^3 - 6a^2 + 7a - 5} \\ & \underline{3a^3 - 9a^2} \\ & \qquad 3a^2 + 7a \\ & \qquad \underline{3a^2 - 9a} \\ & \qquad\qquad 16a - 5 \\ & \qquad\qquad \underline{16a - 48} \\ & \qquad\qquad\qquad 43 \text{ Remainder} \end{array}$$

When dividing in arithmetic, we often have a remainder; thus $63 \div 17$ gives us a quotient of 3 and a remainder of 12. In other words

$$\frac{63}{17} = 3 + \frac{12}{17}$$

In the same manner, the result of the above example means

$$\frac{3a^3 - 6a^2 + 7a - 5}{a - 3} = 3a^2 + 3a + 16 + \frac{43}{a - 3}$$

From these examples, the method used in dividing one multinomial by another may be summarized as follows:

1. Arrange the terms of both dividend and divisor according to descending powers of some common letter.
2. Divide the first term of the dividend by the first term of the divisor to obtain the first term of the quotient.
3. Multiply the divisor by the term of the quotient obtained in step 2 and subtract this product from the dividend to obtain a new dividend.
4. Repeat steps 2 and 3 until a remainder is obtained that is either zero or is of lower degree than the divisor.

Example 3. Divide $5x^2 - 6x^3 - 4$ by $3x - 2x^2 - 2$.

Solution. When certain terms in the descending order of a letter are missing, leave a blank space to indicate where they should be. Thus, following the steps as outlined above, we have

$$\begin{array}{r|l} & 3x + 2 \\ \hline -2x^2 + 3x - 2 & -6x^3 + 5x^2 \qquad\quad - 4 \\ & \underline{-6x^3 + 9x^2 - 6x} \\ & \qquad -4x^2 + 6x - 4 \\ & \qquad \underline{-4x^2 + 6x - 4} \end{array}$$

EXERCISE 22

Divide:

1. $6x^2 + 11x + 3$ by $3x + 1$
2. $a^2 + a - 6$ by $a + 3$
3. $12m^2 + 7mn - 12n^2$ by $4m - 3n$
4. $6x^2 - xy - 2y^2$ by $2x + y$
5. $8a^4 - 8a^2 - 6$ by $2a^2 - 3$
6. $9y^2 + 24y + 16$ by $3y + 4$
7. $5x^2 - 6x - 4$ by $x - 1$
8. $3a^2 + 7a - 7$ by $3a - 2$
9. $8a^2 + 14ab - 15b^2$ by $4a - 3b$
10. $2m^2 - 17mn + 35n^2$ by $m - 5n$
11. $6p^2 - 13pq + 5q^2$ by $3p - 5q$
12. $12x^2 + 6xy - 5y^2$ by $2x + y$
13. $4x^3 - 5x^2 + 2x - 16$ by $x - 2$
14. $a^3 - 4a^2 - 3a + 2$ by $a + 1$
15. $53x^2 - 30x + 15x^3 - 8$ by $3x - 2$

16. $2mn^2 + m^3 + 9m^2n - 48n^3$ by $m - 2n$
17. $2p^3 - 33pq^2 + 20q^3 + 5p^2q$ by $2p - 5q$
18. $5x^3 + 7x^2 - 2$ by $x + 1$
19. $4a^3 - 3a + 1$ by $2a - 1$
20. $4x^3 - 21xy^2 + 4x^2y + 9y^3$ by $2x - 3y$
21. $14a^3 + 17a^2b + 9b^3$ by $2a + 3b$
22. $8p^3 - 27q^3$ by $2p - 3q$
23. $y^3 - y^2 + y - 1$ by $y + 1$
24. $a^3 + 2a^2 + 3a + 4$ by $a + 2$
25. $x^4 - 7x^3 + 8x^2 + 28x - 48$ by $x - 3$
26. $9x^4 - 6x^3 + 13x^2 - 16x + 4$ by $3x - 1$
27. $a^4 - 6a^2b^2 + 8b^4 - a^3b + 4ab^3$ by $a - 2b$
28. $2m^4 - 3m^3n - 16mn^3 + 24n^4$ by $2m - 3n$
29. $24x^4 - 2x + 4x^3 - 1$ by $2x - 1$
30. $x^4 - a^4$ by $x - a$
31. $x^3 - 6x^2 + 7x + 4$ by $x^2 - 2x - 1$
32. $4x^3 - 4x^2 - 5x + 3$ by $2x^2 - x - 3$
33. $a^3 - 3a^2b + 3ab^2 - b^3$ by $a^2 - 2ab + b^2$
34. $m^3 + 8n^3$ by $m^2 - 2mn + 4n^2$
35. $4y^3 - 13y - 6$ by $2y^2 - 2 - 3y$
36. $a + 6 - a^3$ by $3 + 2a + a^2$
37. $x^4 + x^2y^2 + y^4$ by $x^2 + xy + y^2$
38. $m^4 - 2m^2 + 1$ by $m^2 - 2m + 1$
39. $6x^4 + 13x - 11x^3 - 10 - x^2$ by $3x^2 - 5 - x$
40. $4x^4 - 9x^2y^2 - 3y^4 - 8x^3y + 16xy^3$ by $xy + 2x^2 - 3y^2$

REVIEW OF CHAPTER III

A

Evaluate the following numerical expressions:

1. $(-2)(-3)(-4)(+5)$
2. $(+8) \div (-2)$
3. $-(-2)^5$
4. $-(-2)^3(-3)^2$
5. $\dfrac{-24}{(-3)(+2)}$
6 $(-2)(-3) - (-4)(-5)$
7. $3(-2)^2 - 7(-2) - 5$
8. $\dfrac{-6}{+2} - \dfrac{12}{-3}$
9. $[(-1) - (+7)][(-2) + (-3)]$
10. $[(-2) + (+8)] \div [(-3) - (-6)]$

Evaluate the following for the values indicated:

11. $5b^2$ when $b = -3$
12. x^2y^3 when $x = 5,\ y = -2$
13. $2x + 7$ when $x = -5$
14. $a^2 - b^2$ when $a = 5,\ b = -3$

15. $m^2 - m - 2$ when $m = -3$

16. $\frac{1}{x} - \frac{1}{y}$ when $x = -3,\ y = -2$

17. $x - \frac{1}{x}$ when $x = -\frac{2}{3}$

18. $p^2 + 5pq - 6q^2$ when $p = -4,\ q = 2$

19. $(a - b)(a + 2b)$ when $a = 1,\ b = -4$

20. $\frac{x^3 - y^3}{x - y}$ when $x = -5,\ y = -3$

B

Multiply:

1. $\begin{array}{r} -6xy \\ -2x \\ \hline \end{array}$

2. $\begin{array}{r} 4ab \\ -3ab \\ \hline \end{array}$

3. $\begin{array}{r} -5x \\ 2a \\ \hline \end{array}$

4. $(2x)(-3x)$

5. $(-6pq^2)(-5p^2q)$

6. $(ab)(-3ac)$

7. $(3xy)(-2xz)(-5y^2)$

8. $(-4a)(-5b)(-6c)$

9. $(-2uv^2)(-3u^2w)(vw)(-2uvw)$

10. $(5m^2n^3)(m^3n)(-3mn)(-2n^3)$

11. $\begin{array}{r} 6a - 5b \\ 3a \\ \hline \end{array}$

12. $\begin{array}{r} 2ab^2 + 3b \\ 5ab \\ \hline \end{array}$

13. $\begin{array}{r} -a^3 - ab^2 + 5b^2 \\ -a^2b \\ \hline \end{array}$

14. $\begin{array}{r} 2x^2 - 5x + 1 \\ -3x^2 \\ \hline \end{array}$

15. $5a(2a^2 - 3ab - b^2)$

16. $-3mn(3m - 4n + 5)$

17. $ab(5ab^2 - 3a^2b^2 + 2a^2b)$

18. $-2x(5x^2 - 6xy - 3y^2)$

19. $-3pq(4p^2 - 6pq + q^2 - 5)$

20. $2ab^2(3a - 2b^2 - a^2 + 3b - ab)$

21. $\begin{array}{r} 2a + 3b \\ a - b \\ \hline \end{array}$

22. $\begin{array}{r} 5x - y \\ 2x + 3y \\ \hline \end{array}$

23. $\begin{array}{r} x^2 + xy + y^2 \\ x - y \\ \hline \end{array}$

24. $\begin{array}{r} 3a^2 - 2a + 5 \\ a - 3 \\ \hline \end{array}$

25. $\begin{array}{r} 2m^2 - 3m + 5 \\ 2m^2 + 3m - 5 \\ \hline \end{array}$

26. $\begin{array}{r} 5x^2 - 4x - 2 \\ 2x^2 - 3x + 4 \\ \hline \end{array}$

27. $(x - 2y)(x^2 + 2xy - y^2)$

28. $(3a + b)(2a^2 - ab - b^2)$

29. $(3m - 2n)(m^2 - mn + n^2)$

30. $(5p - 2q)(4p^2 - 2pq - 3q^2)$

C

Multiply by inspection:

1. $(x + 2)(x + 3)$

2. $(x - 5)(x + 6)$

3. $(x + 10)^2$

4. $(x - 7)^2$

5. $(x - 5)(2x - 3)$

6. $(x + 2)(2x + 1)$

7. $(2x + 3)^2$

8. $(2x - 5)^2$

9. $(x - 1)(3x + 5)$

10. $(x + 3)(3x - 7)$

11. $(2x + y)(3x + y)$

12. $(3x - 2y)(2x + 3y)$

13. $(5x - y)^2$

14. $(2x + 7y)^2$

15. $(x + 2y)(5x + 3y)$

16. $(4x - 7y)(2x + 5y)$

17. $(3x - 7y)^2$

18. $(2x + y)^2$

19. $(10x - 3y)(10x + 7y)$

20. $(8x - 5y)(6x - 5y)$

D

Divide:

1. $\dfrac{-27a^2b}{3ab}$ **2.** $\dfrac{-12x^3}{-2x^2}$

3. $\dfrac{10m^3n^2}{-5mn}$ **4.** $\dfrac{-15x^2yz}{3xz}$

5. $18p^2q^4 \div (-6pq^2)$

6. $-21abc \div (7abc)$

7. $-14x^2y^2z^2 \div (-2xyz^2)$

8. $3.4mv^2 \div (1.7mv)$

9. $\frac{1}{2}m^5n^4 \div (-\frac{1}{4}m^2n^3)$

10. $-\frac{2}{3}ab^2c \div (-\frac{1}{6}ab)$

11. $\dfrac{8x^5 - 12x^3}{-4x^2}$

12. $\dfrac{-6a^2b^3 + 15a^3b}{-3a^2b}$

13. $\dfrac{10x^3y^4 - 25x^4y^5}{5x^3y^4}$

14. $\dfrac{3m^2n^3 - 9mn^2 - 6m^2n}{-3mn}$

15. $\dfrac{4p^2q^2 + 6p^3q^2 - 8pq^3}{-2pq^2}$

16. $(2.4a^3b - 1.6a^2b^2) \div (0.4a^2b)$

17. $(\frac{5}{6}x^2yz - \frac{2}{3}xy^2z) \div (-\frac{1}{6}yz)$

18. $(15a^2b^3c^4 - 20a^3b^2c^3 + 5a^2b^2c^2) \div (-5a^2b^2c^2)$

19. $(\frac{3}{4}p^5q^4 - \frac{1}{2}p^5q^6 - \frac{1}{4}p^4q^3) \div (-\frac{1}{4}p^3q^3)$

20. $(2.8x^2y^3 + 5.6x^3y^2 - 8.4x^2y^2) \div (1.4x^2y)$

Divide:

21. $x^2 + 3x - 40$ by $x - 5$

22. $m^2 - 7m - 18$ by $m + 2$

23. $4x^2 - 20xy + 25y^2$ by $2x - 5y$

24. $3x^2 + 20x + 12$ by $3x + 2$

25. $5c^2 - 12c + 4$ by $c - 2$

26. $x^3 - 2x^2 - 16x + 5$ by $x - 5$

27. $6a^3 - 7a^2x + x^3$ by $2a - x$

28. $6x^3 + 7x^2y - 4xy^2 - 4y^3$ by $3x + 2y$

29. $a^4 + 2a^3 + 4a - 4$ by $a^2 + 2$

30. $x^4 - 16x^2 + 8x - 1$ by $x^2 - 4x + 1$

CHAPTER IV

EQUATIONS AND STATED PROBLEMS

37. Equivalent Equations. We learned in Chapter I that a *conditional equation* or *equation* differed from an *algebraic identity* in that an equation reduced to a numerical identity only for certain values of the unknown, while an algebraic identity reduced to a numerical identity for *all* values of the unknown.

Illustration.

$x + 1 = 7$ is an equation, since it is true only for $x = 6$.
$7x - 3x = 4x$ is an identity, since it is true for any value of x.

Two equations having the same root (or roots) are said to be *equivalent equations.* Thus $x - 4 = 0$ and $3x = 12$ are equivalent equations, since they have one and the same solution, namely 4. On the other hand, $x = 4$ and $x^2 = 4x$ are not equivalent equations, since the first is true only for the value 4 and the second is true for the two values 4 and 0.

The fundamental law of equalities may be restated as follows:

If both sides of an equation are increased, decreased, multiplied, or divided by the same number (division by zero excluded), an equivalent equation is obtained.

The above principle is applicable regardless of whether the numbers are positive or negative, as is illustrated in the following examples.

Example 1. Solve the equation $x + 7 = 3$ and check your result.

Solution. $x + 7 = 3$, given equation;
$x = 3 - 7$, subtracting 7 from both sides;
$x = -4$, simplifying.

Check. $(-4) + 7 = 3$, substituting (-4) for x in the given equation:
$3 = 3$, simplifying.

Example 2. Solve the equation $-5x = 35$ and check your result.

Solution.

$$-5x = 35, \quad \text{given equation;}$$

$$x = \frac{35}{-5}, \quad \text{dividing both sides by } \mathbf{-5};$$

$$x = -7, \quad \text{simplifying.}$$

Check.

$$-5(-7) = 35, \quad \text{substituting } (-7) \text{ for } x \text{ in the given equation;}$$

$$35 = 35, \quad \text{simplifying.}$$

EXERCISE 23

Solve the following equations and check your results:

1. $x + 10 = 0$	**2.** $x - 42 = 42$	**3.** $\frac{x}{2} = 2\frac{1}{2}$
4. $7x = 0$	**5.** $9\frac{1}{3} + x = 3$	**6.** $\frac{1}{3}x = 4$
7. $-4x = 17$	**8.** $x - 8\frac{3}{4} = 6\frac{1}{2}$	**9.** $\frac{x}{1\frac{1}{2}} = 5$
10. $3\frac{3}{4} + x = 1\frac{1}{2}$	**11.** $x - 17 = -25$	**12.** $-7x = 49$
13. $5x = 4x - 5$	**14.** $x + 14 = -8$	**15.** $\frac{x}{\frac{3}{4}} = 2\frac{2}{3}$
16. $2\frac{1}{3}x = -16\frac{1}{3}$	**17.** $2\frac{1}{3} + x = \frac{1}{2}$	**18.** $\frac{1}{2}x = 5 - \frac{1}{2}x$
19. $x - 1.7 = 2.3$	**20.** $\frac{x}{2.3} = 7$	**21.** $2.7x = 24.3$
22. $x + 11.9 = 2.7$	**23.** $\frac{x}{4.5} = -2.1$	**24.** $x - 3.3 = -12.7$
25. $6.97 + x = 3.2$	**26.** $\frac{x}{-7.7} = -3.6$	**27.** $x - 2.005 = 3.26$
28. $1.21x = 2.904$	**29.** $1.01 + x = -1.1$	**30.** $x - 3.92 = -2.26$
31. $-8.9x = 9.79$	**32.** $\frac{x}{1.01} = -0.02$	**33.** $x - \frac{5}{6} = -12\frac{1}{2}$
34. $191.2 + x = 67.93$	**35.** $\frac{x}{3.11} = 129$	**36.** $5.001x = -555.111$
37. $x + 36.9 = 45.45$	**38.** $x - 23.6 = -161$	**39.** $-2.67x = 0.00534$

38. Transposition. In order to solve equations it is often necessary to add (or subtract) some expression to both sides of an equation in order to eliminate a term from one side of it. Thus, adding 3 to both sides of the equation $x - 3 = 5$ eliminates the 3 on the left side of the equation and gives us $x = 5 + 3$.

The net result of this operation is the same as though we had actually transferred the original 3 from the left member to the right member and changed its sign. In general, we have

Rule of Transposition

Any term may be moved from one member of an equation to the other member, provided that its algebraic sign is changed.

NOTE. The above rule applies to *terms* and *not to factors*. Thus, in the equation $-2x = 6$, we do *not* obtain $x = 6 + 2$ by transposition. It is evident that these two equations do not have the same root and so are not equivalent equations.

Example 1. In the equation $8x - 3 = 5x - 1$ transpose the terms $5x$ and -3.

Solution. By the rule of transposition,

$$8x - 5x = 3 - 1.$$

Example 2. In the equation,

$$5x - 8 - 6x + 2 = 3x - 6 - 2x + 9,$$

transpose terms so that all terms containing x are on the left side of the equation and all terms not containing x are on the right side.

Solution.

$$5x - 6x - 3x + 2x = -6 + 9 + 8 - 2.$$

39. General Solution of Linear Equations. By the process of transposition it is always possible to move all the terms containing the unknown to one side of the equation and all other terms to the opposite side of the equation. Since the terms on each side of the equation are then similar, we may combine them. This reduces to an equivalent equation of the form

$$ax = b,$$

where a and b represent the numbers obtained in collecting terms. The solution of this equation is readily obtained by dividing both sides of it by the number a.

Any equation which may be reduced to the form $ax = b$ is known as an **equation of the first degree** in one unknown or, more commonly, as a **linear equation** in one unknown.

The method to be followed in solving linear equations may be briefly summarized as follows:

1. Remove parentheses, if there are any in the equation.
2. Transpose all terms containing the unknown to one side of the equation and all other terms to the opposite side of the equation.
3. Collect like terms on each side of the equation.
4. Divide both sides of the equation by the coefficient of the unknown.

Example 1. Solve the equation $3(x - 5) = x - 2(x + 7)$ and check your result.

Solution.

$$
\begin{aligned}
3(x - 5) &= x - 2(x + 7), && \text{given equation;}\\
3x - 15 &= x - 2x - 14, && \text{step 1,}\\
3x - x + 2x &= 15 - 14, && \text{step 2,}\\
4x &= 1, && \text{step 3,}\\
x &= \tfrac{1}{4}, && \text{step 4.}
\end{aligned}
$$

Check.

$$
\begin{aligned}
3(\tfrac{1}{4} - 5) &= \tfrac{1}{4} - 2(\tfrac{1}{4} + 7), && \text{substituting } \tfrac{1}{4} \text{ for } x;\\
3(-\tfrac{19}{4}) &= \tfrac{1}{4} - 2(\tfrac{29}{4}), && \text{simplifying parentheses;}\\
-\tfrac{57}{4} &= \tfrac{1}{4} - \tfrac{58}{4}, && \text{removing parentheses;}\\
-\tfrac{57}{4} &= -\tfrac{57}{4}, && \text{simplifying.}
\end{aligned}
$$

Example 2. Solve the equation $(x - 2)(x + 5) - x^2 = 8$ and check your result.

Solution.

$$
\begin{aligned}
(x - 2)(x + 5) - x^2 &= 8, && \text{given equation;}\\
x^2 + 3x - 10 - x^2 &= 8, && \text{step 1,}\\
x^2 - x^2 + 3x &= 8 + 10, && \text{step 2,}\\
3x &= 18, && \text{step 3,}\\
x &= 6, && \text{step 4.}
\end{aligned}
$$

Check.

$$
\begin{aligned}
(6 - 2)(6 + 5) - (6)^2 &= 8, && \text{substituting 6 for } x;\\
(4)(11) - (6)^2 &= 8, && \text{simplifying parentheses;}\\
44 - 36 &= 8, && \text{removing parentheses;}\\
8 &= 8, && \text{simplifying.}
\end{aligned}
$$

Note. If the terms containing the unknown to powers higher than the first do not eliminate one another, the equation is *not* linear. Such equations are discussed later.

Example 3. Solve $5x - (4x - 7)(3x - 5) = 6 - 3(4x - 9)(x - 1)$.

Solution.

$$5x - (4x - 7)(3x - 5) = 6 - 3(4x - 9)(x - 1),$$
$$5x - (12x^2 - 41x + 35) = 6 - 3(4x^2 - 13x + 9),$$
$$5x - 12x^2 + 41x - 35 = 6 - 12x^2 + 39x - 27,$$
$$12x^2 - 12x^2 + 5x + 41x - 39x = 6 - 27 + 35,$$
$$7x = 14,$$
$$x = 2.$$

NOTE. Since the − sign before a parenthesis affects every term within it, in the second line of the above solution we do not remove the parentheses until we have formed the products.

EXERCISE 24

Solve the following equations and check your results:

1. $5x + 3 = 2x + 9$
2. $x - 5 = 11 - 3x$
3. $3x + 2 = 7x - 18$
4. $7x + 10 = 11x + 6$
5. $9x + 32 = x$
6. $5x - 3 - 7x = 15 - x$
7. $3x - 7 - x = 5x + 14$
8. $27 - 2x = x$
9. $1 - x = 2 - 3x$
10. $10 - x - 3 = 5x - 13 - 4x$
11. $(2x - 7) - (x - 5) = 0$
12. $3 - (2 - x) = 2 + 3x$
13. $5 + (x - 11) = 5x - 14$
14. $2 - (5x - 3) - (7 - x) = 6$
15. $2(3x - 5) = 5(x - 2)$
16. $7x - 3(x - 6) = 2$
17. $9 - 5(x + 4) = 3(2 - x) - 1$
18. $3(7 - x) = 2(x - 2)$
19. $3(2x - 7) = 5 - (1 - x)$
20. $4(2x - 5) - 3(x - 5) = 0$
21. $3 - (x - 1) = -1 - 2(5 - x)$
22. $6(5 - x) + 15 = 5(2x - 7)$
23. $9(4 - 5x) + 2 = 7(6x - 7)$
24. $(5x - 7) - 5(7x - 12) + 7 = 0$
25. $7x - 3 = 4x - 1$
26. $1 - (6 - x) = 5 - 3x$
27. $2(3x - 1) = 9(x - 2) - 7$
28. $x - 3(5x - 6) = 3 - 10x$
29. $5 - (6 + 9x) = 9 - (4x - 1)$
30. $3(6x - 5) - 7(3x + 10) = 0$
31. $9(7x - 6) = 5 - 8(3 - 7x)$
32. $5(x + 1) - 3(x - 2) + 2(x - 6) = 0$
33. $6(9 - x) - 5(7 - x) = x$
34. $x - 2(x + 1) + 3(x + 2) = x - 4$
35. $5(x - 9) - 7(3x + 6) = 25$
36. $7(5 + 3x) - 6(x + 5) = 0$
37. $7 - 4(3x - 1) - 6(x + 11) = 2x$
38. $14 - 6(5x - 9) = 7(10x + 4)$
39. $15(62x + 7) = 38(24x + 3)$
40. $8(5x - 3) - 11(3x + 7) = 102$
41. $(x + 1)^2 = x^2 + 5$
42. $(x - 3)(x + 1) = (x - 2)(x - 1)$

43. $(2x + 1)(x - 7) - 2(x^2 - 3) = 25$
44. $(x + 3)^2 - (x + 2)^2 = 1$
45. $(x + 5)(x - 1) = (x + 3)(x - 2) + 13$
46. $x(x - 3) - 2x^2 = 7 - x(x + 2)$
47. $(2x + 3)^2 - x(4x - 1) = 3x - 11$
48. $(2x - 3)(x + 5) - 2(x^2 + 2x - 7) = 5$
49. $(x + 1)(x + 2)(x + 3) = (x - 3)(x + 4)(x + 5) + 84$
50. $(x - 2)(x + 3)(x - 4) = x^2(x - 3) + 34$

40. Stated Problems. We have already learned the general procedure for solving stated problems. There are, however, certain types of problems which occur frequently and whose solution depends on the proper use of a particular formula or idea. We shall discuss some of these types in detail.

41. Mixture Problems. The following examples illustrate this type of problem.

Example 1. A grocer has some tea worth 60 cents per pound and some worth 80 cents per pound. How much of each must he use in order to have a 40-pound mixture worth 75 cents per pound?

Analysis. Let x represent the number of pounds of the tea sold at 60 cents. Then, since the mixture contains 40 pounds in all, we know that $40 - x$ represents the number of pounds of the 80-cent tea used in the mixture.

Now observe that x pounds at 60 cents per pound is worth $60x$ cents, the total value of the cheaper tea used. Also $(40 - x)$ pounds at 80 cents per pound is worth $80(40 - x)$ cents, the total value of the more expensive tea used. Since the total value of the mixture, namely $40 \times 75 = 3000$ cents, must equal the sum of the total values of its parts, we have the equality expressed by the following equation:

$$60x + 80(40 - x) = 3000.$$

The analysis of this problem will probably be clearer if the facts are put in tabulated form.

	No. of Lb.	× Price per Lb. =	Total Value
Cheap tea	x	60	$60x$
Expensive tea . . .	$40 - x$	80	$80(40 - x)$
Mixture	40	75	3000

Solution.

$$\begin{aligned} 60x + 80(40 - x) &= 3000, \\ 60x + 3200 - 80x &= 3000, \\ -20x &= -200, \\ x &= 10. \end{aligned}$$

Answer. The grocer should use 10 pounds of the 60-cent tea and 30 pounds of the 80-cent tea.

Check.

10 lb. at 60¢ = 600¢
30 lb. at 80¢ = 2400¢
40 lb. cost 3000¢

Hence, 1 lb. costs 75 cents checks with the statement.

Example 2. How much cream which is 30% butterfat must be taken with milk which is 3% butterfat in order to have 540 pounds which is 4.5% butterfat?

Analysis. Let x = the number of pounds of cream.

	No. of Lb.	× Percentage	= Lb. of Butterfat
Cream	x	0.30	$0.30x$
Milk	$540 - x$	0.03	$0.03(540 - x)$
Mixture	540	0.045	24.3

Solution.

$$\begin{aligned} 0.30x + 0.03(540 - x) &= 24.3, \\ 0.30x + 16.2 - 0.03x &= 24.3, \\ 0.27x &= 8.1, \\ x &= 30. \end{aligned}$$

Answer. 30 lb. of cream should be mixed with 510 lb. of milk.

Check.

30 lb. at 30% = 9 lb. butterfat
510 lb. at 3% = 15.3 lb. butterfat
24.3 lb. butterfat.

Or 540 lb. at $4\frac{1}{2}$% = 24.3 lb. butterfat.

Example 3. One solution contains 6 parts alcohol to 3 parts water and another contains 3 parts alcohol to 7 parts water. How much of each should be taken in order to obtain 110 ounces of a solution which is half alcohol and half water?

Analysis. Let x = the number of ounces of the first solution.

	Amt. of Sol. $\times$	Frac. Alc. =	Amt. of Alc.
First solution . . .	x	$\frac{2}{3}$	$\frac{2}{3}x$
Second solution . . .	$110 - x$	$\frac{3}{10}$	$\frac{3}{10}(110 - x)$
Mixture	110	$\frac{1}{2}$	55

Solution.

$$\begin{aligned} \tfrac{2}{3}x + \tfrac{3}{10}(110 - x) &= 55, \\ \tfrac{2}{3}x + 33 - \tfrac{3}{10}x &= 55, \\ \tfrac{11}{30}x &= 22, \\ x &= 60. \end{aligned}$$

Answer. 60 ounces of the stronger solution should be added to 50 ounces of the weaker solution.

Check.

60 oz. at $\frac{2}{3}$ alc. = 40 oz. alc.
50 oz. at $\frac{3}{10}$ alc. = 15 oz. alc.
55 oz. alc.

Or 110 oz. at $\frac{1}{2}$ alc. = 55 oz. alc.

EXERCISE 25

Mixture Problems

1. A grocer blended 68-cent coffee with 83-cent coffee so as to obtain 60 pounds worth 75 cents per pound. How much of each did he take?
2. A confectioner mixed chocolates worth \$1.00 a pound with nougats worth 75 cents a pound to obtain a 5-pound box of candy worth \$4.50. How much of each candy did he take?
3. How much coffee worth 85 cents a pound should be blended with 10 pounds of coffee worth 70 cents a pound in order to obtain a mixture worth 75 cents a pound?
4. A candy dealer has 20 pounds of bonbons which sell at 75 cents a pound. He wishes to mix chocolates worth 90 cents a pound with the bonbons in order to obtain a combination worth 80 cents a pound. How many pounds of chocolates should he take?
5. How much cream that is 25% butterfat should be mixed with milk that is 3% butterfat in order to obtain 11 gallons of light cream that is 20% butterfat?
6. A druggist has on hand a 10% and a 15% solution of argyrol. How should he mix these in order to obtain 10 ounces of 12% solution?

7. A grocer has a 50-pound mixture of nuts which sell for 60 cents a pound. He wishes to reduce this price to 55 cents a pound by adding more almonds which sell for 40 cents a pound. How many pounds of almonds should he add to the mixture?

8. How much tea worth 60 cents a pound should be blended with 40 pounds of tea worth 75 cents a pound in order to obtain a mixture worth 68 cents a pound?

9. A goldsmith has two alloys of gold, the first being 90% pure gold and the second being 60% pure gold. How many ounces of each must he take to make 45 ounces of an alloy which will be 72% pure gold?

10. How much pure alcohol should be added to 24 ounces of a 45% solution in order to obtain a 60% solution?

11. How many nuts at 58 cents a pound should be mixed with nuts worth 43 cents a pound in order to obtain a mixture of 60 pounds of nuts worth 50 cents a pound?

12. A perfumer wishes to blend perfume costing $3.75 an ounce with 350 ounces of perfume worth $1.50 an ounce. If the blend is to sell for $2.00 an ounce, how much of the more expensive perfume should he use?

13. How much water should be added to 1 gallon of a full-strength syrup in order to obtain a syrup which is 15% of full strength?

14. How much of an alloy which is 30% copper should be added to 175 pounds of an alloy which is 65% copper in order to obtain an alloy which is 50% copper?

15. One solution contains 2 parts alcohol to 3 parts water and another contains 7 parts alcohol to 3 parts water. How much of each should be taken in order to obtain 30 quarts of a solution which is 1 part alcohol to 1 part water?

16. One alloy is 2 parts silver to 8 parts copper and another is 3 parts silver to 5 parts copper. How much of each should be taken in order to obtain 280 pounds of an alloy which is 3 parts silver to 7 parts copper?

17. A metal worker has 120 pounds of an alloy which is 1 part gold to 7 parts copper. How much of an alloy which is 1 part gold to 15 parts copper should be combined with the first in order to obtain an alloy which is 1 part gold to 9 parts copper?

18. A candy dealer wants to make up 100 one-pound boxes of candy to sell for 66 cents a box. He plans to use 40 pounds of candy worth 75 cents a pound. What should be the price of the other candy he uses?

19. A grocer has 20 pounds of walnuts worth 60 cents a pound and 10 pounds of Brazil nuts worth 72 cents a pound. How many pounds of almonds

at 40 cents a pound must he mix with these in order to have a mixture of the three nuts which he can sell for 56 cents a pound?

20. A confectioner has 20 pounds of chocolates worth 80 cents a pound and 10 pounds of nougats worth 70 cents a pound. How many pounds of bonbons at 62 cents a pound must he add to these in order to have a mixture of the three candies which he can sell at $1.46 for a two-pound box?

42. Motion Problems. The following examples illustrate this type of problem.

Example 1. Two cars leave the same place at the same time and travel in opposite directions, one of them traveling 5 miles per hour faster than the other. After 4 hours they are 220 miles apart; how fast is each car traveling?

Analysis. Let x = the rate of the slower car. We may then tabulate the conditions of the problem as follows:

	Rate $\times$	Time $=$	Distance
Slower car	x	4	$4x$
Faster car	$x + 5$	4	$4(x + 5)$

Since the cars are traveling in opposite directions, their total distance apart is equal to the sum of the distances that each car has traveled; thus

$$4x + 4(x + 5) = 220.$$

It is sometimes clearer to picture the distances traveled. In this case, we have

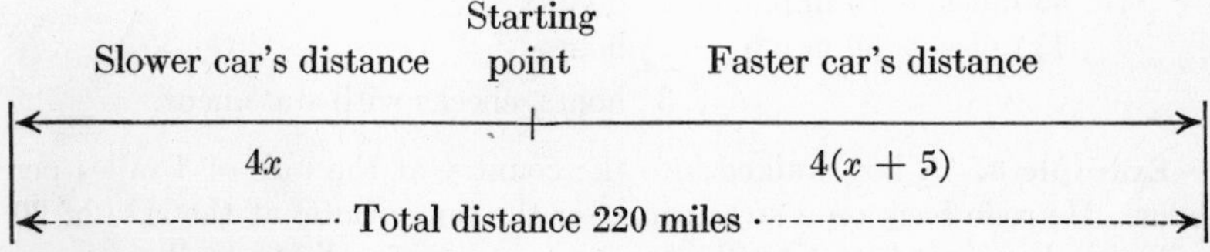

Solution. Hence, as above,

$$\begin{aligned} 4x + 4(x + 5) &= 220, \\ 4x + 4x + 20 &= 220, \\ 8x &= 200, \\ x &= 25. \end{aligned}$$

Answer. One car travels at the rate of 25 m.p.h. (miles per hour) and the other at 30 m.p.h.

Check.

$$\begin{aligned} 25 \text{ m.p.h.} \times 4 \text{ hours} &= 100 \text{ miles} \\ 30 \text{ m.p.h.} \times 4 \text{ hours} &= \underline{120} \text{ miles} \\ & \quad 220 \text{ miles, checks with statement.} \end{aligned}$$

Example 2. A cyclist traveling at an average rate of 10 miles per hour leaves a city and three hours later a motorist traveling at an average rate of 30 miles per hour sets out to overtake him. How far from the city are the two when they meet?

Analysis. Let t = the time in hours that the motorist travels.

	Rate $\times$	Time $=$	Distance
Motorist	30	t	$30t$
Cyclist	10	$t + 3$	$10(t + 3)$

City

Cyclist's distance $10(t + 3)$

Motorist's distance $30t$

Solution.

$$\begin{aligned} 30t &= 10(t + 3), \\ 30t &= 10t + 30, \\ 20t &= 30, \\ t &= 1\tfrac{1}{2}. \end{aligned}$$

Answer. They meet at a distance of 45 miles from the city.

Check.

$$\begin{aligned} 45 \text{ miles} \div 10 \text{ m.p.h.} &= 4\tfrac{1}{2} \text{ hours} \\ 45 \text{ miles} \div 30 \text{ m.p.h} &= \underline{1\tfrac{1}{2}} \text{ hours} \\ & \quad 3 \text{ hours, checks with statement.} \end{aligned}$$

Example 3. A boy walked into the country at the rate of 4 miles per hour. He rode back on a streetcar, over the same route, at the rate of 20 miles per hour. If he returned after 3 hours, how far did he walk?

Analysis. Let t = the time in hours that he walked.

	Rate $\times$	Time $=$	Distance
Walked	4	t	$4t$
Rode	20	$3 - t$	$20(3 - t)$

Distance walked $4t$

Distance rode $20(3-t)$

Solution.

$$4t = 20(3 - t),$$
$$4t = 60 - 20t,$$
$$24t = 60,$$
$$t = 2\tfrac{1}{2}.$$

Answer. He walked a distance of 10 miles.

Check.

10 miles ÷ 4 m.p.h. = $2\frac{1}{2}$ hours
10 miles ÷ 20 m.p.h. = $\frac{1}{2}$ hour
3 hours, checks with statement.

EXERCISE 26

Motion Problems

1. Two cars traveling in opposite directions were 240 miles apart at the end of 4 hours. If one traveled 6 miles per hour faster than the other, find the rate of each car.
2. Two mail planes left St. Louis traveling in opposite directions. One traveled 37 miles per hour faster than the other and at the end of 2 hours they were 722 miles apart. What was the rate of each plane?
3. Two boys, 14 miles apart, start walking toward each other and they meet in 2 hours. If one boy walks 1 mile farther in an hour than the other boy, how fast does each walk?
4. A passenger train and a freight train leave towns 320 miles apart and travel toward each other. If the passenger train travels 30 miles per hour faster than the freight and they meet in 4 hours, what are their respective rates of speed?
5. Two cars travel in the same direction, one going twice as fast as the other. At the end of 4 hours they are 84 miles apart. How fast is each traveling?
6. A cyclist and a hiker travel in the same direction, the cyclist going three times as fast as the hiker. At the end of 3 hours they are 24 miles apart. How fast is each traveling?
7. A freight train leaves Detroit for Chicago, traveling at an average rate of 18 miles per hour. Four hours later, a passenger train leaves Detroit

for Chicago, traveling at an average rate of 42 miles per hour. How far from Detroit do the two trains pass?

8. One hour after a hiker left a camp, a cyclist set out to overtake him. If the hiker travels 3 miles per hour and the cyclist 9 miles per hour, how long will it take the cyclist to catch the hiker?

9. One automobile went 10 miles farther when traveling 50 miles per hour than a second one which traveled 2 hours longer at the rate of 40 miles per hour. How long did each travel?

10. Two automobiles travel at the rate of 60 miles per hour and 45 miles per hour respectively. If the faster automobile travels 2 hours longer than the slower automobile and goes twice as far, what distance is traveled by each?

11. A boy starts to walk to a town 21 miles away at the rate of 4 miles per hour. After walking for a time, he is invited to ride by a motorist and is taken to the town at the rate of 30 miles per hour. If the boy reached his destination 2 hours after he started, how far did he walk?

12. A cyclist starts to ride to a village 10 miles away at the rate of 9 miles per hour. After riding for a time his bicycle breaks down and he walks the remainder of the distance to town at the rate of 3 miles per hour. If he arrived at the village an hour and forty minutes after he started, how far did he walk?

13. A boy ran to a store at the rate of 6 miles per hour and he walked back at the rate of 4 miles per hour. If it took 10 minutes for the round trip, how far away is the store?

14. A cyclist rides into the country at the rate of 10 miles per hour. His bicycle breaks down and he walks back at the rate of 3 miles per hour. If he returned 3 hours and 15 minutes after he started, how far into the country did he go?

15. A motorist traveling at a certain rate can go from A to B in nine hours. If he travels 5 miles an hour faster he can go from A to B in one hour less. How far is it from A to B?

16. Two trains travel different routes, one 14 miles longer than the other. The train traveling the longer route travels 2 miles per hour slower than the other train and takes 6 hours for the trip. If the train traveling the shorter route takes 5 hours for its trip, find the length of each route.

17. A motorist drives to a distant city in 4 hours and returns by another route which is 26 miles longer. On the return trip he traveled 5 miles per hour faster and took 4 hours and 12 minutes. Find the length of the shorter route.

18. A motorist drove at an average rate of 45 miles per hour outside the city limits and 20 miles per hour inside the city limits. If a trip of 25 miles took him 50 minutes, how much of it was within the city limits?

19. A train averages a speed of 40 miles per hour across the plains and 25 miles per hour through the mountains. If a trip of 70 miles took an hour and 54 minutes, how much of it was through the mountains?

20. At 7 A.M. a fast train traveling 60 miles per hour leaves Los Angeles for San Francisco, a distance of 400 miles. At 9 A.M. a train traveling 40 miles per hour leaves San Francisco for Los Angeles. At what time will these two trains pass each other?

43. Age Problems. The following examples illustrate this type of problem.

Example 1. A man is seven times as old as his son, and in two years he will be five times as old as his son is then. How old is each now?

Analysis. Let x = the son's present age in years. The conditions of the problem may then be tabulated as follows:

Time	Now	2 Years Hence
Son	x	$x + 2$
Man	$7x$	$7x + 2$

Solution. Since at the future time the man's age will be five times that of his son, we have the equation

$$7x + 2 = 5(x + 2),$$
$$7x + 2 = 5x + 10,$$
$$2x = 8,$$
$$x = 4.$$

Answer. The boy's present age is 4 years and his father's age is 28 years.

Check. In two years their respective ages will be 30 and 6. Since 30 is five times 6, the above answer satisfies the conditions of the problem.

Example 2. Dora is one-third as old as her mother and in four years she will be as old as her mother was twenty years ago. Find Dora's age.

Analysis. Let x = Dora's present age in years.

Time	Now	4 Years Hence	20 Years Ago
Dora	x	$x + 4$	
Mother	$3x$		$3x - 20$

Solution.

$$3x - 20 = x + 4,$$
$$2x = 24,$$
$$x = 12.$$

Answer. Dora is 12 years old.

Check. Dora's mother is three times as old, or 36 years. Twenty years ago she was 16 years old. In four years Dora will be 16 years old.

EXERCISE 27

Age Problems

1. A father is three times as old as his son, and in twelve years he will be just twice as old as his son. Find their present ages.
2. Roy is twice as old as John, and in three years the sum of their ages will be 21 years. Find their present ages.
3. Mary is two years older than Edna, and three years ago Mary was twice as old as Edna. Find their present ages.
4. An uncle is twice as old as his nephew, and eight years ago he was three times as old as his nephew. Find their present ages.
5. A man is 38 years old and his son is 14 years old. How many years will it be before the man's age is just twice that of his son?
6. A man is 36 years old and his brother is 29 years old. How many years ago was it when the man was twice as old as his brother?
7. Harry is three times as old as Joel, and in four years he will be twice as old as Joel. How old was Harry when Joel was born?
8. Joan is 5 years older than Ellen, and three years ago the sum of their ages was 17 years. In how many years will Joan be 21 years old?
9. The sum of the present ages of George and his father is 60 years. In six years his father will be twice as old as George will be. Find their present ages.
10. Earl is twice as old as Sally, and next year the sum of their ages will be four times as much as Sally's age last year. Find their present ages.
11. Boyd is 2 years older than Jimmy, and in three years the sum of their ages will be twice as much as the sum of their ages three years ago. Find their present ages.
12. Dick is one year older than Guy, and Dick's age in two years will be twice what Guy's age was two years ago. Find their present ages.
13. Doris is three years older than her sister, and in two years her age will be three times what her sister's age was last year. Find their present ages.

14. Three years ago Tom was twice as old as Dean, and in two years the sum of their ages will be 28 years. Find their present ages.

15. A man 35 years old has two sons, one of whom is twice as old as the other. In three years the sum of all their ages will be 59 years. How old are the boys at present?

16. A mother is thirty years older than her twin daughters. In two years her age will be three times the combined ages of the twins. How old are the twins now?

17. A man has two sons, one twice as old as the other. The man is now four times as old as the older boy, and in three years he will be five times as old as the younger boy. Find their present ages.

18. A mother has two daughters, one three times as old as the other. The mother is five times as old as her older daughter, and in five years she will be five times as old as her younger daughter. Find their present ages.

19. A is twice as old as B, and in four years he will be three times as old as C, who is 6 years younger than B. Find their ages.

20. A man being asked his age replied, "If you take one year from my present age the result will be three times my son's age, and three years ago my age was twice what his will be in four years." Find his age.

44. Investment Problems. The following examples illustrate this type of problem.

Example 1. A man invested part of \$3000 at $5\frac{1}{2}\%$ and the remainder at 4%. If his total yearly income from these investments is \$139.50, how much does he have invested at each rate?

Analysis. Let x = the amount in dollars invested at $5\frac{1}{2}\%$. The problem may then be set up as follows:

	Amount $\times$	Rate $=$	Interest
First investment. .	x	0.055	$0.055x$
Second investment .	$3000 - x$	0.04	$0.04(3000 - x)$

Solution. Since the total yearly interest is \$139.50, we have the equation:

$$
\begin{aligned}
0.055x + 0.04(3000 - x) &= 139.50,\\
0.055x + 120 - 0.04x &= 139.50,\\
0.015x &= 19.50,\\
x &= 1300.
\end{aligned}
$$

Answer. The man has \$1300 invested at $5\frac{1}{2}\%$, and \$1700 at 4%.

Check.

$$5\tfrac{1}{2}\% \text{ of } \$1300 = \$\ 71.50$$
$$4\% \text{ of } \$1700 = \$\ 68.00$$
$$\$139.50, \text{ checks with statement.}$$

Example 2. A man has \$4500 more invested at 3% than he has at 6%. His annual income from these two investments is equivalent to a return of 4% on the entire sum invested. How much money has he invested at each rate?

Analysis. Let x = the amount in dollars invested at 6%.

	Amount $\times$	Rate $=$	Interest
6% investment	x	0.06	$0.06x$
3% investment	$x + 4500$	0.03	$0.03(x + 4500)$
Combined . .	$2x + 4500$	0.04	$0.04(2x + 4500)$

Solution.

$$0.06x + 0.03(x + 4500) = 0.04(2x + 4500),$$
$$0.06x + 0.03x + 135 = 0.08x + 180,$$
$$0.01x = 45,$$
$$x = 4500.$$

Answer. The man has \$4500 invested at 6%, and \$9000 at 3%.

Check.

$$6\% \text{ of } \$4500 = \$270$$
$$3\% \text{ of } \$9000 = \$270$$
$$\$540$$

Or

$$4\% \text{ of } \$13{,}500 = \$540.$$

Example 3. A man has \$4000 invested at $r\%$ and \$2500 at $(r + 1)\%$. If his annual interest on the \$4000 investment is \$27.50 more than the annual interest on the \$2500 investment, what are the respective rates of interest?

Analysis.

	Amount $\times$	Rate $=$	Interest
First investment . . .	4000	$\frac{r}{100}$	$40r$
Second investment . .	2500	$\frac{r+1}{100}$	$25(r + 1)$

Solution.

$$40r - 25(r + 1) = 27.50,$$
$$40r - 25r - 25 = 27.50,$$
$$15r = 52.50,$$
$$r = 3.5.$$

Answer. The \$4000 is invested at $3\frac{1}{2}\%$, and the \$2500 at $4\frac{1}{2}\%$.

Check.

$3\frac{1}{2}\%$ of \$4000 = \$140.00
$4\frac{1}{2}\%$ of \$2500 = \$112.50
Difference = \$ 27.50, checks with statement.

EXERCISE 28

Investment Problems

1. A sum of \$3800 is invested, part at 4% and the remainder at 5%. The yearly income from the two investments is \$169. Find the amount invested at each rate.
2. A man invested \$4000, part of it at 5% and the remainder at 6%. If his yearly income is \$225, what is the amount invested at each rate?
3. A man has \$4000 invested, part at 4% and the remainder at 6%. If the annual return on the 6% investment is \$10 more than that on the 4% investment, find how much is invested at each rate.
4. A man has \$3150 invested, part at 3% and the remainder at 4%. If the yearly interest on each investment is the same, how much interest does he receive each year?
5. A man has \$4300 invested, part at $2\frac{1}{2}\%$ and the remainder at 4%. If the yearly interest on these investments is \$142, find how much he has invested at each rate.
6. Mr. Drew has twice as much money invested in 4% bonds as he has in stocks paying 6%. If his annual income from the stocks and bonds is \$1190, how much has he invested in bonds?
7. Mr. Baldwin has \$9300 invested, part at $3\frac{1}{2}\%$ and the remainder at 5%. If his annual income on the 5% investment is \$6 more than on the $3\frac{1}{2}\%$ investment, find his total annual income.
8. A man has \$2000 more invested in 3% bonds than he has invested in 7% oil stock. If his total annual income from these is \$580, how much does he have invested in oil stock?
9. A man invested part of \$21,000 at 4% and the remainder at 7%. His annual income from these two investments was equivalent to a return of 6% on the entire sum invested. How much money was invested at each rate?

10. Mr. Smith invested \$18,000 in two business enterprises which paid interest at the rate of 3% and 6% respectively. If his annual income from these two enterprises was equivalent to a return of 5% on the entire sum invested, find how much he had invested at each rate.

11. A man has \$1200 more invested at 4% than he has at 5%. If his annual income from these two investments is \$147, how much does he have invested at each rate?

12. A man has \$10,000 less invested at 6% than he has at 4%. If his annual income from these two investments is \$1800, how much does he have invested at each rate?

13. Mr. Cox has \$4000 invested at one rate of interest and \$2500 invested at a rate twice as great. If his annual interest on these two investments is \$225, at what rate of interest is the \$4000 invested?

14. A man has \$2200 invested at r% and \$1800 invested at $(r + 1)$%. If his annual income on these two investments is \$158, find r.

15. A man has part of \$4500 invested at 4% and the remainder at 6%. His annual return on each investment is the same. What average rate of interest does he realize on his money?

16. A man has twice as much money invested at 4% as he has at 3%, and he has \$100 more invested at 5% than he has at 3%. If his annual income from these three investments is \$405, how much does he have invested at each rate?

17. A man has \$2500 invested in three bonds which pay different rates of interest. He has twice as much invested in a 3% bond as he has in a 4% bond, and the remainder is invested in a 6% bond. If his annual dividends from these three bonds are \$110, what is the value of each bond?

18. A man has \$10,000 to invest. He invests \$4000 at 6% and \$3500 at 4%. At what rate should he invest the remainder in order to have a yearly income of \$500?

19. A man has \$7500 invested, part at $3\frac{1}{2}$% and the remainder at 5%. If the amounts invested at each rate had been interchanged, the yearly interest would have been increased by \$7.50. Find the amount invested at each rate.

20. Mr. Thomas has \$$x$ invested at 4% and \$$(2x - 500)$ invested at 5%, while Mr. Mills has \$$(x + 300)$ invested at 3% and \$$2x$ at $4\frac{1}{2}$%. If they both have the same yearly income, find the value of each of their holdings.

45. Value Problems. The following examples illustrate this type of problem.

Example 1. A purse contains \$2.10 in nickels and dimes. If there are 29 coins in all, how many of each kind are there?

Analysis. Let x = the number of nickels. The problem may then be tabulated as follows:

	Number	× Value of Each =	Total Value
Nickels	x	5	$5x$
Dimes	$29 - x$	10	$10(29 - x)$

Solution. Since the combined value of the nickels and dimes is \$2.10, we have:

$$\begin{aligned} 5x + 10(29 - x) &= 210, \\ 5x + 290 - 10x &= 210, \\ -5x &= -80, \\ x &= 16. \end{aligned}$$

Answer. The purse contains 16 nickels and 13 dimes.

Check.

16 nickels = \$ .80
13 dimes = \$1.30
\$2.10, checks with statement.

Example 2. A confectioner buys some candy bars at 2 for a nickel and sells them at 3 for a dime. If he makes \$2.50, how many candy bars did he buy?

Analysis. Let x = the number of candy bars.

	Number ×	Price =	Value
Bought.	x	$2\frac{1}{2}$	$\frac{5}{2}x$
Sold	x	$3\frac{1}{3}$	$\frac{10}{3}x$

Solution.

$$\begin{aligned} \tfrac{10}{3}x - \tfrac{5}{2}x &= 250, \\ \tfrac{5}{6}x &= 250, \\ x &= 300. \end{aligned}$$

Answer. He bought 300 candy bars.

Check.

300 bars at 3 for 10¢ = \$10.00
300 bars at 2 for 5¢ = \$ 7.50
Profit = \$ 2.50, checks with statement.

Example 3. If linen costs three times as much as calico, and a woman spends \$11.25 for 10 yards of linen and 15 yards of calico, find the price of each.

Analysis. Let x = the price in cents per yard of the calico.

	Amount $\times$	Price $=$	Value
Calico.	15	x	$15x$
Linen	10	$3x$	$30x$

Solution.

$$15x + 30x = 1125,$$
$$45x = 1125,$$
$$x = 25.$$

Answer. The calico costs 25 cents per yard and the linen costs 75 cents per yard.

Check.

15 yards calico at 25¢ per yard = \$ 3.75
10 yards linen at 75¢ per yard = \$ 7.50
\$11.25, checks with statement.

EXERCISE 29

Value Problems

1. A pile of 55 coins, consisting of nickels and dimes, is worth \$3.90. Find the number of each.
2. A man has \$3.35 in his purse, consisting of dimes and quarters. If there are 23 coins in all, how many of each has he?
3. A pile of coins, consisting of quarters and half dollars, is worth \$11.75. If there are 2 more quarters than half dollars, how many of each are there?
4. A boy has 5 more than twice as many pennies as he has dimes. Altogether he has \$3.05. How many coins does he have?
5. A pile of coins, consisting of nickels, dimes, and quarters, is worth \$4.55. There are 4 more dimes than nickels and 3 more quarters than dimes. How many of each are there?
6. A man has twice as many nickels as he has pennies, and he has 5 less dimes than nickels. If he has \$4.46 in all, how many coins of each kind does he have?

7. A man has twice as many dimes as quarters in his purse. If the dimes were quarters and the quarters were dimes, he would have \$1.20 more than he now has. How many of each has he?

8. A boy was sent to the post office for some stamps of the following denominations: 2¢, 3¢, 6¢. He paid \$1.90 for them, and received twice as many 3-cent stamps as 2-cent stamps and 25 fewer 6-cent stamps than 3-cent stamps. Find the number of each kind that he bought.

9. The gate receipts at a baseball game were \$1708.45 from 2858 paid admissions. If grandstand tickets sold for 65 cents each and bleacher tickets sold for 40 cents each, how many tickets of each kind were sold?

10. A fair charged 60 cents admission for adults and 25 cents for children. The receipts for one day were \$343.45 on 650 paid admissions. How many adults and children attended the fair that day?

11. A lady bought 100 Christmas cards. For some she paid 5 cents each and for the remainder she paid 3 cents each. If the cards cost \$3.76 in all, how many of each kind did she buy?

12. An ice cream vendor sells both 5-cent and 10-cent ice cream bars. On a certain day 328 bars were sold and the receipts were \$22.20. How many of each kind were sold?

13. If silk costs four times as much as linen, and I spend \$84.32 for 12 yards of silk and 20 yards of linen, find the cost of each per yard.

14. A woman spends \$89.70 for 20 yards of satin and 30 yards of cotton. If the satin costs as many dimes per yard as the cotton costs cents per yard, find the price of each.

15. A boy bought some apricots at three for a nickel. He ate three and sold the remainder at 30 cents a dozen. If he made \$1.15, how many apricots did he buy?

16. A dealer buys some horses at \$100 each; through disease he loses 10% of the horses. By selling the remainder of them at \$125 each, he makes \$7000 on the transaction. How many horses did he buy?

17. A boy has 25 cents more than he needs to buy 6 tennis balls. If the tennis balls each cost 5 cents more, he would have 80 cents too little to buy 7 tennis balls. How much money has he?

18. A woman has 45 cents more than she needs to purchase 50 Christmas cards. If she should purchase cards each of which costs 1 cent more, she would only have 25 cents more than enough to purchase 40 cards. How much money has she?

19. The admission charges for a football game are as follows: \$1.00 for general admission, \$2.00 for reserved seats, \$3.00 for box seats. There

are twice as many reserved seats as there are seats for general admission and the stadium has a total seating capacity of 25,000. If the receipts for a capacity crowd are $44,600, how many seats of each kind are there?

20. The admission charges at a dance were 40 cents for men, 25 cents for women, and 10 cents for children. 140 people attended of which there were half as many children as there were men. If the receipts were $38.90, how many women attended the dance?

46. Digit Problems. In arithmetic a two-figured number such as 48 indicates the sum of 4 tens and 8 ones, that is, $48 = 4 \times 10 + 8$. When letters are used to represent digits this latter representation must be used. Thus, the number whose tens' digit is x and whose units' digit is y is represented by $10x + y$. More generally, the number whose hundreds' digit is a, whose tens' digit is b, and whose units' digit is c, is represented by $100a + 10b + c$; and so on.

The following examples illustrate this type of problem.

Example 1. The sum of the digits of a two-figured number is 13. If the digits are reversed, the resulting number is 45 more than the original number. Find the number.

Analysis. Let $x =$ the tens' digit; then $13 - x =$ the units' digit.

Hence, the number is represented algebraically by $10(x) + (13 - x)$ and the number with the digits reversed is represented by $10(13 - x) + (x)$.

Solution. Thus, we have the equation:

$$\begin{aligned} 10(13 - x) + (x) &= 10(x) + (13 - x) + 45, \\ 130 - 10x + x &= 10x + 13 - x + 45, \\ -18x &= -72, \\ x &= 4. \end{aligned}$$

Answer. The number is 49.

Check. To verify this answer we note that $94 = 49 + 45$.

Example 2. In a two-figured number the units' digit is 2 more than the tens' digit. The number itself is equal to four times the sum of the digits. Find the number.

Analysis. Let $x =$ the tens' digit; then $x + 2 =$ the units' digit.

Hence, $10x + (x + 2)$ or $11x + 2$ represents the number.

Solution.

$$\begin{aligned} 11x + 2 &= 4(x + x + 2), \\ 3x &= 6, \\ x &= 2. \end{aligned}$$

Answer. The number is 24.

Check. 4 times the sum of the digits, $2 + 4$, equals 24, which is the number.

Example 3. In a three-figured number the units' digit is twice the hundreds' digit, and the tens' digit is 1 more than the hundreds' digit. The number obtained by reversing the digits is 49 less than twice the number. What is the number?

Analysis. Let $x =$ the hundreds' digit; then $x + 1 =$ the tens' digit, and $2x =$ the units' digit.

Hence, $100(x) + 10(x + 1) + (2x) = 112x + 10 =$ the number,
and $100(2x) + 10(x + 1) + (x) = 211x + 10 =$ the number with digits reversed.

Solution.

$$211x + 10 = 2(112x + 10) - 49,$$
$$211x + 10 = 224x + 20 - 49,$$
$$-13x = -39,$$
$$x = 3.$$

Answer. The number is 346.

Check. $2 \times 346 - 49 = 643$, which is the number with digits reversed.

EXERCISE 30

Digit Problems

1. The sum of the digits of a two-figured number is 5. If 9 is subtracted from the number, the digits will be interchanged. Find the number.
2. The sum of the digits of a two-figured number is 13. If 27 is added to the number, the digits will be interchanged. Find the number.
3. In a two-figured number the tens' digit is 3 more than the units' digit. The number itself is 17 times the units' digit. Find the number.
4. In a two-figured number the tens' digit is 1 more than the units' digit. The number itself is six times the sum of the digits. Find the number.
5. The sum of the digits of a two-figured number is 9. If the digits are interchanged, the number obtained is only $\frac{3}{8}$ as large as the original number. Find the number.
6. The units' digit of a two-digit number is 5 more than the tens' digit, and the number is three times as great as the sum of the digits. What is the number?
7. In a two-figured number the tens' digit is 1 less than the units' digit. If the digits are interchanged, the sum of the number obtained and the original number is 143. Find the number.

8. The sum of the digits of a two-figured number is 9. If the digits are interchanged, the resulting number equals the original number plus the sum of the digits. Find the number.

9. In a two-figured number the units' digit is 7 more than the tens' digit. The number with digits reversed is three times as large as the sum of the original number and the two digits. Find the number.

10. The units' digit of a two-digit number is 7 more than the tens' digit. If 26 is added to the number, the result obtained is five times the sum of the digits. Find the number.

11. In a two-figured number the tens' digit is 1 more than twice the units' digit, and the number itself is 2 more than twice the number obtained by interchanging the digits. Find the number.

12. The sum of the digits of a two-figured number is 15. The number itself equals the sum of six times the tens' digit and seven times the units' digit. Find the number.

13. In a two-figured number the tens' digit is 2 less than the units' digit. The number itself equals the sum of five times the tens' digit and four times the units' digit. Find the number.

14. The sum of the digits of a two-figured number is 13. If 29 is subtracted from the number with its digits reversed, the result is a number which is one-half the original number. Find the number.

15. In a two-figured number the tens' digit is twice the units' digit. If 5 is subtracted from the number with its digits reversed, the result is a number which is one-third the original number. Find the number.

16. The sum of the digits of a two-digit number is 8. If the number decreased by the number with digits reversed is 18, find the number.

17. In a three-figured number the tens' digit is twice the units' digit, and the sum of the digits is 7. If the number with digits reversed is 99 more than the original number, find the number.

18. In a three-figured number the tens' digit is twice the hundreds' digit, and the units' digit is twice the tens' digit. If the number with digits reversed is 594 more than the original number, find the number.

19. In a three-figured number the tens' digit is the same as the units' digit, and the hundreds' digit is 3 less than the units' digit. If the number itself is equal to 28 times the sum of the digits, what is the number?

20. A three-figured number is symmetrical; here this means that the hundreds' digit equals the units' digit. If the tens' digit is 3 more than the units' digit and the number itself is equal to 28 times the sum of the digits, what is the number?

REVIEW OF CHAPTER IV

A

Solve the following equations and check your results:

1. $6x - 2 = 3x + 13$
2. $7x + 11 = 2x + 21$
3. $6 - 7x = 5x + 54$
4. $x - 2 + 3x = 4x - 5 + 6x$
5. $6 - (x - 5) = x + 9$
6. $7x - 5 + 3x + 6 - 6x + 7 = 0$
7. $5 - (x - 3) = x$
8. $x - (1 - x) = 6$
9. $1 - (2 - x) = 5 - 2x$
10. $(6x - 4) - (3x + 2) = (2x - 3)$
11. $3(x + 6) = 4(2x - 3)$
12. $5(x - 3) + 2(9 - 2x) = 3(x + 3)$
13. $5x - 2(x + 3) = 12 - 3(x - 2)$
14. $4(x - 3) - 5(x - 2) = 0$
15. $53x - 6(9x - 4) = 6x$
16. $3(x - 3) + 2(x - 2) = 13 - 5x$
17. $2(29x + 6) - 15(3x + 6) = 0$
18. $9(9x - 6) - 1 = 5 - 3(x + 6)$
19. $4[5x - 3(x + 1)] = 7[2(x - 5) + 4]$
20. $5[x - 4(3 + 2x)] + 6(x - 4) = 3$
21. $(2x + 5)^2 - (2x - 5)^2 = 60$
22. $(2x - 1)(3x + 4) = (x + 2)(6x - 5)$
23. $(3x - 2)^2 - x(9x - 7) = 7 - 8x$
24. $(x + 1)^3 = x^2(x + 3)$

B

1. Two motorists leave towns which are 237 miles apart and travel toward each other. If one motorist averages 7 miles per hour more than the other and they meet in 3 hours, find their respective rates of travel.
2. A pile of 36 coins, consisting of half dollars, quarters, and dimes, is worth \$7.40. If there are seven more dimes than quarters, how many of each kind of coin are there?
3. The sum of two numbers is 19, and one of them with 11 added to it is five times the other. Find the numbers.
4. A grocer wishes to mix sugar at 11 cents a pound with another sort at 8 cents a pound to make 42 pounds to be sold at 9 cents a pound. What quantity of each must he take?
5. The sum of the digits of a two-figured number is 10. If the digits are reversed, the resulting number is 36 less than the original number. Find the number.
6. The length of a rectangle is 5 feet more than twice its width. If the perimeter of the rectangle is 88 feet, find its dimensions.
7. A part of \$4100 is invested at 4% and the remainder at 5%. If the annual income from the two amounts is \$186.50, find the amount invested at each rate.

8. Divide \$358 among A, B, and C, so that B will have \$30 more than A, and C will have \$40 more than B.

9. A chemist has two acid solutions; one is 65% acid and the other is 20% acid. How much of each should he take in order to have 100 cc. of acid solution which is 47% acid?

10. A is four times as old as B, and in 10 years A will be twice as old as B. Find their ages.

11. Two sides of a triangle are equal, and the third side is 2 more than one-third the sum of the equal sides. If the perimeter of the triangle is 42, how long are its sides?

12. A boy starts to walk to a town 10 miles away at the rate of 3 miles per hour. After walking part of the way he obtains a ride from a passing motorist who is traveling at an average rate of 24 miles per hour. If the boy reached his destination one hour after he started, how far did he walk?

13. A certain number between 10 and 100 is 45 more than the number obtained by reversing its digits. If the sum of its digits is 9, what is the number?

14. A man has a \$100 bill changed into \$1, \$2, and \$5 bills. If he received 4 more \$1 bills than \$5 bills and 40 bills altogether, how many \$2 bills did he receive?

15. George's age in 8 years will be four times what it was last year. How old is George?

16. How much tea at 74 cents a pound should a grocer mix with 20 pounds of tea at 62 cents a pound in order to obtain a mixture worth 69 cents a pound?

17. A hiker can climb a mountain trail at the rate of 2 miles per hour and he can descend the same trail at the rate of 5 miles per hour. If he started up the trail and then returned after 42 minutes, how far up the trail had he gone?

18. A man had \$3450 invested, part of it at 4% and the remainder at $5\frac{1}{2}$%. If the total yearly income was \$159, find how much was invested at each rate.

19. The length of a rectangle is 5 more than its width. If its width were doubled and its length were halved, the perimeter of the rectangle would be increased by 12. Find the dimensions of the rectangle.

20. John is 5 years older than his sister, and next year he will be twice as old as his sister was three years ago. Find their ages.

21. At a school entertainment the admission charges were 25 cents for adults and 10 cents for students. If 332 tickets were sold and the receipts were $46.25, how many adults attended?

22. In a certain examination the number of successful candidates was five times the number of those who failed. If 2 less had failed, the number of those who passed would have been six times the number of those who failed. Find the number of candidates.

23. A man has $400 more invested at $3\frac{1}{2}\%$ than he has at 5%. If the total yearly interest is $209.50, find how much money he has invested altogether.

24. The sum of the digits of a two-figured number is 12. If the sum of four times the tens' digit and three times the units' digit is equal to the number, what is the number?

C

1. At 7 A.M. a motorist leaves for a distant city and travels at an average rate of 30 miles per hour. At 9 A.M. a second motorist leaves the same place and travels the same road at a rate of 45 miles per hour. At what time will the second motorist overtake the first?

2. A transportation company charges $2.25 per hundred pounds for crated goods and $2.75 per hundred pounds for uncrated goods. If a man sends 500 pounds and is billed for $12.57, how much of his order was crated?

3. In a bag containing black balls and white balls, there are 16 more white balls than there are black balls. If twice the total number of balls exceeds the number of black balls by 200, find how many balls the bag contains.

4. A grocer has 15 pounds of black walnuts worth 46 cents a pound and 25 pounds of hickory nuts worth 42 cents a pound. How many pounds of filberts at 47 cents a pound must be mixed with these in order to have a mixture of the three nuts which he can sell for 45 cents a pound?

5. In a two-figured number the units' digit is 2 more than the tens' digit. If the product of the digits is 12 more than the square of the tens' digit, what is the number?

6. The width of a rectangle is 3 more than one-half its length. If the width were tripled and the length decreased by 6, the perimeter of the rectangle would be increased by 20. Find the dimensions of the rectangle.

7. A man has $5000 invested, part at 4% and the remainder at 5%. If the amounts invested at each rate had been interchanged, the yearly interest would have been increased $4. Find the amount invested at each rate.

8. The difference of the squares of two consecutive integers is 27. What are the integers?

9. Two vessels contain mixtures of wine and water; in one there is three times as much wine as water, in the other twice as much water as wine. Find how much must be drawn off from each to fill a third vessel which holds 5 gallons, so that its contents will be half wine and half water.

10. Helen is twice as old as Stella, and the product of their ages next year will exceed the product of their present ages by 13. How old are they now?

11. The area of a square is the same as that of a rectangle whose length is 10 feet more and whose width is 6 feet less than a side of the square. Find the area of the square.

12. A and B start together from the same place, walking at different rates. When A has walked 8 miles B doubles his pace and 3 hours later passes A. If A walks at the rate of 4 miles an hour, what is B's rate at first?

13. In a three-figured number the units' digit is zero, and the tens' digit is twice as large as the hundreds' digit. If the tens' digit and the hundreds' digit are interchanged, the resulting number is 270 larger. What is the number?

14. A man receives \$6 and board for every day that he works. He has to pay \$2 for board on each day that he does not work. After 60 days he has accumulated \$272. How many days was he idle?

15. In two years a father will be five times as old as his son, and in four years his age will be four times that of his son. How many times older than his son is he now?

16. A metallurgist has 45 grams of an alloy which is 1 part silver to 8 parts copper. How much of an alloy which is half silver and half copper must be added to this in order to have an alloy which is 1 part silver to 3 parts copper?

17. At 7 A.M. a boy scout started hiking from Camp A to Camp B, a distance of 12 miles. At 8 A.M. he doubled his pace, and at 8:30 A.M. he reduced his pace by 2 miles per hour. If he arrived at Camp B at 10 A.M., at what speed did he start his hike?

18. A man has \$6000 invested; part is at 4%, twice this amount at 5%, and the remainder at 6%. If the total annual interest on these three investments is \$296, find how much is invested at each rate.

19. When the sides of a square are increased 3 inches, the area is increased by 39 square inches. What is the side of the square?

20. Ray's age in four years will be twice what Gary's was three years ago; and Gary's age in five years will be twice what Ray's age was last year. Find their present ages.

21. A dealer bought some pictures at $1.00 each. Two of the pictures were accidentally destroyed, but he sold the remainder at $1.49 each and made $31.32 on the deal. How many pictures did he buy?

22. The sum of two numbers is 29, and the square of the larger exceeds the square of the smaller by 203. Find the numbers.

23. A man has $100 more than twice as much invested at 4% than he has invested at 6%. If the amounts invested at each rate had been interchanged, the annual interest would have been increased $22. Find the amount invested at each rate.

24. In a two-figured number the units' digit is 1 more than the tens' digit. If 495 is added to the square of the number, the result is the square of the number with its digits reversed. Find the number.

United They Stand

The united age of a man and his wife is four times the united age of their children. Last year their united age was five times the united age of their children, and six years hence their united age will be twice the united age of the children. How many children have they?

1000 to 1?

The whole numbers are arranged in five columns as follows:

1	2	3	4	5
9	8	7	6	
	10	11	12	13
17	16	15	14	
. . .	. . .	. . .	. . .	. . .

In which column will the number 1000 fall?

How Old Is Ann?

The combined age of Mary and Ann is 60 years, and Ann is as old as Mary was when Ann was a fourth the age that Mary will be when Ann is as old as Mary is now. How old is Ann?

CHAPTER V

FACTORING

47. Factoring. When an algebraic expression is the product of two or more quantities, each of these quantities is called a *factor* of the expression, and the determination of these factors is called the *factoring* of the expression.

48. Common Monomial Factors. We have already learned that to multiply a multinomial by a monomial we apply the distributive law for multiplication. Algebraically this rule may be expressed as follows:

$$a(x - y + z) = ax - ay + az.$$

Conversely, if we have an expression such as $ax - ay + az$, it may be written in the *factored* form $a(x - y + z)$.

In general, if every term of an algebraic expression contains a common factor, the expression may be represented as the product of that common factor and the quotient obtained by dividing the original expression by the common factor.

Example 1. Factor the expression $4x - 6$.

Solution. Each of the two terms contains the common factor 2; hence,

$$4x - 6 = 2(2x - 3).$$

Example 2. Factor the expression $x^3 - 2x^2 + 5x$.

Solution. Each of the three terms contains the common factor x; hence,

$$x^3 - 2x^2 + 5x = x(x^2 - 2x + 5).$$

NOTE. In factoring exercises the student should always verify his results by *mentally* forming the product of the factors he has chosen.

Example 3. Factor the expression $3a^4b - 3a^3b^2 + 6a^2b^2$.

Solution. Each of the three terms contains the common factor $3a^2b$; hence,

$$3a^4b - 3a^3b^2 + 6a^2b^2 = 3a^2b(a^2 - ab + 2b).$$

NOTE. In seeking the factor that is common to all terms, the student should remember that he is seeking the *greatest* factor that is common to all terms. In other words, when the original expression is divided by the common factor, the terms of the quotient obtained should *not* contain any other common factor.

Example 4. Factor the expression $a(x + y) + b(x + y)$.

Solution. Thinking of $(x + y)$ as a single quantity, say p, we may factor as follows:

$$a(x + y) + b(x + y) = ap + bp = p(a + b) = (x + y)(a + b).$$

NOTE. The above substitution was made only to indicate that the expression $(x + y)$ is considered as a single quantity. When you can factor without actually making such a substitution, do so, as it saves time.

EXERCISE 31

Factor:

1. $3x + 12y$
2. $7a^2 - 49ab$
3. $8x - 2x^2$
4. $x^3y^2 - x^2y^3$
5. $4a^2 - 16a$
6. $18a^3x^5 - 21a^4x^2$
7. $81a^2x^2 - 9ax$
8. $38xy + 57y^2$
9. $4x^3 - x^2 + x$
10. $3x^2 - 6xy + 9y^2$
11. $2a^3b + 4a^2b^2 + 8ab^3$
12. $x^5 - 2x^3 + 4x^2$
13. $x^3y^2z^4 + x^2y^3z^3 + x^4y^3z^4$
14. $15x^3 + 10ax^3 - 15a^2x^3$
15. $6bx^2 + 15bx - 18b$
16. $a^2b^2 - 3a^3b^3 + 6a^4b^4$
17. $2x^4 - 4x^3 + 10x^2 - 8x$
18. $x^4y^2 + x^3y^3 + x^2y^4 + xy^5$
19. $9ax^2 - 12a^2x^2 - 6ax + 3a$
20. $4a^4b - 6a^3b + 12a^2b - 10ab$
21. $x(a + b) - y(a + b)$
22. $m(2x + y) + n(2x + y)$
23. $2a(x - y) + 3b(x - y)$
24. $5x(2a - b) - 3y(2a - b)$
25. $x(x - y) + y(x - y)$
26. $a(2a - 3b) - 3b(2a - 3b)$
27. $7m(a - b) - n(a - b)$
28. $2x(2x - y) - y(2x - y)$
29. $a(x - y) + b(x - y)$
30. $3x(a - b) - y(a - b)$
31. $2x(3x - y) - y(3x - y)$
32. $x(2a + b) + y(2a + b)$
33. $a(x - y) + b(x - y) - c(x - y)$
34. $3x(2x + y) - y(2x + y) + (2x + y)$
35. $a(a + b) - 3b(a + b) + 5(a + b)$
36. $x^2(a - b) + y^2(a - b) + z^2(a - b)$
37. $ab(x^2 + y^2) + ac(x^2 + y^2) + bc(x^2 + y^2)$
38. $x^2(x + 1) + 2x(x + 1) + 7(x + 1)$
39. $3x^2(x - 2) - x(x - 2) + 4(x - 2)$
40. $x^2(2x - 3) - x(2x - 3) + (2x - 3)$

49. Common Binomial Factors. Although all the terms of an expression may not contain a common factor, it is sometimes possible to group the terms so that each group will have a common factor.

Example 1. Factor $ac + bc + ad + bd$.

Solution. Grouping together the first two terms and the last two terms, we have

$$(ac + bc) + (ad + bd).$$

Factoring the expressions within each set of parentheses, we have

$$c(a + b) + d(a + b).$$

We see here that the quantity $(a + b)$ is common to both terms; hence we have

$$(a + b)(c + d).$$

Example 2. Factor $x^2 - ax + bx - ab$.

Solution. Following the procedure described in Example 1, we have

$$\begin{aligned} x^2 - ax + bx - ab &= (x^2 - ax) + (bx - ab), \\ &= x(x - a) + b(x - a), \\ &= (x - a)(x + b). \end{aligned}$$

Example 3. Factor $12a^2 - 4ab - 3ac + bc$.

Solution.

$$\begin{aligned} 12a^2 - 4ab - 3ac + bc &= (12a^2 - 4ab) - (3ac - bc), \\ &= 4a(3a - b) - c(3a - b), \\ &= (3a - b)(4a - c). \end{aligned}$$

NOTE. In the first line of work it is usually sufficient to see that each pair contains some common factor. Thus, in Example 3, by a different grouping, we have

$$\begin{aligned} 12a^2 - 4ab - 3ac + bc &= (12a^2 - 3ac) - (4ab - bc), \\ &= 3a(4a - c) - b(4a - c), \\ &= (4a - c)(3a - b). \end{aligned}$$

This is the same result as before, because the order of factors is immaterial.

EXERCISE 32

Factor:

1. $a^2 + ab + ac + bc$

2. $3mx - nx - 3my + ny$

3. $as - bs - at + bt$

4. $5x + xy + 5y + y^2$

5. $6x^2 + 3xy - 2ax - ay$

6. $ab - 3a - 2b + 6$

7. $x^4 + a^2x^2 + b^2x^2 + a^2b^2$

8. $acx^2 + adx + bcx + bd$

9. $x - y - xy + x^2$

10. $3a + b - 6a^2 - 2ab$

11. $5ac - 5bc - a + b$

12. $x^3 - y^3 - x^2y + xy^2$

13. $a^3 + 2ab^2 - 4b^3 - 2a^2b$

14. $ax + 9by - 3ay - 3bx$

15. $2x^3 - 3x^2 + 2x - 3$

16. $6a^3 - 4a^2 + 9a - 6$

17. $5y^3 + 5y^2 - y - 1$

18. $x^3 - x^2y + 2xy^2 - 2y^3$

19. $12b^3 - 15b^2 - 16b + 20$

20. $6x^3 - 3x^2 - 14x + 7$

21. $12 - 4a + 3a^2 - a^3$

22. $9 - 6x + 12x^2 - 8x^3$

23. $x^3 + 1 + x^2 + x$

24. $6a^3 - 2 + 3a - 4a^2$

25. $ax + ay + bx + by - cx - cy$

50. Factoring of Simple Trinomials. Before considering the factoring of trinomials, let us recall how the product of two binomials gives rise to a trinomial. Thus, by Article 33, p. 58, we have:

$$\begin{aligned}(x + 5)(x + 2) &= x^2 + 7x + 10,\\ (x - 5)(x - 2) &= x^2 - 7x + 10,\\ (x + 5)(x - 2) &= x^2 + 3x - 10,\\ (x - 5)(x + 2) &= x^2 - 3x - 10.\end{aligned}$$

In each of the above products:

1. The first term of the trinomial is the product of the first terms of the binomials.
2. The second term of the trinomial is obtained by finding the algebraic sum of the product of the two *outer terms* and the two *inner terms*.
3. The third term of the trinomial is the product of the second terms of the binomials.

We now consider the converse problem, that is, to resolve trinomial expressions, such as $x^2 + 2x - 15$, into their component binomial factors. We observe from the above results that the binomial factors must satisfy the three requirements:

1. The first terms must be factors of x^2.
2. The second terms must be factors of -15.
3. These factors must be chosen in such a manner that when the product of the outer terms of the binomials is combined with the product of the inner terms, the result will be $+2x$.

The totality of binomial factors which satisfy the first two conditions is:

$$(x + 15)(x - 1),\ (x - 15)(x + 1),\ (x - 5)(x + 3),\ (x + 5)(x - 3).$$

However, only in the last of these *trial factors* do we obtain $+2x$, the required middle term. Hence,

$$x^2 + 2x - 15 = (x + 5)(x - 3).$$

Example 1. Factor $a^2 + 7a + 12$.

Solution. The second terms of the factors must be such that their product is $+12$ and their sum is $+7$. It is clear that they must be $+3$ and $+4$; hence

$$a^2 + 7a + 12 = (a + 3)(a + 4).$$

Example 2. Factor $x^2 - x - 6$.

Solution. The second terms of the factors must be such that their product is -6 and their *algebraic sum* is -1. Hence, they must have *opposite* signs, and the greater of them must be *negative* in order to give its sign to their sum. Hence, the required factors are -3 and $+2$.

$$x^2 - x - 6 = (x - 3)(x + 2).$$

When the third term of the trinomial is negative, the following method may be adopted, if desired.

Example 3. Factor $x^4 + 5x^2 - 104$.

Solution. Find two numbers whose product is 104 and whose *difference* is 5. These are 13 and 8; hence, we insert signs so that the positive will predominate:

$$x^4 + 5x^2 - 104 = (x^2 + 13)(x^2 - 8).$$

NOTE. Always verify results by forming the product *mentally*.

EXERCISE 33

Give the products:

1. $(x + 7)(x - 5)$
2. $(a + 8)(a - 1)$
3. $(x - 4)(x + 11)$
4. $(y - 3)(y + 10)$
5. $(x - 2y)(x - 3y)$
6. $(x - 3y)(x + 4y)$
7. $(x^2 - 6)(x^2 + 4)$
8. $(ab + 1)(ab - 5)$
9. $(x + \frac{1}{2})(x - \frac{1}{4})$
10. $(a - 0.7)(a + 0.4)$

Factor:

11. $x^2 + 3x + 2$
12. $a^2 + 5a - 14$
13. $b^2 - b - 2$
14. $x^2 - 7x + 12$
15. $c^2 - 6c + 9$
16. $y^2 + 2y - 8$
17. $x^2 - 5x - 24$
18. $x^2 - 8x + 7$
19. $m^2 + 10m + 21$
20. $a^2 - 11a - 12$
21. $y^2 - 7y - 30$
22. $x^2 + 10x - 75$
23. $a^2 - 3a - 18$
24. $c^2 - 11c + 24$
25. $x^2 - 9x - 10$
26. $m^2 - 13m + 36$
27. $y^2 - y - 20$
28. $x^2 - 19x + 84$

29. $x^2 + 21x + 90$	**30.** $a^2 - 12a - 85$	**31.** $a^2 + 16a - 260$
32. $x^2 + ax - 42a^2$	**33.** $x^2 - 18xy + 45y^2$	**34.** $m^2 - 26mn + 169n^2$
35. $36 - 9x - x^2$	**36.** $56 - 15x + x^2$	**37.** $84 + 20b + b^2$
38. $110 - a - a^2$	**39.** $x^2y^2 - 28xy - 60$	**40.** $a^4 - 14a^2 - 51$
41. $x^4 - 11x^2 - 152$	**42.** $a^2 - 23ab + 132b^2$	**43.** $361 - 38m^2 + m^4$
44. $a^2b^2 + 23ab + 76$	**45.** $x^2 + \frac{3}{2}x + \frac{1}{2}$	**46.** $a^2 - \frac{4}{3}a + \frac{1}{3}$
47. $x^2 + 0.2x - 0.15$	**48.** $x^2 - 0.7x + 0.12$	

51. Factoring of General Trinomials. When the coefficient of the term of the trinomial having the highest power is not unity, the number of possible trial factors is considerably greater.

It will be helpful in particular to note that

1. If the third term of the trinomial is *positive*, the second terms of its factors both have the *same sign*, and this sign is the same as that of the middle term of the trinomial.
2. If the third term of the trinomial is *negative*, the second terms of its factors have *opposite signs*.

Example 1. Factor $5x^2 + 7x - 6$.

Solution. Write $(5x \quad 2)(x \quad 3)$ for a first trial. Since 2 and 3 must have opposite signs, the *difference* between the inner and the outer products must be $7x$. However, since $(5 \times 3) - (2 \times 1) = 13$, this combination fails to give the correct coefficient of the middle term.

Next try $(5x \quad 3)(x \quad 2)$. Since $(5 \times 2) - (3 \times 1) = 7$, these factors will be correct if we insert signs so that the *positive* predominates. Thus,

$$5x^2 + 7x - 6 = (5x - 3)(x + 2).$$

Example 2. Factor $7x^2 - 19x + 10$.

Solution. We note that the factors which give 10 are both negative, and so we write $(7x - \quad)(x - \quad)$. We now must place factors of 10 in the parentheses, so that the *sum* of the inner and outer products yields 19. The possible factors of 10 are 10 and 1, or 5 and 2; and since $(7 \times 2) + (5 \times 1) = 19$,

$$7x^2 - 19x + 10 = (7x - 5)(x - 2).$$

Example 3. Factor $4x^2 - 12xy + 9y^2$.

Solution.

$$4x^2 - 12xy + 9y^2 = (2x - 3y)(2x - 3y) = (2x - 3y)^2.$$

Example 4. Factor $8 + 6x - 5x^2$.

Solution. $8 + 6x - 5x^2 = (4 + 5x)(2 - x).$

EXERCISE 34

Give the products:

1. $(2x + 3)(2x + 7)$
2. $(2x - 3)(x + 6)$
3. $(3x - 1)(3x + 1)$
4. $(4x + 1)(x - 3)$
5. $(2x - 1)(3x - 2)$
6. $(2ax + 3)(3ax - 1)$
7. $(4a + b)(3a - b)$
8. $(5x - y)(5x - 2y)$
9. $(5 - x)(3 + 2x)$
10. $(3 + 2x)(1 + 4x)$

Factor:

11. $3x^2 + 5x + 2$
12. $3x^2 + x - 2$
13. $3x^2 + 7x - 6$
14. $2x^2 - x - 15$
15. $2x^2 - x - 1$
16. $3x^2 + 11x + 6$
17. $4x^2 + x - 14$
18. $4x^2 + 11x - 3$
19. $3x^2 - 10x + 3$
20. $3x^2 + 10x + 7$
21. $3a^2 + a - 4$
22. $2a^2 - a - 28$
23. $5a^2 - 12a + 4$
24. $3a^2 - 2a - 5$
25. $4y^2 - y - 5$
26. $2b^2 - 5b + 2$
27. $3c^2 + 13c + 12$
28. $2m^2 + 9m + 10$
29. $6a^2 + 17a - 14$
30. $5b^2 - 12b + 4$
31. $27a^2 - 3ab - 4b^2$
32. $5x^2 - 11xy + 2y^2$
33. $4m^2 - 8mn + 3n^2$
34. $15a^2 - 22ab + 8b^2$
35. $6x^2y^2 - 5xy - 4$
36. $2a^2b^2 + 11ab + 15$
37. $6a^4 - 13a^2 + 6$
38. $2x^4 + 11x^2y^2 - 21y^4$
39. $9x^2y^2 - 24xyz + 7z^2$
40. $12a^2b^2 - 11abcd + 2c^2d^2$
41. $6 - 11x + 4x^2$
42. $6 - x - 2x^2$
43. $5 - 9x - 2x^2$
44. $9 + 35x - 4x^2$
45. $5 + 13x - 6x^2$
46. $8x^2 - 38xy + 35y^2$
47. $24m^2 + 22mn - 21n^2$
48. $5 + 32a - 21a^2$
49. $14 + 23x + 3x^2$
50. $16x^2 - 6x - 27$

52. Difference of Two Squares. Multiplying $(a + b)$ by $(a - b)$, we obtain the identity:

$$(a + b)(a - b) = a^2 - b^2,$$

a result which may be stated as follows:

The product of the sum and the difference of any two quantities is equal to the difference of their squares.

Conversely, we also have:

The difference of the squares of any two quantities is equal to the product of their sum and their difference.

Example 1. Factor $9x^2 - 16y^2$.

Solution. $9x^2 - 16y^2 = (3x)^2 - (4y)^2$.

Hence, the first factor is the sum of $3x$ and $4y$, and the second factor is the difference of $3x$ and $4y$; thus,

$$9x^2 - 16y^2 = (3x + 4y)(3x - 4y).$$

Example 2. Factor $1 - 36m^6$.

Solution.

$$\begin{aligned} 1 - 36m^6 &= (1)^2 - (6m^3)^2, \\ &= (1 + 6m^3)(1 - 6m^3). \end{aligned}$$

NOTE. The intermediate step may be omitted when the principle is understood.

EXERCISE 35

Give the products:

1. $(2a + b)(2a - b)$
2. $(2y - 5)(2y + 5)$
3. $(7m - 4n)(7m + 4n)$
4. $(3a^2 - 5)(3a^2 + 5)$
5. $(1 + x^3)(1 - x^3)$
6. $(4y + 3x)(3x - 4y)$
7. $\left(\frac{2a}{3} - \frac{3b}{2}\right)\left(\frac{2a}{3} + \frac{3b}{2}\right)$
8. $(\frac{1}{3} - a)(\frac{1}{3} + a)$
9. $(2y + 0.5)(2y - 0.5)$
10. $(3x^2y + 2z^3)(3x^2y - 2z^3)$

Factor:

11. $x^2 - 16$
12. $25 - a^2$
13. $n^4 - 64$
14. $16x^2 - 49y^2$
15. $9a^2 - 25b^2$
16. $1 - 81x^2$
17. $\frac{x^2}{4} - y^2$
18. $121a^2 - 36b^2$
19. $a^2b^2 - 1$
20. $m^2 - \frac{1}{9}$
21. $36p^2 - 49q^2$
22. $25c^2 - 144$
23. $1 - 100x^2$
24. $81a^4 - 4x^4$
25. $x^6 - 25$
26. $1 - a^2b^2$
27. $a^2x^4 - 49$
28. $4a^2b^2 - 9c^2d^2$
29. $a^2 - 64x^6$
30. $x^4 - 25$
31. $16x^{16} - 25y^{24}$
32. $a^6b^6 - 4$
33. $a^2b^4 - c^4d^6$
34. $1 - 100x^4y^6z^8$
35. $64a^{16} - 49b^{14}$
36. $81a^4 - 100x^4$
37. $\frac{1}{4}p^2 - \frac{1}{9}q^2$
38. $\frac{1}{16}a^2 - \frac{1}{49}b^2$
39. $0.01x^2 - 1$
40. $0.64m^2 - 0.25n^2$

Find, by factoring, the value of each of the following:

41. $(525)^2 - (475)^2$
42. $(272)^2 - (271)^2$
43. $(101)^2 - (99)^2$
44. $(349)^2 - (339)^2$
45. $(750)^2 - (250)^2$
46. $(732)^2 - (268)^2$
47. $(666)^2 - (166)^2$
48. $(1693)^2 - (307)^2$
49. $(1532)^2 - (1468)^2$
50. $(1267)^2 - (967)^2$

53. Difference of Two Squares When One or Both Are Binomials. When one or both of the squares is a binomial (or, in fact, any multiple expression), we employ the method described in the preceding article.

Example 1. Factor $(a + 2b)^2 - 9x^2$.

Solution. The sum of $a + 2b$ and $3x$ is $a + 2b + 3x$, and their difference is $a + 2b - 3x$; hence,

$$(a + 2b)^2 - 9x^2 = (a + 2b + 3x)(a + 2b - 3x).$$

Example 2. Factor $m^2 - (2a - 3b)^2$.

Solution. The sum of m and $2a - 3b$ is $m + 2a - 3b$, and their difference is $m - (2a - 3b) = m - 2a + 3b$. Hence,

$$m^2 - (2a - 3b)^2 = (m + 2a - 3b)(m - 2a + 3b).$$

If the factors contain similar terms, they should be collected so as to give the result in its simplest form.

Example 3. Factor $(3x + 5y)^2 - (2x - y)^2$.

Solution.

$$\begin{aligned}(3x + 5y)^2 - (2x - y)^2 &= [(3x + 5y) + (2x - y)][(3x + 5y) - (2x - y)],\\ &= (3x + 5y + 2x - y)(3x + 5y - 2x + y),\\ &= (5x + 4y)(x + 6y).\end{aligned}$$

EXERCISE 36

Factor:

1. $(a + b)^2 - c^2$
2. $a^2 - (b + c)^2$
3. $(a - b)^2 - c^2$
4. $a^2 - (b - c)^2$
5. $x^2 - (2y - z)^2$
6. $(x - y)^2 - 4z^2$
7. $(a + b)^2 - 4b^2$
8. $9x^2 - (2x + y)^2$
9. $9x^2 - (2y - 1)^2$
10. $(a - 2b)^2 - 4$
11. $(3m - n)^2 - 16$
12. $1 - (3x - 2y)^2$
13. $(x + y)^2 - x^2$
14. $x^2 - (y - x)^2$
15. $(a + b)^2 - (c + d)^2$
16. $(a - b)^2 - (c - d)^2$
17. $(a - b)^2 - (c + d)^2$
18. $(a + b)^2 - (c - d)^2$
19. $(3a + 5b)^2 - (a - 2b)^2$
20. $(4a - 3b)^2 - (a + 2b)^2$
21. $(5x - 3y)^2 - (4x + y)^2$
22. $(2x - y)^2 - (x - y)^2$
23. $(7x - 2)^2 - (3x + 5)^2$
24. $(5x + 3)^2 - (4x - 5)^2$
25. $(3a + 2)^2 - (2a + 3)^2$
26. $16a^2 - (3a - 5)^2$
27. $(2m + n)^2 - (m - 3n)^2$
28. $(4p - q)^2 - (4p + q)^2$
29. $(x - 5y)^2 - (3x + y)^2$
30. $(2x - 3y)^2 - (3x + 7y)^2$

31. $(5a - x)^2 - (4a - x)^2$
32. $(5a + 3)^2 - (6a - 5)^2$
33. $16 - (2a + 7)^2$
34. $(5x - 7)^2 - 1$
35. $(2x - 1)^2 - (2x - 3)^2$
36. $(3y + 5)^2 - (2y - 9)^2$

54. Summary of Factoring. In many problems of factoring it is possible to resolve the given expression into more than two factors. Thus, to factor $2a^2 - 2b^2$ we first remove the common factor 2, and then factor the remainder since it is the difference of two squares.

$$2a^2 - 2b^2 = 2(a^2 - b^2) = 2(a + b)(a - b).$$

In general, to factor more complex expressions we proceed as follows:

1. In *all* problems of factoring, remove monomial factors before attempting any other method of factoring.
2. Depending on the number of terms in the multinomial that remains, attempt to factor as follows:
   (*a*) If it has two terms, try to factor as the difference of two squares.
   (*b*) If it has three terms, try to factor as a trinomial.
   (*c*) If it has four or more terms, try to factor by grouping.

This process is applied to any new factors obtained until every remaining factor can be factored no further.

Example 1. Factor $x^3 - 2x^2 - 3x$.
Solution. This expression contains the common factor x; hence,

$$x^3 - 2x^2 - 3x = x(x^2 - 2x - 3).$$

Since three terms remain in the second factor, we attempt to factor as a trinomial; and since

$$x^2 - 2x - 3 = (x + 1)(x - 3),$$

we have finally;

$$x^3 - 2x^2 - 3x = x(x + 1)(x - 3).$$

Example 2. Factor $a^4 - x^4$.
Solution.

$$\begin{aligned} a^4 - x^4 &= (a^2 + x^2)(a^2 - x^2), \\ &= (a^2 + x^2)(a + x)(a - x). \end{aligned}$$

Example 3. Factor $x^4 - 7x^2y^2 + 12y^4$.
Solution.

$$\begin{aligned} x^4 - 7x^2y^2 + 12y^4 &= (x^2 - 3y^2)(x^2 - 4y^2), \\ &= (x^2 - 3y^2)(x + 2y)(x - 2y). \end{aligned}$$

Example 4. Factor $a^2x^2 - a^2y^2 - b^2x^2 + b^2y^2$.

Solution.

$$\begin{aligned} a^2x^2 - a^2y^2 - b^2x^2 + b^2y^2 &= (a^2x^2 - a^2y^2) - (b^2x^2 - b^2y^2), \\ &= a^2(x^2 - y^2) - b^2(x^2 - y^2), \\ &= (x^2 - y^2)(a^2 - b^2), \\ &= (x + y)(x - y)(a + b)(a - b). \end{aligned}$$

Example 5. Factor $a^2 - b^2 + a + b$.

Solution.

$$\begin{aligned} a^2 - b^2 + a + b &= (a^2 - b^2) + (a + b), \\ &= (a + b)(a - b) + (a + b)1, \\ &= (a + b)(a - b + 1). \end{aligned}$$

EXERCISE 37

Express each of the following as a product of three factors:

1. $x^4 - x^2y^2$ **2.** $7a^3 - 63a$
3. $2x^2 - 18y^2z^2$ **4.** $a^2x^3 - 64x^3y^2$
5. $6x^3 + 34x^2 + 20x$ **6.** $8x^2 - 4x - 24$
7. $24x^2y^2 - 30xy^3 - 36y^4$ **8.** $2ax^2 + 19ax + 35a$
9. $m^4 - 16$ **10.** $x^4y^4 - 1$
11. $x^3 + x^2y - xy^2 - y^3$ **12.** $a^2x + a^2y + abx + aby$

Express each of the following as a product of four factors:

13. $a^4 - 5a^2b^2 + 4b^4$ **14.** $ax^4 - ay^4$
15. $a^2x^2 - a^2y^2 + abx^2 - aby^2$ **16.** $y^8 - c^8$

Express each of the following as a product of as many factors as you can obtain:

17. $1 - (x - y)^2$ **18.** $a^4b^4 - 256$
19. $28x^4y + 64x^3y - 60x^2y$ **20.** $6a^2b - 46ab - 72b$
21. $a^2 - 6a - 247$ **22.** $(2x - 3)^2 - (x - 1)^2$
23. $x^8 - 1$ **24.** $4(x - y)^3 - (x - y)$
25. $x^2 - 4y^2 + x - 2y$ **26.** $6x^3y + 10x^2y - 4xy$
27. $9(x + y)^2 - (2x - y)^2$ **28.** $3x^3 - 3x^2 + x - 1$
29. $abc - a^7b^5c^3$ **30.** $x^4y^4 - 9y^4$
31. $40 - 3x - x^2$ **32.** $3bx^2 + 19bx + 26b$
33. $axy + bcxy - az - bcz$ **34.** $(a - 2b)^2 - 4$
35. $60x^2 - 46xy - 28y^2$ **36.** $ax^2 - ay^2 + bx^2 - by^2$
37. $a^3 - 2a^2 - 4a + 8$ **38.** $x^3y^3 + x^2y^4 + xy^5$
39. $(x^2 - 4y^2) + x(x + 2y)$ **40.** $10 - 18x - 4x^2$
41. $x^2(2x + 1) - (4x^2 - 1)$
42. $(m + n)(2x^2 - 3y^2) - (m + n)(x^2 - 2y^2)$

43. $acx^2 + adx + bcx + bd$
44. $(x^2 - y^2) - (x - y)^2$
45. $21x^2 + 82x - 39$
46. $y^3 - 2y^2 - y + 2$
47. $(a + b)^2 + a + b$
48. $(3a + 1)^2 - (2a - 1)^2$
49. $3x^4 - 2x^3 + 3x^2 - 2x$
50. $(c - 2d)^2 - 1$

55. The Sum and Difference of Two Cubes. (Optional.) By direct multiplication we can easily show that

$$A^3 + B^3 = (A + B)(A^2 - AB + B^2),$$

and

$$A^3 - B^3 = (A - B)(A^2 + AB + B^2).$$

Using these two relations as formulas, we can factor any binomial that can be written as the sum or the difference of two cubes.

Example 1. Factor $8x^3 + 27y^3$.

Solution. The given expression may be represented as the sum of two cubes in the following manner:

$$8x^3 + 27y^3 = (2x)^3 + (3y)^3.$$

Comparing the new expression with the first formula above, we see that $A = 2x$ and $B = 3y$. Hence substituting these values in the formula, we have

$$\begin{aligned}(2x)^3 + (3y)^3 &= [(2x) + (3y)][(2x)^2 - (2x)(3y) + (3y)^2],\\ &= (2x + 3y)(4x^2 - 6xy + 9y^2).\end{aligned}$$

Example 2. Factor $(2a - 1)^3 - a^3$.

Solution. Comparing the given expression with the second formula above, we see that $A = 2a - 1$ and $B = a$. Hence, substituting these values in the formula, we have

$$\begin{aligned}(2a - 1)^3 - a^3 &= [(2a - 1) - a][(2a - 1)^2 + (2a - 1)a + a^2],\\ &= (a - 1)(7a^2 - 5a + 1).\end{aligned}$$

EXERCISE 38

Factor:

1. $a^3 - 1$
2. $m^3 + 8n^3$
3. $27x^3 + y^3$
4. $z^3 - 8$
5. $y^3 + \frac{1}{64}$
6. $8c^3 - 27$
7. $x^3y^3 + 1$
8. $125 - p^3$
9. $a^3b^3 + c^3$
10. $\dfrac{x^3}{27} - \dfrac{y^3}{64}$
11. $m^6 - n^3$
12. $8p^3 + \frac{1}{8}q^3$
13. $(a + b)^3 - c^3$
14. $(a - b)^3 + c^3$
15. $a^3 - (b + c)^3$
16. $a^3 + (b - c)^3$
17. $(x + 1)^3 - 8x^3$
18. $x^3 + (2x - 3)^3$
19. $(2a + 1)^3 + (a + 1)^3$
20. $(q + 1)^3 - 8(q - 1)^3$

Factor completely:

21. $2ax^3 - 54a^4$ **22.** $x^4 + xy^3$ **23.** $x^6 - 64$

24. $a^6 - 7a^3 - 8$ **25.** $x^4 + x^3 + x + 1$ **26.** $2ab^4 - 16abc^3$

27. $a^4b^4 + ab$ **28.** $p^4 - p^3q + pq^3 - q^4$ **29.** $8x^6 + 7x^3 - 1$

30. $a^6 + x^6$ **31.** $c^6 + c^3 - 2$ **32.** $a^4b - ab^4$

33. $x^3(a + b) - 8(a + b)$ **34.** $16x^2y^3z - 2x^2z^4$ **35.** $a^6 - 2a^3 + 1$

56. The Factor Theorem. (Optional.) One of the most useful methods of factoring depends on the following theorem:

The Factor Theorem

If any polynomial in x becomes equal to 0 when a is written for x, then the polynomial is exactly divisible by x − a.

NOTE. A **polynomial** in x is an expression of the form $ax^n + bx^{n-1} + \cdots + px + q$, in which the coefficients are constants and n is a positive integer.

Proof. Let P stand for the polynomial. Divide P by $x - a$ until the remainder no longer contains x. Let R denote this remainder, and Q the quotient obtained. Then

$$\frac{P}{x-a} = Q + \frac{R}{x-a},$$

or

$$P = Q(x - a) + R.$$

Since this equation is true for all values of x, we will assume that x equals a. By hypothesis, the substitution of a for x makes P equal to 0; thus

$$0 = Q \cdot 0 + R.$$

Hence

$$R = 0.$$

Since the remainder is zero, the given polynomial is exactly divisible by $x - a$.

The following examples illustrate the application of this theorem.

Example 1. Factor $x^3 - x - 6$.

Solution. By trial we find that this expression reduces to 0 when 2 is substituted for x. Hence, by the factor theorem $x - 2$ is a factor. Dividing the given expression by $x - 2$, we obtain the quotient $x^2 + 2x + 3$. Thus, in factored form we have

$$x^3 - x - 6 = (x - 2)(x^2 + 2x + 3).$$

NOTE. In seeking values which will make the given polynomial equal to zero, observe that we test only those numerical values which divide evenly into the constant term of the polynomial. This is true only if the highest power of x in the polynomial has a coefficient equal to ± 1 and the other coefficients are integers. We shall restrict ourselves to such polynomials. Thus, in Example 1, we would have tried the values ± 1, ± 2, ± 3, and ± 6. If none of these make the polynomial equal to zero, then we may conclude that the polynomial cannot be factored by means of the factor theorem.

Example 2. Factor $x^4 + 5x - 6$.

Solution. The only numbers which divide evenly into the constant term are ± 1, ± 2, ± 3, and ± 6; and since $(+1)^4 + 5(+1) - 6$ equals zero we know that $x - 1$ is a factor. Hence, dividing $x^4 + 5x - 6$ by $x - 1$, we have

$$x^4 + 5x - 6 = (x - 1)(x^3 + x^2 + x + 6).$$

For the new factor, $x^3 + x^2 + x + 6$, we again seek a zero; and we find that $x = -2$ makes the expression zero. Thus $x - (-2)$ or $x + 2$ is a factor; and again by division we find the other factor to be $x^2 - x + 3$. Thus far then,

$$x^4 + 5x - 6 = (x - 1)(x + 2)(x^2 - x + 3).$$

Having reduced the remaining factor to a trinomial, we attempt to factor it as a trinomial; and since it cannot be factored, we conclude that the original expression can be factored no further.

EXERCISE 39

By the factor theorem show that:

1. $x - 1$ is a factor of $x^2 - 5x + 4$.
2. $x + 2$ is a factor of $x^3 + 2x^2 + x + 2$.
3. $x + 1$ is a factor of $x^3 + 1$.
4. $x - 2$ is a factor of $x^3 - 5x^2 + 12$.
5. $x + 1$ and $x + 2$ are factors of $x^3 + 2x^2 - x - 2$.
6. $x - 1$ and $x + 1$ are factors of $x^4 - 1$.

Express each of the following as a product of two factors:

7. $x^3 + 2x^2 - 1$
8. $x^3 + x^2 + x - 3$
9. $x^3 + x + 10$
10. $x^3 - 3x^2 - 16$
11. $x^3 - 2x^2 + x - 12$
12. $x^3 + 2x^2 - 7x - 8$
13. $x^5 + x - 2$
14. $x^5 + 2x^3 - 2x + 1$

Express each of the following as a product of three factors:

15. $x^3 + 6x^2 + 11x + 6$
16. $x^3 + 4x^2 + x - 6$
17. $x^4 - 3x^3 + 6x - 4$
18. $x^4 + x^3 - x - 1$
19. $x^5 - x^4 - 2x^3 + 2x^2 - 2x - 4$
20. $x^5 + x^2 - x - 1$

Factor completely:

21. $x^3 + 2x^2 - 5x - 6$

22. $x^3 - 3x^2 + 4x - 4$

23. $x^3 - 3x - 2$

24. $x^3 - 7x + 6$

25. $x^3 - 2x^2 + 3$

26. $x^3 + 3x^2 - 4$

27. $x^3 + 4x^2 + 7x + 6$

28. $x^3 - 5x^2 + 8x - 4$

29. $x^3 + 3x^2y + 3xy^2 + y^3$

30. $x^3 - 7x^2y + 11xy^2 - 5y^3$

31. $x^4 - 11x^2 + 18x - 8$

32. $x^4 + x^3 - 2$

33. $x^4 - 2x^3 + x^2 - 4$

34. $x^4 + 5x^3 + 5x^2 - 5x - 6$

35. $x^4 - x^3 - 3x^2 + 5x - 2$

36. $x^4 + x^3 - x^2 - 1$

37. $x^5 + x^4 - x^2 - 7x + 6$

38. $x^5 + x^4 - 2x^3 - 2x^2 + x + 1$

39. $x^5 - 10x^3 - 20x^2 - 15x - 4$

40. $x^5 - 10x^2 + 15x - 6$

REVIEW OF CHAPTER V

A

Factor:

1. $2x^3 - 8x^2$

2. $5ab^2 + 25a^2b$

3. $p^3q - pq^2$

4. $9x^2y^3 + 3x^2y^2$

5. $3x^3 + 3x^2 + 3x$

6. $5m^3n^2 - 10m^2n^3 + 5m^2n^2$

7. $a^4b^5 + a^6b^3 - a^5b^4$

8. $6x^2y^3 - 9x^4y - 3x^2y$

9. $4u^3v^2 - 2u^4v^3 - 8u^3v^3 + 10u^2v^5$

10. $5pq^3 - 10p^3q^2 - 5pq + 15p^2q^2$

11. $x(2x - y) + y(2x - y)$

12. $3x(a + b) - y(a + b)$

13. $7a(a - 3) - 4(a - 3)$

14. $a(a - x) + x(a - x)$

15. $3m(m - 3n) - n(m - 3n)$

16. $a(2a - b) - 3b(2a - b)$

17. $x(a + b) + y(a + b) + z(a + b)$

18. $xy(m - n) + xz(m - n) + yz(m - n)$

19. $x^2(x - 3) - 3x(x - 3) + (x - 3)$

20. $a^2(2a + b) - 5ab(2a + b) - b^2(2a + b)$

B

Factor by grouping:

1. $ax + ay + bx + by$

2. $ma + mb - na - nb$

3. $3a^3 + 6a^2 - 2a - 4$

4. $x^3 - 6x^2 + 2x - 12$

5. $2m^3 - 3m^2 - 4m + 6$

6. $3p^3 + 9p^2 + p + 3$

7. $x^3 + x^2 - x - 1$

8. $a^3 - a^2x + ax^2 - x^3$

9. $3a^3 - a^2 - 3a + 1$

10. $2x^3 + 5x^2 - 4x - 10$

11. $a^2 + ac - ab - bc$

12. $m^2 - mx - my + xy$

13. $ab + 3a + 2b + 6$

14. $xy - x - y + 1$

15. $6mn - 2m + 3n - 1$

16. $6pq + 9p + 4q + 6$

17. $2x^3 + 4x^2y - 3xy^2 - 6y^3$

18. $3a^3 - 9a^2b - 2ab^2 + 6b^3$

19. $a^3 - 4a^2y - 2ay^2 + 8y^3$

20. $3p^3 + 6p^2q + 2pq^2 + 4q^3$

C

Factor:

1. $x^2 - x - 42$
2. $a^2 + 2a - 48$
3. $m^2 + 7mn + 10n^2$
4. $x^2 + 3xy - 18y^2$
5. $35 - 2c - c^2$
6. $20 + 8x - x^2$
7. $x^2y^2 - 17xy + 42$
8. $p^2 + 20p + 64$
9. $x^2 - 4x - 192$
10. $y^4 + 16y^2 - 36$
11. $3x^2 - 5x - 12$
12. $8x^2 - 2xy - 3y^2$
13. $4a^2 + 8a + 3$
14. $10m^2 - 3mn - 4n^2$
15. $4x^2y^2 - 11xy - 3$
16. $6p^2 + 13pq - 5q^2$
17. $25a^2 + 10ab - 3b^2$
18. $15y^2 + 8y - 12$
19. $50x^2 - 15x - 104$
20. $20m^2 + 63mn - 45n^2$

D

Factor:

1. $x^2 - 9y^2$
2. $25a^2 - x^2$
3. $x^2y^2 - a^2b^2$
4. $16m^2 - 49n^2$
5. $100 - 81x^4$
6. $1 - a^2b^2c^2$
7. $\frac{4}{9}x^2 - \frac{9}{16}$
8. $\frac{25}{36}p^2q^2 - 1$
9. $0.81a^2 - 0.49b^2$
10. $0.01x^2 - 0.0001$
11. $(x + y)^2 - 9$
12. $4 - (x - y)^2$
13. $(2a - 3b)^2 - a^2$
14. $(m + n)^2 - m^2$
15. $(2x + 5)^2 - (2x + 3)^2$
16. $(x - 3y)^2 - (3x - 2y)^2$
17. $(5a - 2b)^2 - (4a + 7b)^2$
18. $(2m - n)^2 - (3m - 2n)^2$
19. $(2x - 3)^2 - (3x + 5)^2$
20. $(9a + 4)^2 - (5a - 9)^2$

E

Factor completely:

1. $9a^2x^2 - 49x^2$
2. $x^4 - 4x^3 + 4x^2$
3. $2x^3 + 4x^2 - 30x$
4. $a^4 - x^4$
5. $4y^4 - 4y^2$
6. $6m^2 - 36m - 96$
7. $5a^3b - 5a^2b^2 - 60ab^3$
8. $x^4 - x^2 - 12$
9. $x(x^2 - 1) + 3(x^2 - 1)$
10. $3ax^3 - 18ax^2 - 21ax$
11. $x^5 - 8x^3 + 16x$
12. $4c^3 - 100cy^4$
13. $81x^5y^4 - x$
14. $a^2(x + 1) - b^2(x + 1)$
15. $a^3 - a^2 - 4a + 4$
16. $(2x + 1)^2 - 5(2x + 1)$
17. $(3a + 5)^2 - (a - 3)^2$
18. $(a + b)^2 + 4(a + b) + 3$
19. $x^2 - (3x^2 - 2)^2$
20. $(x^2 + y^2)^2 - 4x^2y^2$
21. $(x + 5)^2 - (x + 5) - 12$
22. $2a^4 + 6a^3 - 8a^2 - 24a$
23. $(a^2 + 2ab + b^2) - c^2$
24. $2(x - y)^2 + 5(x - y) - 12$

25. $9x^2(4x^2 - 1) - (4x^2 - 1)$
26. $ax^5 + 3ax^3 - 4ax$
27. $64a^3x^5 + 24a^2x^4 - 54ax^3$
28. $(x^2 + 6)^2 - 25x^2$
29. $x^2 + 2xy + y^2 - z^2$
30. $a^2 - b^2 - 2a + 2b$

F (Optional)

Factor:

1. $p^3 - 27$
2. $\frac{x^3}{8} + 1$
3. $a^3x^3 + 1$
4. $64 - y^3$
5. $(x - 1)^3 - 8$
6. $1 - (1 - 2x)^3$
7. $a^2x^4 + a^5x$
8. $x^6 + 9x^3 + 8$
9. $54x^2y^4 - 2x^2y$
10. $x^6 - 1$
11. $x^3 - 2x^2 - 9$
12. $x^3 - 2x + 1$
13. $x^4 - 2x^3 + 1$
14. $x^4 + x^3 + 2x - 4$
15. $x^5 + x^4 - 9x + 7$
16. $x^5 - x^4 - 6x - 4$
17. $x^4 - 4x^3 - 25x$
18. $x^4 - x^3 + 4x - 16$
19. $x^4 - 2x^3 - 3x^2 + 4x + 4$
20. $x^4 - 2x^3 + 2x - 1$

What's the Difference?

Show that all numbers divisible by 4, that is 4, 8, 12, 16, and so on, can be written as the difference of two squares.

Chattanooga Choo Choo

A fast train leaves Chattanooga at 2 P.M. and reaches Memphis at 6 P.M. A slower train leaves Memphis at 12 noon and arrives in Chattanooga at 6 P.M. If both trains travel uniformly, at what time do they meet?

A Perfect Square

Show that the sum of 1 and the product of any four consecutive integers is a perfect square.

CHAPTER VI

FRACTIONS

57. Notations. When the sum of N equal numbers is 1, we define the quantity $\frac{1}{N}$ to represent the magnitude of each of these numbers. The sum of M such numbers is represented by the **fraction** $\frac{M}{N}$. In this fraction, M is called the **numerator** and N the **denominator;** and it is read "M over N" or "M divided by N."

58. Reduction of Fractions. The rules governing the use of fractions in algebra are precisely the same as the corresponding rules used in arithmetic.

Fractions. Rule 1

The value of a fraction is not changed if we multiply or divide both the numerator and denominator by the same quantity (zero excluded).

Thus, just as $\frac{6}{9} = \frac{2}{3}$ by dividing both terms by 3, so $\frac{ax}{bx} = \frac{a}{b}$ by dividing both terms by x.

If we divide both the numerator and the denominator by the *highest common factor*, the resulting fraction is said to be **reduced to its lowest terms.**

Example 1. Reduce $\frac{21x^2y}{28xy^2}$ to its lowest terms.

Solution. Dividing numerator and denominator by $7xy$, we have

$$\frac{21x^2y}{28xy^2} = \frac{3x}{4y}.$$

Example 2. Reduce $\dfrac{24a^3x^3}{6a^3x - 6a^2x^2}$ to its lowest terms.

Solution. $$\frac{24a^3x^3}{6a^3x - 6a^2x^2} = \frac{24a^3x^3}{6a^2x(a - x)} = \frac{4ax^2}{a - x}.$$

NOTE. It is advisable, in general, not to perform any division until both numerator and denominator are expressed in factored form.

Example 3. Reduce $\dfrac{6x^2 - 8xy}{9xy - 12y^2}$ to its lowest terms.

Solution. $$\frac{6x^2 - 8xy}{9xy - 12y^2} = \frac{2x(3x - 4y)}{3y(3x - 4y)} = \frac{2x}{3y}.$$

EXERCISE 40

Reduce to the lowest terms:

1. $\dfrac{36}{63}$ 2. $\dfrac{143}{187}$ 3. $\dfrac{15a^2b}{9ab}$ 4. $\dfrac{10x^3y^2}{12x^2y^3}$

5. $\dfrac{32x^3y^2z}{56x^2y^2z^2}$ 6. $\dfrac{46a^3b^4c^5}{69a^2b^3c^4}$ 7. $\dfrac{24a^3c^2x^3}{36a^5x^2}$ 8. $\dfrac{21m^7n^6}{63m^6n^7}$

9. $\dfrac{-51abcx^3}{-85a^3bcx}$ 10. $\dfrac{38bx^5y^4}{95ax^3y^7}$ 11. $\dfrac{x^2 - x}{x^2 + x}$ 12. $\dfrac{4a + 12}{6a^2 + 18a}$

13. $\dfrac{7a + 7b}{a^2 + ab}$ 14. $\dfrac{ax + bx}{a^2 - b^2}$ 15. $\dfrac{ax}{a^2x^2 - ax}$ 16. $\dfrac{3x^2 - 6xy}{2x^2y - 4xy^2}$

17. $\dfrac{2ab + 2b}{a^2 - 1}$ 18. $\dfrac{mx + my}{4m}$ 19. $\dfrac{mx - my}{mx + my}$ 20. $\dfrac{x^2 - y^2}{x^2 + xy}$

21. $\dfrac{abx + bx^2}{acx + cx^2}$ 22. $\dfrac{4a^2 - 9b^2}{4a^2 + 6ab}$ 23. $\dfrac{x^2 - 2x + 1}{x^2 - 1}$

24. $\dfrac{12a^3b^2 + 18a^2b^3}{12a^3b^2 + 8a^4b}$ 25. $\dfrac{y(2c^2 - 3cy)}{c(4c^2y - 9y^3)}$ 26. $\dfrac{m^2 - 5m}{m^2 - 4m - 5}$

27. $\dfrac{a^2 + 3ab + 2b^2}{a^2 - 4b^2}$ 28. $\dfrac{p^2 - q^2}{(p - q)^2}$ 29. $\dfrac{x^2 - 6x + 9}{x^2 - 4x + 3}$

30. $\dfrac{x^2 - 2x - 15}{x^2 - 8x + 15}$ 31. $\dfrac{2x^2 - 5x + 3}{2x^2 - x - 3}$ 32. $\dfrac{3x^2 + 8x + 5}{3x^2 - x - 10}$

33. $\dfrac{9x^2 - 25}{3x^2 - 11x + 10}$ 34. $\dfrac{x^2 + 4x - 21}{x^2 + 16x + 63}$

35. $\dfrac{a^2x^2 - 16a^2}{ax^2 + 9ax + 20a}$ 36. $\dfrac{3x^4 + 9x^3y + 6x^2y^2}{x^4 + x^3y - 2x^2y^2}$

37. $\dfrac{x^2y^2 - xy}{x^2y^2 - 1}$

38. $\dfrac{ax^4 - 3ax^3 - 4ax^2}{a^2x^3 + 5a^2x^2 + 4a^2x}$

39. $\dfrac{18m^2 + 3mn - 10n^2}{21m^2 - 26mn + 8n^2}$

40. $\dfrac{a^3b^4 + a^2b^3}{a^5b^4 - a^3b^2}$

41. $\dfrac{ma + mb + na + nb}{ma - mb + na - nb}$

42. $\dfrac{x^3 - 3x^2 + x - 3}{x^2 - 9}$

43. $\dfrac{x^3 + 2x^2 + x + 2}{x^3 - 5x^2 + x - 5}$

44. $\dfrac{acx^2 + adx + bcx + bd}{c^2x^2 + 2cdx + d^2}$

45. $\dfrac{2a^3 + 2a^2 - 3a - 3}{2a^3 - 2a^2 - 3a + 3}$

46. $\dfrac{6a^3 + 3a^2b + 2ab^2 + b^3}{3a^3 + 6a^2b + ab^2 + 2b^3}$

59. Multiplication of Fractions. For the multiplication of fractions, we have:

Fractions. Rule 2

To multiply two or more fractions, multiply the numerators to obtain the numerator of the product, and multiply the denominators to obtain the denominator of the product.

Example 1. Find the product $\dfrac{2a}{3b} \times \dfrac{9b^2}{4a^2}$, and simplify.

Solution.

$$\frac{2a}{3b} \times \frac{9b^2}{4a^2} = \frac{18ab^2}{12a^2b} = \frac{3b}{2a}.$$

Example 2. Find the product $\dfrac{x+1}{x-3} \times \dfrac{x-3}{x^2-1}$, and simplify.

Solution.

$$\frac{x+1}{x-3} \times \frac{x-3}{x^2-1} = \frac{(x+1)(x-3)}{(x-3)(x+1)(x-1)} = \frac{1}{x-1}.$$

Example 3. Simplify $\dfrac{6x^2 - ax - 2a^2}{ax - a^2} \times \dfrac{x-a}{9x^2 - 4a^2}$.

Solution.

$$\frac{6x^2 - ax - 2a^2}{ax - a^2} \times \frac{x-a}{9x^2 - 4a^2} = \frac{\cancel{(3x - 2a)}(2x+a)}{a\cancel{(x-a)}} \times \frac{\cancel{(x-a)}}{(3x+2a)\cancel{(3x-2a)}},$$

$$= \frac{2x+a}{a(3x+2a)}.$$

NOTE. The elimination, by division, of a factor from the numerator and denominator of a fraction is often referred to as the **canceling** of the factor. Observe especially that if all the factors of the numerator just cancel those of the denominator, the value of the fraction is 1.

Example 4. Simplify $\frac{x^2 + y^2}{a^2 + 2ab} \times \frac{a^2 - 4b^2}{x^3 + xy^2} \times \frac{ax}{a - 2b}$.

Solution.

$$\frac{x^2+y^2}{a^2+2ab} \times \frac{a^2-4b^2}{x^3+xy^2} \times \frac{ax}{a-2b} = \frac{(x^2+y^2)}{a(a+2b)} \times \frac{(a + 2b)(a - 2b)}{x(x^2+y^2)} \times \frac{ax}{(a-2b)} = 1$$

since all the factors cancel each other.

EXERCISE 41

Find the products:

1. $\frac{5}{8} \times \frac{4}{15}$
2. $\frac{34}{39} \times \frac{91}{119}$
3. $\frac{10x}{9y} \times \frac{3y}{5x}$
4. $\frac{4ax}{15by} \times \frac{3ay}{8bx}$
5. $\frac{12a}{35b} \times \frac{21bx}{4a^2}$
6. $\frac{6m}{5n} \times \frac{10mn}{9}$
7. $\frac{18ab}{25cd} \times \frac{10ac}{27bd}$
8. $\frac{2x^2}{3y^2} \times \frac{9y}{8x^2}$
9. $\frac{100pq}{63m^2} \times \frac{7pm}{16q}$
10. $\frac{28st^2}{91r} \times \frac{65r^3t}{4s}$
11. $\frac{21a^2b^3c}{13x^2yz^4} \times \frac{39x^3yz^2}{28ab^3c^3}$
12. $\frac{36a^2bx}{81by^3z} \times \frac{27ay^3z^2}{12a^3xz}$
13. $\frac{2x - 4}{9} \times \frac{6}{5x - 10}$
14. $\frac{x + 1}{x} \times \frac{x^2}{x^2 + x}$
15. $\frac{c}{c + d} \times \frac{c^2 - d^2}{2c}$
16. $\frac{4xy}{x + 3} \times \frac{3x^2 + 9x}{16y^2}$
17. $\frac{a + b}{2a + b} \times \frac{4a^2 - b^2}{a^2 - b^2}$
18. $\frac{10x^2 - 5x}{12x^3 + 24x^2} \times \frac{x^2 + 2x}{2x - 1}$
19. $\frac{3ax^2 - 9ax}{10x^2 + 5x} \times \frac{2x^3 + x^2}{a^2x - 3a^2}$
20. $\frac{x^2y^3 - xy^4}{x^4y^3 + x^3y^4} \times \frac{x^3y^3 + x^2y^4}{x^3y - x^2y^2}$
21. $\frac{x^2 - 6x - 16}{x^2 + 4x - 21} \times \frac{x^2 - 8x + 15}{x^2 + 9x + 14}$
22. $\frac{2x^2 + 5x + 2}{x^2 - 4} \times \frac{x^2 + 4x}{2x^2 + 9x + 4}$
23. $\frac{x^2 - 4x + 3}{42x^2} \times \frac{7x^2 - 7x}{x^2 - 2x + 1}$
24. $\frac{a^2 + 6a + 9}{a^2 - 1} \times \frac{a^2 - 2a - 3}{a^2 - 9}$
25. $\frac{2b + 1}{2b + 2} \times \frac{2b^2 - b - 3}{4b^2 - 4b - 3}$

26. $\dfrac{x^2 - 14xy - 15y^2}{x^2 - 4xy - 45y^2} \times \dfrac{x^2 - 6xy - 27y^2}{x^2 - 12xy - 45y^2}$

27. $\dfrac{a^2 - 4}{2a^2 + 5a + 2} \times \dfrac{4a^2 - 1}{2a^2 - 5a + 2}$

28. $\dfrac{m^2 - 5m - 14}{m^2 - 4} \times \dfrac{(m - 2)^2}{m^2 - 6m - 7}$

29. $\dfrac{x^2 - x - 20}{x^2 + 7x + 12} \times \dfrac{x^2 + 9x + 18}{x^2 - 7x + 10}$

30. $\dfrac{4x^2 - 8x - 5}{3x^2 + 4x + 1} \times \dfrac{3x^2 - 20x - 7}{2x^2 - 3x - 5}$

31. $\dfrac{a^2 - b^2}{a^2b^2} \times \dfrac{a^4}{a + b} \times \dfrac{b^2}{a^2 - ab}$

32. $\dfrac{x^2 + x}{2x + 1} \times \dfrac{10x + 5}{xy - y} \times \dfrac{x^2y - xy}{10x + 10}$

33. $\dfrac{27a^2x^3}{4bc} \times \dfrac{5ab - 5bx}{a^2x - x^3} \times \dfrac{8ac + 8ax}{45ax^2}$

34. $\dfrac{1 - x^2}{4x^2} \times \dfrac{3ax}{2 - 2x} \times \dfrac{8}{ax + a}$

35. $\dfrac{2x + 10}{3x - 9} \times \dfrac{12x - 36}{5x + 5} \times \dfrac{ax + a}{4x + 20}$

36. $\dfrac{x^2 - 1}{x^2 - 4} \times \dfrac{x^2 - 2x}{x^2 + x} \times \dfrac{x^2 + 2x}{x^2 - x}$

37. $\dfrac{x^2 + x - 2}{x^2 + 6x + 9} \times \dfrac{x^2 - x - 12}{x^2 - 3x + 2} \times \dfrac{x^2 + x - 6}{x^2 - 6x + 8}$

38. $\dfrac{2x^2 + 9x - 5}{2x^2 + 3x + 1} \times \dfrac{3x^2 + x - 2}{4x^2 - 8x + 3} \times \dfrac{4x^2 - 4x - 3}{3x^2 + 4x - 4}$

39. $\dfrac{x^2 + y^2}{3x^2y - 3xy^2} \times \dfrac{x^2 + xy}{x^4 - y^4} \times \dfrac{x^2 - 2xy + y^2}{y}$

40. $\dfrac{x^3 - 3x^2 + 2x - 6}{x + 5} \times \dfrac{x^2 + 7x + 10}{x^2 + 2}$

60. Division of Fractions. For the division of fractions the following rule applies:

Fractions. Rule 3

To divide one fraction by another, invert the divisor and proceed as in multiplication.

Example 1. Divide $\frac{5}{8}$ by $\frac{3}{4}$.

Solution. Inverting the divisor and multiplying, we have

$$\tfrac{5}{8} \div \tfrac{3}{4} = \tfrac{5}{8} \times \tfrac{4}{3} = \tfrac{5}{6}.$$

Example 2. Divide $\dfrac{3a^2b}{5cd}$ by $\dfrac{9ab^2}{20c^2}$.

Solution.

$$\frac{3a^2b}{5cd} \div \frac{9ab^2}{20c^2} = \frac{3a^2b}{5cd} \times \frac{20c^2}{9ab^2} = \frac{4ac}{3bd}.$$

Example 3. Simplify $\dfrac{x^2 + 4x - 5}{x^2 - 3x + 2} \div \dfrac{x^2 + 7x + 10}{x^2 - 4}$.

Solution.

$$\begin{aligned}\frac{x^2 + 4x - 5}{x^2 - 3x + 2} \div \frac{x^2 + 7x + 10}{x^2 - 4} &= \frac{x^2 + 4x - 5}{x^2 - 3x + 2} \times \frac{x^2 - 4}{x^2 + 7x + 10},\\ &= \frac{(x + 5)(x - 1)}{(x - 2)(x - 1)} \times \frac{(x + 2)(x - 2)}{(x + 2)(x + 5)},\\ &= 1.\end{aligned}$$

Example 4. Simplify $\dfrac{x^2 - y^2}{2ax + 2ay} \times \dfrac{ax}{2x - y} \div \dfrac{x^2 - xy}{4ax - 2ay}$.

Solution.

$$\begin{aligned}\frac{x^2 - y^2}{2ax + 2ay} \times \frac{ax}{2x - y} \div \frac{x^2 - xy}{4ax - 2ay}\\ &= \frac{x^2 - y^2}{2ax + 2ay} \times \frac{ax}{2x - y} \times \frac{4ax - 2ay}{x^2 - xy},\\ &= \frac{(x + y)(x - y)}{2a(x + y)} \times \frac{ax}{(2x - y)} \times \frac{2a(2x - y)}{x(x - y)},\\ &= a.\end{aligned}$$

EXERCISE 42

Find the quotients:

1. $\dfrac{3a}{4b} \div \dfrac{9a}{8b}$
2. $\dfrac{24a^2}{33b} \div \dfrac{12a}{55b^2}$
3. $\dfrac{27ab}{32cd} \div \dfrac{45ac}{56bd}$
4. $\dfrac{81ax}{82by} \div \dfrac{27cx}{123dy}$
5. $\dfrac{42x^2y}{5z} \div \dfrac{63xy^2}{10z^3}$
6. $\dfrac{38a^2b^2}{39cd} \div \dfrac{57ab^2}{26c^2d}$
7. $\dfrac{8m^3n}{9p^2q^2} \div \dfrac{64mn}{63p^2q}$
8. $\dfrac{18a^2x^3}{25by} \div \dfrac{81a^3x^2}{125b^2y}$

9. $\frac{86a^2b^3c}{17m^2n^2p^3} \div \frac{129abc^2}{34mn^2p}$

10. $\frac{57abx^3}{87cdy} \div \frac{76a^2bx^2}{58cdy^2}$

11. $\frac{ax - a}{bx - 3b} \div \frac{cx - c}{dx - 3d}$

12. $\frac{a^2 - 49}{a^2 - 4} \div \frac{a + 7}{a + 2}$

13. $\frac{x - 4}{x^2 - 4} \div \frac{x^2 - 16}{x - 2}$

14. $\frac{xy + 3y}{x^2 - x} \div \frac{x + 3}{x}$

15. $\frac{3x - 15}{5} \div \frac{2x - 10}{3}$

16. $\frac{10x - 15}{6} \div \frac{14x - 21}{9}$

17. $\frac{5x + 5}{9x - 9} \div \frac{2x + 2}{3x - 3}$

18. $\frac{35x - 25}{12x + 44} \div \frac{28x - 20}{36x + 132}$

19. $\frac{x^2 - 3x - 10}{x^2 + 2x - 35} \div \frac{x^2 + 9x + 14}{x^2 + 4x - 21}$

20. $\frac{x^2 + 6x + 9}{x^2 - 2x - 15} \div \frac{x^2 + 2x - 3}{x^2 - 4x - 5}$

21. $\frac{x^2 + 4x + 4}{x^2 - 9} \div \frac{x^2 - 4}{x^2 - 6x + 9}$

22. $\frac{4a^2 - 6ax}{4a^2 - 16x^2} \div \frac{4a^2 - 9x^2}{4a^2 - 8ax}$

23. $\frac{1 - x^2}{3 - 3x} \div \frac{2x^2 - x - 3}{4x^2 - 9}$

24. $\frac{2x^2 - 3x - 2}{2x^2 + 5x + 2} \div \frac{16x^2 - 49}{4x^2 + x - 14}$

25. $\frac{a^2 - x^2}{ax - 2x^2} \div \frac{a^2 - 2ax + x^2}{2a^2 - 4ax}$

26. $\frac{x^3 + x^2 - 20x}{x^2 + 3x - 28} \div \frac{x^2 + 5x}{ax + 7a}$

27. $\frac{6x^2 - x - 12}{2x^2 + 15x - 27} \div \frac{6x^2 + 11x + 4}{2x^2 + 17x - 9}$

28. $\frac{10x^2 - 29x + 10}{6x^2 - 29x + 20} \div \frac{10x^2 - 19x + 6}{12x^2 - 28x + 15}$

Simplify:

29. $\frac{3x + 9}{2x + 14} \times \frac{6x - 30}{7x + 21} \div \frac{3x - 15}{7x + 49}$

30. $\frac{20x + 10}{2x - 18} \times \frac{4x - 36}{6x - 10} \div \frac{10x + 5}{18x - 30}$

31. $\frac{a^2 + ax}{3ax + 2x^2} \times \frac{2ax - x^2}{ax + x^2} \div \frac{4a^2 - 2ax}{9a + 6x}$

32. $\frac{2ax + x^2}{3a^2 + ax} \times \frac{6a + 2x}{6a - 12x} \div \frac{2a^2 + ax}{ax^2 - 2x^3}$

33. $\frac{4x^2 - 9}{4x - 4} \times \frac{4x - 7}{2x^2 - x - 3} \div \frac{2x + 3}{4x^2 - 4}$

34. $\dfrac{x^2 - x - 20}{x^2 - 25} \times \dfrac{x^2 - x - 2}{x^2 + 2x - 8} \div \dfrac{x + 1}{x^2 + 5x}$

35. $\dfrac{x^4 - y^4}{4x^2 - 8xy + 3y^2} \times \dfrac{2x - y}{x^2 - xy} \div \dfrac{x^2 + y^2}{2x^2 - 3xy}$

36. $\dfrac{x^3 + x^2 + x + 1}{2x^2 - 5x + 3} \times \dfrac{2x^2 - 3x}{x^2 - 1} \div \dfrac{x^2 + 1}{x - 1}$

61. Addition and Subtraction of Fractions. We learned in arithmetic that to add or subtract fractions it is first necessary to express each of the fractions in terms of some common integral unit, which we call a common denominator. To avoid working with fractions that are not in their lowest terms, we use the *lowest common denominator*, which is the lowest common multiple of the denominators of the given fractions.

Fractions. Rule 4

To express fractions in terms of their lowest common denominator, (1) find the lowest common multiple of the given denominators, and take it for the common denominator; (2) divide it by the denominator of the first fraction, and multiply the numerator of this fraction by the quotient so obtained; (3) do the same with all the other given fractions.

Example 1. Express the fractions $\frac{1}{2}$, $\frac{2}{3}$, $\frac{3}{4}$ in terms of their lowest common denominator.

Solution. The lowest common multiple of 2, 3, and 4 is 12. Dividing this number by each of the denominators in turn, and multiplying the corresponding numerators by the respective quotients, we have the equivalent fractions

$$\tfrac{6}{12}, \tfrac{8}{12}, \tfrac{9}{12}.$$

Example 2. Express the following fractions in terms of their lowest common denominator:

$$\frac{a}{3xy}, \frac{b}{6xz}, \frac{c}{2yz}.$$

Solution. The lowest common multiple of the denominators is $6xyz$. Thus, we have the equivalent fractions

$$\frac{2az}{6xyz}, \frac{by}{6xyz}, \frac{3cx}{6xyz}.$$

We may now state the rule for the addition and subtraction of fractions.

Fractions. Rule 5

To add or subtract fractions, (1) express them in terms of their lowest common denominator; (2) add or subtract the numerators, and retain the common denominator.

Example 3. Add $\dfrac{4}{3a^2} + \dfrac{5}{6a}$.

Solution. The lowest common denominator is $6a^2$; hence

$$\frac{4}{3a^2} + \frac{5}{6a} = \frac{8}{6a^2} + \frac{5a}{6a^2} = \frac{8 + 5a}{6a^2}.$$

Example 4. Combine $\dfrac{a - 2b}{ab} + \dfrac{3b - x}{bx} - \dfrac{3a - 2x}{ax}$ into one fraction.

Solution. The lowest common denominator is abx; hence the expression

$$= \frac{x(a - 2b) + a(3b - x) - b(3a - 2x)}{abx},$$

$$= \frac{ax - 2bx + 3ab - ax - 3ab + 2bx}{abx} = 0$$

since the terms in the numerator eliminate one another.

EXERCISE 43

Combine and simplify:

1. $\dfrac{x}{2} + \dfrac{y}{6} + \dfrac{z}{3}$ **2.** $\dfrac{3x}{4} + \dfrac{2x}{3} - \dfrac{5x}{12}$ **3.** $\dfrac{ab}{7} + \dfrac{5ab}{8} - \dfrac{9ab}{14}$

4. $\dfrac{4m}{5} - \dfrac{m}{6} - \dfrac{3m}{4}$ **5.** $\dfrac{3x}{5} - \dfrac{3x}{10} - \dfrac{x}{4}$ **6.** $\dfrac{6a}{7} + \dfrac{a}{3} + \dfrac{7a}{9}$

7. $\dfrac{1}{x} + \dfrac{1}{y} + \dfrac{1}{z}$ **8.** $\dfrac{2}{a} + \dfrac{3}{b} - \dfrac{4}{c}$ **9.** $\dfrac{2}{3m} - \dfrac{3}{2m} + \dfrac{4}{5m}$

10. $\dfrac{a}{x} - \dfrac{b}{y} - \dfrac{c}{z}$ **11.** $\dfrac{a}{bc} + \dfrac{b}{ac} + \dfrac{c}{ab}$ **12.** $\dfrac{5}{3x} + \dfrac{3}{4x} - \dfrac{1}{2x}$

13. $\dfrac{a + 2}{2} + \dfrac{a + 4}{4}$ **14.** $\dfrac{a - 3}{3} - \dfrac{a - 1}{4}$ **15.** $\dfrac{m}{3} - \dfrac{2 - m}{2}$

16. $\frac{x}{2} - \frac{x-4}{4}$ **17.** $\frac{a-1}{a} - \frac{b+4}{b}$ **18.** $\frac{x-y}{x} - \frac{x+y}{y}$

19. $\frac{3}{x} - \frac{2}{x^2} + \frac{1}{x^3}$ **20.** $\frac{5}{4x} + \frac{3}{5x^2} - \frac{1}{10x^3}$

21. $\frac{3x-2}{3} + \frac{2x-1}{6} + \frac{x-4}{4}$ **22.** $\frac{5x+2}{9} + \frac{x+2}{6} - \frac{3x-7}{12}$

23. $\frac{x-7}{15} + \frac{x-3}{25} - \frac{x+4}{45}$ **24.** $\frac{2x-1}{7} + \frac{x+2}{3} - \frac{3x-1}{9}$

25. $\frac{2x+1}{5} - \frac{2x-1}{7} - \frac{x-3}{10}$ **26.** $\frac{2x+3}{4} - \frac{5x+2}{5} + \frac{3x-7}{6}$

27. $\frac{2x-1}{x} - \frac{x+4}{2x} - \frac{9}{8x^2}$ **28.** $\frac{a+1}{a} - \frac{a+2}{a^2} + \frac{a+3}{a^3}$

29. $\frac{x-y}{xy} + \frac{y-z}{yz} + \frac{z-x}{xz}$ **30.** $\frac{a-x}{x} + \frac{a+x}{a} - \frac{a^2-x^2}{ax}$

31. $\frac{x+3}{5x} + \frac{2x^2-5}{10x^2} - \frac{6x^3+1}{15x^3}$ **32.** $\frac{2x-3y}{xy} + \frac{5x-z}{xz} + \frac{4}{x}$

33. $\frac{3a+4b}{ab} - \frac{a-5c}{ac} - \frac{9}{a}$ **34.** $\frac{2x-a}{ax} + \frac{x-2b}{bx} + \frac{3}{x}$

35. $\frac{y-m}{my} + \frac{y-n}{ny} + \frac{2}{y}$ **36.** $\frac{x+a}{ax} - \frac{a+1}{a} + \frac{2x-1}{x}$

62. Fractions with Binomial Denominators. The following examples illustrate the general procedure for combining fractions.

Example 1. Add $\frac{4a}{3x-6}$ and $\frac{3a}{2x-4}$.

Solution. The lowest common denominator is $6(x-2)$; hence,

$$\frac{4a}{3x-6} + \frac{3a}{2x-4} = \frac{4a}{3(x-2)} + \frac{3a}{2(x-2)} = \frac{8a+9a}{6(x-2)} = \frac{17a}{6(x-2)}.$$

Example 2. Subtract $\frac{3}{2a-5}$ from $\frac{4}{a-3}$.

Solution. The lowest common denominator is $(a-3)(2a-5)$; hence,

$$\frac{4}{a-3} - \frac{3}{2a-5} = \frac{4(2a-5) - 3(a-3)}{(a-3)(2a-5)},$$

$$= \frac{8a - 20 - 3a + 9}{(a-3)(2a-5)},$$

$$= \frac{5a-11}{(a-3)(2a-5)}.$$

Example 3. Combine and simplify $\frac{1}{x+1} + \frac{2}{x+2} - \frac{3}{x+3}$.

Solution. The lowest common denominator is $(x + 1)(x + 2)(x + 3)$. Therefore the expression

$$= \frac{(x+2)(x+3) + 2(x+1)(x+3) - 3(x+1)(x+2)}{(x+1)(x+2)(x+3)},$$

$$= \frac{x^2 + 5x + 6 + 2(x^2 + 4x + 3) - 3(x^2 + 3x + 2)^*}{(x+1)(x+2)(x+3)},$$

$$= \frac{x^2 + 5x + 6 + 2x^2 + 8x + 6 - 3x^2 - 9x - 6}{(x+1)(x+2)(x+3)},$$

$$= \frac{4x+6}{(x+1)(x+2)(x+3)}.$$

EXERCISE 44

Simplify:

1. $\frac{5}{2a+6} + \frac{2}{a+3}$
2. $\frac{7}{5x-10} - \frac{4}{3x-6}$
3. $\frac{1}{2x-8} - \frac{5}{7x-28}$
4. $\frac{5}{a+1} + \frac{3}{2a+2}$
5. $\frac{3}{ax-x^2} + \frac{4}{2a-2x}$
6. $\frac{a}{2x-4y} - \frac{b}{3x-6y}$
7. $\frac{3}{x+1} + \frac{5}{2x+2} - \frac{7}{3x+3}$
8. $\frac{2}{a-5} - \frac{7}{5a-25} + \frac{3}{2a-10}$
9. $\frac{5}{2a-b} - \frac{4}{6a-3b} - \frac{1}{4a-2b}$
10. $\frac{1}{a^2-ax} + \frac{1}{ax-x^2} + \frac{1}{a-x}$
11. $\frac{4}{2x-3} + \frac{3}{3x-1}$
12. $\frac{4}{3a+b} - \frac{5}{4a-3b}$
13. $\frac{7}{7a-2} - \frac{5}{5a+4}$
14. $\frac{3}{2y+3} + \frac{2}{3y-2}$
15. $\frac{x}{x-y} - \frac{y}{x+y}$
16. $\frac{a}{x+a} - \frac{b}{x+b}$
17. $\frac{1}{x-1} + \frac{1}{2x-1} - \frac{1}{2x-2}$
18. $\frac{1}{3x-6} - \frac{1}{2x+1} + \frac{1}{6x-12}$

* See Note, p. 72.

19. $\dfrac{10}{2x+4} + \dfrac{1}{x-1} + \dfrac{3}{2x-2}$

20. $\dfrac{1}{x+3} - \dfrac{2}{3x+9} - \dfrac{1}{3x-9}$

21. $\dfrac{x-2}{x-1} + \dfrac{x+2}{x+1}$

22. $\dfrac{x+2}{x-2} - \dfrac{x-2}{x+2}$

23. $\dfrac{a-4}{a-2} - \dfrac{a-7}{a-5}$

24. $\dfrac{b+3}{b+2} - \dfrac{b+5}{b+4}$

25. $\dfrac{2y-3}{y-2} + \dfrac{y+3}{3y-2}$

26. $\dfrac{a-2b}{a-6b} - \dfrac{a+b}{a-3b}$

27. $\dfrac{3x-y}{x-2y} + \dfrac{2x-y}{x+2y}$

28. $\dfrac{3a-x}{a-3x} - \dfrac{2a-x}{a-2x}$

29. $\dfrac{3x-2}{x+1} - \dfrac{x-3}{2x+2} + \dfrac{x}{x-1}$

30. $\dfrac{2x+5}{2x-4} - \dfrac{x-1}{2x+1} - \dfrac{2x-1}{4x+2}$

31. $\dfrac{1}{x-1} + \dfrac{1}{x-2} + \dfrac{1}{x-3}$

32. $\dfrac{4}{2x+1} + \dfrac{1}{x-1} - \dfrac{3}{x+1}$

33. $\dfrac{2}{a+b} - \dfrac{5}{a-b} + \dfrac{3}{a+2b}$

34. $\dfrac{4}{a-2} + \dfrac{2}{a-1} - \dfrac{6}{3a+1}$

35. $\dfrac{x}{x+1} - \dfrac{2x}{x+2} + \dfrac{x}{x+3}$

36. $\dfrac{2}{x-y} - \dfrac{3}{x-2y} - \dfrac{2}{x-4y}$

63. Fractions with Multinomial Denominators. The following examples illustrate further the method for combining fractions.

Example 1. Combine $\dfrac{1}{2x-3y} - \dfrac{6y}{4x^2-9y^2}$.

Solution. Since $4x^2 - 9y^2 = (2x-3y)(2x+3y)$, the lowest common denominator is $(2x-3y)(2x+3y)$. Hence,

$$\begin{aligned}\frac{1}{2x-3y} - \frac{6y}{4x^2-9y^2} &= \frac{(2x+3y)-6y}{(2x-3y)(2x+3y)},\\ &= \frac{2x-3y}{(2x-3y)(2x+3y)},\\ &= \frac{1}{2x+3y}.\end{aligned}$$

Example 2. Combine $\dfrac{13}{2x^2+7x-15} - \dfrac{4}{x^2+6x+5}$.

Solution.

$$\frac{13}{2x^2+7x-15}-\frac{4}{x^2+6x+5}=\frac{13}{(2x-3)(x+5)}-\frac{4}{(x+5)(x+1)},$$
$$=\frac{13(x+1)-4(2x-3)}{(2x-3)(x+5)(x+1)},$$
$$=\frac{5x+25}{(2x-3)(x+5)(x+1)},$$
$$=\frac{5(x+5)}{(2x-3)(x+5)(x+1)},$$
$$=\frac{5}{(2x-3)(x+1)}.$$

When possible it is always advisable to reduce fractions to lowest terms before combining them.

Example 3. Combine $\dfrac{a^2-2ax-8x^2}{a^2-16x^2}-\dfrac{2ax}{2a^2+4ax}$.

Solution.

$$\frac{a^2-2ax-8x^2}{a^2-16x^2}-\frac{2ax}{2a^2+4ax}=\frac{(a+2x)(a-4x)}{(a+4x)(a-4x)}-\frac{2ax}{2a(a+2x)},$$
$$=\frac{a+2x}{a+4x}-\frac{x}{a+2x},$$
$$=\frac{(a+2x)^2-x(a+4x)}{(a+4x)(a+2x)},$$
$$=\frac{a^2+4ax+4x^2-ax-4x^2}{(a+4x)(a+2x)},$$
$$=\frac{a(a+3x)}{(a+4x)(a+2x)}.$$

Occasionally some modification of the preceding general methods may be used with advantage. One is illustrated in the following example.

Example 4. Combine $\dfrac{x+3}{x-2}-\dfrac{x+2}{x-3}+\dfrac{6}{x^2-4}$.

Solution. We combine the first two fractions before considering the third fraction:

$$= \frac{x^2 - 9 - (x^2 - 4)}{(x-2)(x-3)} + \frac{6}{x^2 - 4},$$

$$= \frac{-5}{(x-2)(x-3)} + \frac{6}{(x-2)(x+2)},$$

$$= \frac{-5(x+2) + 6(x-3)}{(x-2)(x-3)(x+2)},$$

$$= \frac{x - 28}{(x-2)(x-3)(x+2)}.$$

EXERCISE 45

Combine and simplify:

1. $\dfrac{a}{x-a} - \dfrac{a^2}{x^2 - a^2}$

2. $\dfrac{x+2}{x-2} - \dfrac{x^2+4}{x^2-4}$

3. $\dfrac{1}{(x+y)^2} + \dfrac{1}{x^2 - y^2}$

4. $\dfrac{4}{x^2-4} - \dfrac{3}{x^2 - x - 2}$

5. $\dfrac{a}{a+b} + \dfrac{2ab}{a^2 - b^2}$

6. $\dfrac{xy}{16x^2 - y^2} + \dfrac{x}{4x+y}$

7. $\dfrac{3}{y+1} + \dfrac{4}{y-1} - \dfrac{6y}{y^2-1}$

8. $\dfrac{1}{a+b} - \dfrac{1}{a-b} + \dfrac{2a}{a^2 - b^2}$

9. $\dfrac{5}{1+2x} - \dfrac{2x}{1-2x} - \dfrac{4(1-3x)}{1-4x^2}$

10. $\dfrac{1}{x+1} + \dfrac{6}{(x+1)^2} - \dfrac{5}{x^2-1}$

11. $\dfrac{1}{x^2 - 3x + 2} + \dfrac{1}{x^2 - 5x + 6}$

12. $\dfrac{5}{2x^2 + 3x - 2} - \dfrac{3}{2x^2 + x - 1}$

13. $\dfrac{3}{2 + a - 6a^2} - \dfrac{1}{1 + a - 2a^2}$

14. $\dfrac{4}{x^2 - 4y^2} + \dfrac{1}{x^2 - 5xy + 6y^2}$

15. $\dfrac{1}{x^2 - 2x + 1} - \dfrac{1}{x^2 + 2x + 1}$

16. $\dfrac{4}{4m^2 - 1} - \dfrac{5}{6m^2 + m - 1}$

17. $\dfrac{1}{(x-y)(x+y)} - \dfrac{1}{(x-y)(x+2y)} + \dfrac{1}{(x+y)(x+2y)}$

18. $\dfrac{1}{(x-1)(x-2)} + \dfrac{1}{(x-2)(x-3)} - \dfrac{1}{(x-1)(x-3)}$

19. $\dfrac{1}{x^2 - 3x + 2} + \dfrac{3}{x^2 - x - 2} - \dfrac{2}{x^2 - 1}$

20. $\dfrac{x}{x^2 + 5x + 6} + \dfrac{4}{x^2 + 6x + 8} - \dfrac{2}{x^2 + 7x + 12}$

21. $\dfrac{5}{x^2 - x - 6} + \dfrac{1}{x^2 - 5x + 6} + \dfrac{2x}{x^2 - 4}$

22. $\dfrac{4}{x^2 - 6x + 5} + \dfrac{3x}{x^2 + x - 2} + \dfrac{2}{x^2 - 3x - 10}$

23. $\dfrac{x - 1}{x^2 - 1} - \dfrac{x - 2}{x^2 - 4}$

24. $\dfrac{xy}{x^2y^2 - xy} + \dfrac{xy}{x^2y^2 + xy}$

25. $\dfrac{2x + 4}{2x^2 - x - 10} - \dfrac{3x}{3x^2 + 5x}$

26. $\dfrac{2x + 2}{x^2 + 3x + 2} + \dfrac{3x - 6}{x^2 - 3x + 2}$

27. $\dfrac{x^2 + 1}{x^3 - x^2 + x - 1} - \dfrac{x^2 - 1}{x^3 + x^2 - x - 1}$

28. $\dfrac{x^2 - 3}{x^3 + 2x^2 - 3x - 6} - \dfrac{x^2 - 2}{x^3 + 2x^2 - 2x - 4}$

29. $\dfrac{x + 1}{x - 1} - \dfrac{x - 1}{x + 1} - \dfrac{4x}{(x + 1)^2}$

30. $\dfrac{x + 1}{x + 2} - \dfrac{x - 2}{x - 1} - \dfrac{2}{x^2 - 1}$

31. $\dfrac{x + 3}{x - 3} - \dfrac{x + 4}{x - 2} + \dfrac{6}{x^2 - 3x + 2}$

32. $\dfrac{x + 1}{2x + 1} - \dfrac{x - 1}{2x - 1} - \dfrac{1}{2x}$

64. The Signs of a Fraction. We learned in division that $\dfrac{-a}{-b}$ is the quotient resulting from the division of $-a$ by $-b$, and that this is obtained by dividing a by b and, according to the rule of signs, prefixing $+$. Thus,

$$\frac{-a}{-b} = +\frac{a}{b} = \frac{a}{b}.$$

Similarly, $$\frac{-a}{b} = -\frac{a}{b}, \text{ and } \frac{a}{-b} = -\frac{a}{b}.$$

These results may be stated as follows:

Fractions. Rule 6

1. ***If the signs of both the numerator and the denominator of a fraction are changed, the sign of the whole fraction will be unchanged.***
2. ***If the sign of either the numerator or the denominator alone is changed, the sign of the whole fraction will be changed.***

Illustrations. (1) $\dfrac{b-a}{y-x} = \dfrac{-b+a}{-y+x} = \dfrac{a-b}{x-y}.$

(2) $\dfrac{x-x^2}{2} = -\dfrac{-x+x^2}{2} = -\dfrac{x^2-x}{2}.$

(3) $\dfrac{3}{4-x^2} = -\dfrac{3}{-4+x^2} = -\dfrac{3}{x^2-4}.$

NOTE. The intermediate step may be omitted when the principle is understood.

Example 1. Combine $\dfrac{2+x}{x-1} + \dfrac{x-3}{1-x}.$

Solution. It is evident that $1 - x$ is the negative of $x - 1$; hence it is convenient to alter the sign of one of the denominators. Thus,

$$\frac{2+x}{x-1} + \frac{x-3}{1-x} = \frac{2+x}{x-1} - \frac{x-3}{x-1} = \frac{2+x-x+3}{x-1} = \frac{5}{x-1}.$$

Example 2. Combine $\dfrac{a}{x+a} - \dfrac{x}{x-a} - \dfrac{2x^2}{a^2-x^2}.$

Solution. Here the lowest common denominator of the first two fractions is $x^2 - a^2$. Hence we alter the sign of the denominator in the third fraction; thus,

$$\begin{aligned}\frac{a}{x+a} - \frac{x}{x-a} - \frac{2x^2}{a^2-x^2} &= \frac{a}{x+a} - \frac{x}{x-a} + \frac{2x^2}{x^2-a^2},\\ &= \frac{a(x-a) - x(x+a) + 2x^2}{x^2-a^2},\\ &= \frac{ax - a^2 - x^2 - ax + 2x^2}{x^2-a^2},\\ &= \frac{x^2-a^2}{x^2-a^2} = 1.\end{aligned}$$

EXERCISE 46

Combine and simplify:

1. $\dfrac{1}{4x-4}-\dfrac{1}{5x-5}+\dfrac{1}{1-x}$

2. $\dfrac{2}{6-3x}+\dfrac{5}{x-2}-\dfrac{3}{4-2x}$

3. $\dfrac{1}{2x+4}-\dfrac{2}{2-x}-\dfrac{4}{3x-6}$

4. $\dfrac{5}{2x-3}-\dfrac{2}{3+2x}+\dfrac{3}{3-2x}$

5. $\dfrac{3}{1+c}-\dfrac{2}{1-c}-\dfrac{5c}{c^2-1}$

6. $\dfrac{m}{1+m}-\dfrac{m}{1-m}-\dfrac{m^2}{m^2-1}$

7. $\dfrac{1}{3+x}+\dfrac{1}{3-x}+\dfrac{6}{x^2-9}$

8. $\dfrac{1}{1-2x}+\dfrac{2}{1+2x}+\dfrac{2}{4x^2-1}$

9. $\dfrac{10}{a^2-25}+\dfrac{1}{5-a}-\dfrac{1}{5+a}$

10. $\dfrac{3}{x+3}-\dfrac{x}{x-3}-\dfrac{2x^2}{9-x^2}$

11. $\dfrac{1}{x^2-3x+2}+\dfrac{5}{6-x-x^2}$

12. $\dfrac{7}{x^2-3x-10}+\dfrac{5}{6+x-x^2}$

13. $\dfrac{3}{2x^2+x-1}-\dfrac{2}{2-3x-2x^2}$

14. $\dfrac{7}{2x^2-x-6}+\dfrac{9}{9+3x-2x^2}$

15. $\dfrac{a^2-b^2}{ab}-\dfrac{ab-b^2}{ab-a^2}$

16. $\dfrac{a-x}{x^2-a^2}-\dfrac{b-x}{x^2-b^2}$

17. $\dfrac{x^2+x+1}{x+1}-\dfrac{x^2-x+1}{x-1}$

18. $\dfrac{x+z}{(x-y)(a-x)}+\dfrac{y+z}{(y-x)(a-y)}$

19. $\dfrac{1}{(a-b)(a-c)}+\dfrac{1}{(b-c)(b-a)}+\dfrac{1}{(c-a)(c-b)}$

20. $\dfrac{a}{(a-b)(a-c)}+\dfrac{b}{(b-c)(b-a)}+\dfrac{c}{(c-a)(c-b)}$

65. Mixed Expressions. An algebraic expression which is partly integral and partly fractional is called a *mixed expression.*

Illustration. $1+\dfrac{1}{x}$ and $\dfrac{x^2}{x-y}-x-y$ are mixed expressions.

Before attempting to apply the operations of algebra to such expressions it is advisable to reduce the expressions to fractions.

Example 1. Reduce $x-\dfrac{x}{1+x}$ to a fraction.

Solution. Consider the denominator of the integral term to be 1 and combine the two fractions.

$$x - \frac{x}{1+x} = \frac{x}{1} - \frac{x}{1+x},$$
$$= \frac{x(1+x) - x}{1+x},$$
$$= \frac{x + x^2 - x}{1+x},$$
$$= \frac{x^2}{1+x}.$$

Example 2. Divide $x - 6 + \dfrac{20}{x+3}$ by $x + \dfrac{2}{x-3}$.

Solution. First reducing each mixed expression to a fraction, we have

$$\left(x - 6 + \frac{20}{x+3}\right) \div \left(x + \frac{2}{x-3}\right) = \left(\frac{x^2 - 3x - 18 + 20}{x+3}\right) \div \left(\frac{x^2 - 3x + 2}{x-3}\right),$$
$$= \frac{(x^2 - 3x + 2)}{(x+3)} \times \frac{(x-3)}{(x^2 - 3x + 2)},$$
$$= \frac{x-3}{x+3}.$$

EXERCISE 47

Reduce the following to fractions:

1. $1 + \dfrac{1}{x}$
2. $1 + \dfrac{a}{b} + \dfrac{a^2}{b^2}$
3. $\dfrac{a^2}{a-b} - a$
4. $\dfrac{3x}{x+3} - 2$
5. $y + 1 - \dfrac{y^2}{y-1}$
6. $1 - \dfrac{a(b-a)}{1+ab}$
7. $\dfrac{5x-3}{x+1} - 3$
8. $6 - \dfrac{13x - 15}{3x - 5}$
9. $\dfrac{2x-5}{x-2} + x - 3$
10. $\dfrac{x^2 - 9}{x+4} - x + 7$
11. $1 - \dfrac{a}{a-b} + \dfrac{b}{a+b}$
12. $3y + \dfrac{x^2 - 2xy}{x+y} - x$

Perform the indicated operation and simplify:

13. $\left(4x - 16 + \dfrac{49}{x+3}\right)\left(x + 2 - \dfrac{7}{2x-1}\right)$

14. $\left(x - 2 - \dfrac{4}{x+1}\right)\left(x - 1 - \dfrac{8}{x-3}\right)$

15. $\left(\dfrac{x}{y} - \dfrac{y}{x}\right)\left(x - \dfrac{x^2}{x+y}\right)$

16. $\left(\dfrac{x^2+y^2}{x} - 2y\right)\left(1 + \dfrac{y}{x-y}\right)$

17. $\left(x - \dfrac{6}{x+1}\right) \div \left(x - 3 - \dfrac{12}{x+1}\right)$

18. $\left(2x - 3 + \dfrac{3}{x+2}\right) \div \left(x - 2 + \dfrac{5}{2x+3}\right)$

19. $\left(\dfrac{x+y}{1-xy} - y\right) \div \left(1 + \dfrac{y(x+y)}{1-xy}\right)$

20. $\left(x - \dfrac{1}{x}\right) \div \left(x + \dfrac{1}{x} - 2\right)$

66. Complex Fractions. A *complex fraction* is one that has a fraction in the numerator or in the denominator, or in both. Thus,

$$\frac{\frac{a}{b}}{c}, \quad \frac{a}{\frac{b}{c}}, \quad \frac{\frac{a}{b}}{\frac{c}{d}}$$

are complex fractions.

By definition, $\dfrac{\frac{a}{b}}{\frac{c}{d}}$ is the quotient resulting from the division of $\dfrac{a}{b}$ by $\dfrac{c}{d}$ and hence equals $\dfrac{a}{b} \times \dfrac{d}{c}$ or $\dfrac{ad}{bc}$. In general, to simplify complex fractions, (1) *reduce the numerator and the denominator to fractions* and (2) *invert the denominator and multiply by the numerator.*

Example 1. Simplify $\dfrac{\frac{a}{b} - \frac{b}{a}}{\frac{1}{a} + \frac{1}{b}}$.

Solution.

$$\frac{\frac{a}{b}-\frac{b}{a}}{\frac{1}{a}+\frac{1}{b}}=\frac{\frac{a^2-b^2}{ab}}{\frac{b+a}{ab}}=\frac{(a+b)(a-b)}{ab}\times\frac{ab}{(a+b)}=a-b.$$

NOTE. Many complex fractions may be partially simplified very readily by multiplying the numerator and denominator by some common factor. Thus in the above example, if we multiply the numerator and the denominator by ab, we obtain immediately

$$\frac{a^2-b^2}{b+a},$$

which reduces to the answer $a - b$.

Example 2. Simplify $\dfrac{\frac{x+y}{x-y}-\frac{x-y}{x+y}}{1-\frac{x^2+y^2}{(x+y)^2}}$.

Solution.

$$\frac{\frac{x+y}{x-y}-\frac{x-y}{x+y}}{1-\frac{x^2+y^2}{(x+y)^2}}=\frac{\frac{(x^2+2xy+y^2)-(x^2-2xy+y^2)}{(x-y)(x+y)}}{\frac{(x^2+2xy+y^2)-(x^2+y^2)}{(x+y)^2}},$$

$$=\frac{\frac{4xy}{(x-y)(x+y)}}{\frac{2xy}{(x+y)^2}},$$

$$=\frac{4xy}{(x-y)(x+y)}\times\frac{(x+y)^2}{2xy},$$

$$=\frac{2(x+y)}{x-y}.$$

EXERCISE 48

Simplify:

1. $\dfrac{\frac{x}{y}}{x}$

2. $\dfrac{x}{\frac{y}{x}}$

3. $\dfrac{\frac{a^3}{b^3}}{\frac{a^2}{b}}$

4. $\dfrac{\frac{2a^2b^3}{9mn}}{\frac{4ab^3}{3m^2n}}$

5. $\dfrac{\dfrac{1}{x}+\dfrac{1}{y}}{\dfrac{1}{x}-\dfrac{1}{y}}$ 6. $\dfrac{\dfrac{a}{2}+\dfrac{b}{3}}{\dfrac{a}{3}+\dfrac{b}{2}}$ 7. $\dfrac{y+1}{1+\dfrac{1}{y}}$ 8. $\dfrac{x-1}{x-\dfrac{1}{x}}$

9. $\dfrac{a+\dfrac{a}{b}}{1+\dfrac{1}{b}}$ 10. $\dfrac{2+\dfrac{3x}{4y}}{x+\dfrac{8y}{3}}$ 11. $\dfrac{\dfrac{a}{b}-1}{1-\dfrac{b}{a}}$ 12. $\dfrac{x-\dfrac{1}{y}}{\dfrac{1}{x}-y}$

13. $\dfrac{1-\dfrac{4}{x}+\dfrac{3}{x^2}}{x-\dfrac{9}{x}}$ 14. $\dfrac{2x-5+\dfrac{2}{x}}{\dfrac{4}{x^2}-1}$ 15. $\dfrac{x+3+\dfrac{2}{x}}{x+4+\dfrac{3}{x}}$ 16. $\dfrac{2x+5-\dfrac{3}{x}}{2x+3-\dfrac{2}{x}}$

17. $\dfrac{1-\dfrac{b^2}{a^2}}{1-\dfrac{b}{a}}$ 18. $\dfrac{1-\dfrac{b^2}{a^2}}{1-\dfrac{2b}{a}+\dfrac{b^2}{a^2}}$

19. $\dfrac{x}{1-\dfrac{1-x}{1+x}}$ 20. $\dfrac{\dfrac{a}{1+a}-\dfrac{1-a}{a}}{\dfrac{a}{1+a}+\dfrac{1-a}{a}}$

21. $\dfrac{\dfrac{1}{x-y}-\dfrac{1}{x+y}}{\dfrac{2}{x^2-y^2}}$ 22. $\dfrac{y+\dfrac{xy}{y-x}}{\dfrac{x^2}{x^2-y^2}-1}$

23. $\dfrac{\dfrac{a+x}{a-x}-\dfrac{a-x}{a+x}}{\dfrac{a}{a-x}-\dfrac{a}{a+x}}$ 24. $\dfrac{\dfrac{x-a}{x-b}-\dfrac{x-b}{x-a}}{\dfrac{1}{x-a}-\dfrac{1}{x-b}}$

67. Fractional Equations. We have learned that if we multiply both sides of an equation by the same number, we obtain a new equation that is equivalent to the original equation. We may use this fact to simplify fractional equations prior to solving the equations. Thus, to solve the equation

$$\frac{x}{2}+\frac{x}{3}=5,$$

we multiply both sides by the number 6, which is the common denominator of the fractions. This gives us

$$3x + 2x = 30,$$

an equation which we easily solve by the usual methods.

This process of eliminating fractions is referred to as *clearing an equation of fractions.*

Example 1. Solve the equation $4 - \dfrac{x-5}{8} = \dfrac{7x}{26} - \dfrac{1}{2}$.

Solution. To clear of fractions, multiply both sides of the equation by 104, the lowest common denominator of the fractions:

$$416 - 13(x - 5) = 28x - 52.$$

Solve this equation in the usual manner:

$$\begin{aligned} 416 - 13x + 65 &= 28x - 52, \\ -13x - 28x &= -416 - 65 - 52, \\ -41x &= -533, \\ x &= 13. \end{aligned}$$

Note. In the above equation, $-\dfrac{x-5}{8}$ is regarded as a single term. It is in fact equivalent to $-\frac{1}{8}(x - 5)$.

Example 2. Solve $\dfrac{x-3}{x-1} = \dfrac{x-5}{x-2}$.

Solution. Multiplying both sides by the common denominator $(x - 1)(x - 2)$, we have

$$\begin{aligned} (x - 2)(x - 3) &= (x - 1)(x - 5), \\ x^2 - 5x + 6 &= x^2 - 6x + 5, \\ x &= -1. \end{aligned}$$

Example 3. Solve $\dfrac{3}{2x-2} + \dfrac{5}{x+1} = \dfrac{4}{x-1}$.

Solution. We clear the equation of fractions by multiplying each term by $2(x - 1)(x + 1)$; hence we obtain

$$\begin{aligned} 3(x + 1) + 10(x - 1) &= 8(x + 1), \\ 3x + 3 + 10x - 10 &= 8x + 8, \\ 5x &= 15, \\ x &= 3. \end{aligned}$$

EXERCISE 49

Solve:

1. $\dfrac{x}{2} + \dfrac{x}{3} + \dfrac{x}{4} = 78$

2. $\dfrac{x}{2} - \dfrac{2x}{3} + \dfrac{3x}{5} = 26$

3. $\dfrac{x-5}{4}+\dfrac{x+5}{14}=2$

4. $\dfrac{x+1}{4}-\dfrac{2x-9}{10}=\dfrac{3}{2}$

5. $\dfrac{2(x+3)}{3}-\dfrac{4(2x+5)}{5}+\dfrac{16}{15}=0$

6. $\dfrac{5(x-1)}{4}-\dfrac{3(2x+1)}{7}+\dfrac{1}{2}=0$

7. $\dfrac{3x}{4}-\dfrac{4}{13}(x+8)=\dfrac{x-5}{2}-\dfrac{7}{52}$

8. $x-\dfrac{1}{3}(x+1)+\dfrac{1}{5}(x+3)=\dfrac{3x-2}{3}$

9. $\dfrac{4}{x}=\dfrac{5}{x}-1$

10. $\dfrac{5}{2x}+\dfrac{1}{3}=\dfrac{1}{x}+\dfrac{13}{12}$

11. $\dfrac{x+2}{x}=\dfrac{5}{3}$

12. $\dfrac{1}{x}-\dfrac{4}{x^2}=\dfrac{1}{3x}+\dfrac{8}{x^2}$

13. $\dfrac{10}{x}+\dfrac{x-5}{3x}=2$

14. $\dfrac{2}{3}+\dfrac{2x-5}{x}=\dfrac{3}{x}$

15. $\dfrac{x+3}{x}=\dfrac{x+9}{x+4}$

16. $\dfrac{x-2}{x-3}=\dfrac{x+1}{x-1}$

17. $\dfrac{2x+5}{5x-1}-\dfrac{2x+3}{5x-3}=0$

18. $\dfrac{4}{x+1}-\dfrac{5}{x+4}=0$

19. $\dfrac{1}{x-4}=\dfrac{3}{x+6}$

20. $\dfrac{2x+1}{3x+2}=\dfrac{2x-3}{3x-2}$

21. $\dfrac{3}{5x+5}-\dfrac{1}{2x+2}=\dfrac{1}{10}$

22. $\dfrac{4}{2x-1}+\dfrac{1}{2}=\dfrac{3}{4x-2}$

23. $\dfrac{x}{x+1}-\dfrac{2x}{x+2}+1=0$

24. $\dfrac{5}{x}-\dfrac{3}{x-1}-\dfrac{2}{x+1}=0$

25. $\dfrac{5x+1}{5}-\dfrac{2\frac{4}{5}}{x+2}=\dfrac{2x-1}{2}$

26. $\dfrac{1}{x+1}-\dfrac{1}{x+2}+\dfrac{1}{2x+2}=0$

27. $\dfrac{5}{x-1}-\dfrac{7}{x-2}+\dfrac{2}{x-3}=0$

28. $\dfrac{3x-1}{x-3}-\dfrac{x+5}{x+2}=2$

29. $\dfrac{2}{2x^2+x+2}=\dfrac{3}{3x^2+x+5}$

30. $\dfrac{4}{(2x-3)(3x-2)}=\dfrac{5}{(4x-3)(2x-3)}$

31. $\dfrac{1}{x^2-4}=\dfrac{1}{2x^2+5x+2}$

32. $\dfrac{x-2}{3x^2-2x}=\dfrac{x+1}{3x^2+4x-4}$

68. Literal Equations. An equation in which at least one of the coefficients or terms is represented by a letter (other than the unknown) is called a *literal equation.*

To solve such equations we apply precisely the same processes used in solving numerical equations. That is:

1. Clear of fractions (if fractions occur in the equation).
2. Remove parentheses (if parentheses occur in the equation).
3. Transpose terms so that all and only terms containing the unknown are on one side of the equation.
4. Divide both sides of the equation by the coefficient of the unknown.

Example 1. Solve the equation $\frac{x - 8a}{6} = 3a - 2x$ for x and check your result.

Solution.
$$\begin{aligned} \frac{x - 8a}{6} &= 3a - 2x, && \text{given equation.} \\ x - 8a &= 18a - 12x, && \text{step 1,} \\ 13x &= 26a, && \text{step 3,} \\ x &= 2a, && \text{step 4.} \end{aligned}$$

Check. Substituting $2a$ for x in the given equation, we have

$$\frac{(2a) - 8a}{6} = 3a - 2(2a),$$

$$\frac{-6a}{6} = 3a - 4a,$$

$$-a = -a.$$

Example 2. Solve the equation $\frac{p(x + q)}{9} = \frac{px}{3} - pq$ for x.

Solution.
$$\begin{aligned} \frac{p(x + q)}{9} &= \frac{px}{3} - pq, && \text{given equation.} \\ p(x + q) &= 3px - 9pq, && \text{step 1,} \\ px + pq &= 3px - 9pq, && \text{step 2,} \\ -2px &= -10pq, && \text{step 3,} \\ x &= 5q, && \text{step 4.} \end{aligned}$$

Example 3. Solve $(x - a)^2 = (x - b)^2$ for x.

Solution. Multiplying out gives

$$x^2 - 2ax + a^2 = x^2 - 2bx + b^2;$$

hence,

$$\begin{aligned} 2bx - 2ax &= b^2 - a^2, \\ 2(b - a)x &= b^2 - a^2, \\ x &= \frac{(b + a)(b - a)}{2(b - a)}, \\ x &= \frac{a + b}{2}. \end{aligned}$$

EXERCISE 50

Solve the following equations for x:

1. $3x - 2a = 7a$
2. $4x - 7a = a + 4b$
3. $2a - x = x - 2b$
4. $4ax - 9ab = 3ab$
5. $6ax - 3ab = 8ax - 9ab$
6. $\dfrac{2x + 3b}{9} = \dfrac{x - b}{2}$
7. $3(2x - 3a) + 2(3x - a) = a$
8. $5(x + 4a) - 2(a - x) = x$
9. $\frac{1}{3}(4a - x) = \frac{1}{2}(x - 4a)$
10. $\frac{1}{2}(x - b) = \frac{1}{3}(x + 3a)$
11. $a(5x + a) = 7a^2 + 9ax$
12. $3(bx - 2ab) = b(x - 7a) + 3ab$
13. $\dfrac{2(ax + 2ab)}{3a} = \dfrac{x - 2b}{2}$
14. $ab - \frac{5}{2}(ab - bx) = bx$
15. $2(ax + ab) - a(3b - x) = ab$
16. $3(x - 6b) = 5(b - 2x) + 16b$
17. $3(x + a) + 2(x + 2a) + (x + 3a) = 0$
18. $6a(bx - 3bc) = 7b(ax - 9ac)$
19. $\dfrac{6x - 7a}{7} = \dfrac{5x + 7a}{14} + x$
20. $\frac{1}{2}x - \frac{1}{3}a = \frac{1}{4}x - \frac{1}{5}a$
21. $a(x - 3) + 2x = 8 + a$
22. $a + 2 = 2(x + 4) - a(x + 2)$
23. $ax + b^2 = bx + a^2$
24. $ax + 6b = 2bx + 3a$
25. $\dfrac{x}{a} - \dfrac{x}{b} = \dfrac{b}{a} - \dfrac{a}{b}$
26. $\dfrac{1}{a} - \dfrac{1}{x} = \dfrac{1}{x} - \dfrac{1}{b}$
27. $\dfrac{a}{x} = c(a - b) + \dfrac{b}{x}$
28. $\dfrac{2}{3}\left(\dfrac{x}{ab} + 1\right) = \dfrac{3}{4}\left(\dfrac{x}{ab} - 1\right)$
29. $m(m - x) + 2mn = n(x - n)$
30. $\dfrac{a - x - 2}{4} = \dfrac{2 - x}{a}$
31. $a(x - a) + 3b(x - 2b) = 5ab$
32. $3a(3a - x) = 2b(2b + x)$
33. $(x + a)(2x - a) = (x + b)(2x - b)$
34. $(x + a)(x + b) = x(x - c)$
35. $(x - a)^2 + 2b(b - a) = (x - b)^2$
36. $(x - a)^2 + a = (x - 3)^2 + 3$
37. $x(x - a) + a = x(x - b) + b$
38. $\dfrac{x - a}{2a + 5b} = \dfrac{x - b}{3a + 4b}$
39. $\dfrac{1}{x - a} + \dfrac{1}{x - b} = 0$
40. $\dfrac{1}{x} + \dfrac{1}{x - a} - \dfrac{2}{x + a} = 0$

69. Solving Formulas. A formula may be solved for a particular letter by considering that letter to be the unknown.

Example 1. Solve the formula, $F = m(g - a)$, for a.

Solution. Multiplying out, we have

$$F = mg - ma;$$

transposing, $$ma = mg - F;$$

hence, $$a = \frac{mg - F}{m}.$$

Example 2. Solve $S = \frac{1}{2}n(a + l)$ for l.

Solution.

$S = \frac{1}{2}n(a + l)$, given formula;

$2S = n(a + l)$, clearing of fractions;

$2S = na + nl$, removing parentheses;

$2S - na = nl$, transposing;

$nl = 2S - na$, interchanging members;

$l = \dfrac{2S - na}{n}$, dividing by coefficient of l.

Example 3. Solve $\dfrac{1}{a} - \dfrac{1}{H} = \dfrac{1}{H} - \dfrac{1}{b}$ for H.

Solution.

$\dfrac{1}{a} - \dfrac{1}{H} = \dfrac{1}{H} - \dfrac{1}{b}$, given formula;

$bH - ab = ab - aH$, clearing of fractions;

$aH + bH = 2ab$, transposing;

$(a + b)H = 2ab$, factoring;

$H = \dfrac{2ab}{a + b}$, dividing by the coefficient of H.

EXERCISE 51

Solve each of the following formulas for the letter indicated:

1. $I = prt$, for t
2. $C = 2\pi r$, for r
3. $A + B = 90$, for B
4. $P = a + b + c$, for c
5. $V = \frac{1}{3}Bh$, for h
6. $v = \frac{1}{2}gt^2$, for g
7. $p = 2(l + w)$, for l
8. $V = v + gt$, for g
9. $A = p(1 + rt)$, for t
10. $s = \frac{1}{2}(a + b + c)$, for b
11. $F = \frac{9}{5}C + 32$, for C
12. $V = \frac{1}{3}\pi r^2 h$, for h
13. $A = \frac{1}{2}h(B + b)$, for B
14. $E = I(R + r)$, for r
15. $V = \frac{1}{4}\pi d^2 h$, for h
16. $l = \dfrac{L - M}{Mt}$, for M
17. $F = \dfrac{h}{l}W$, for h
18. $K = \frac{1}{2}mv^2 + mh$, for m

19. $V = \frac{h}{6}(B + 4M + b)$, for M

20. $D = \frac{n + 2}{q}$, for q

21. $b^2 - 4ac = 0$, for c

22. $a = s(1 - r)$, for r

23. $A = \frac{\pi r^2 s}{360}$, for s

24. $V = \frac{1}{3}\pi h^2(3r - h)$, for r

25. $F = k\frac{mM}{r^2}$, for m

26. $s = v + \frac{1}{2}a(2n - 1)$, for n

27. $M = \frac{mgl^3}{4sa^3b}$, for g

28. $H = sm(T - t)$, for t

29. $nE = I(R + nr)$, for r

30. $E = I\left(R + \frac{r}{n}\right)$, for r

31. $\frac{1}{f} = \frac{1}{p} + \frac{1}{q}$, for p

32. $\frac{n}{v} = \frac{m}{M + m}$, for m

33. $S = \frac{a - rl}{1 - r}$, for r

34. $I = \frac{E}{R + \frac{r}{n}}$, for n

35. $(P - p)(V - v) = K$, for p

36. $S = \frac{n}{2}[2a + (n - 1)d]$, for d

70. Work Problems. If it takes a person 8 hours to complete a piece of work, he does $\frac{1}{8}$ of the work each hour. This fraction we call his rate of working. Thus, if he can do the work in x hours, his rate of working is $\frac{1}{x}$.

Work problems and other problems related to them depend on the formula

$$\text{Rate of working} \times \text{Time} = \text{Fraction of work done.}$$

Example 1. A boy can do a piece of work in 15 days and his father can do it in 10 days. How long will it take them when working together?

Analysis. Let x = time in days it takes them working together. The problem may then be tabulated as follows:

	Rate	× Time	= Frac. of Work Done
Boy	$\frac{1}{15}$	x	$\frac{x}{15}$
Father	$\frac{1}{10}$	x	$\frac{x}{10}$

The fraction of the work done by the boy combined with the fraction of the work done by the father must equal the whole piece of work, or 1; hence

Solution.

$$\frac{x}{15} + \frac{x}{10} = 1.$$
$$2x + 3x = 30,$$
$$5x = 30,$$
$$x = 6.$$

Answer. It takes them 6 days when working together.

Check. To verify this answer we note that the boy does $\frac{6}{15} = \frac{2}{5}$ of the whole job in 6 days and the father does $\frac{6}{10} = \frac{3}{5}$ of the whole job in the same time. Since $\frac{2}{5} + \frac{3}{5} = 1$, it is evident that they complete the job.

Example 2. A tank is fitted with two pipes. One pipe can fill it in 15 hours. After it has been open 3 hours, the second pipe is opened and the tank is filled in 4 hours more. How long would it take the second pipe alone to fill the tank?

Analysis. Let x = time in hours for second pipe alone to fill the tank; then

	Rate	× Time	= Fraction of Tank
First pipe	$\frac{1}{15}$	7	$\frac{7}{15}$
Second pipe	$\frac{1}{x}$	4	$\frac{4}{x}$

Solution.

$$\frac{7}{15} + \frac{4}{x} = 1,$$
$$\frac{4}{x} = \frac{8}{15},$$
$$8x = 60,$$
$$x = 7\tfrac{1}{2}.$$

Answer. The second pipe alone can fill the tank in $7\frac{1}{2}$ hours.

Example 3. A tank can be filled by one pipe in 30 minutes and emptied by another in 50 minutes. If both pipes are open, how long will it take to fill the tank?

Analysis. Let x = time in minutes to fill tank; then

	Rate	× Time	= Fraction of Tank
Intake	$\frac{1}{30}$	x	$\frac{x}{30}$
Drain	$\frac{1}{50}$	x	$\frac{x}{50}$

Solution. Since the drain is nullifying part of the intake,

$$\frac{x}{30} - \frac{x}{50} = 1,$$
$$5x - 3x = 150,$$
$$2x = 150,$$
$$x = 75.$$

Answer. It will take 1 hour and 15 minutes to fill the tank.

EXERCISE 52

Work Problems

1. One boy can mow a lawn in 2 hours and another can mow it in 1 hour and 45 minutes. If both work together, how many minutes will it take?
2. A painter can paint a certain house in 36 hours; it will take his apprentice 45 hours to do the same job. If they work together, how long will it take?
3. One faucet can fill a tub in 28 minutes and another faucet can fill it in 21 minutes. How long will it take to fill one-half of the tub if both faucets are open?
4. A tank can be emptied by one pipe in 30 minutes and by another in 20 minutes. If the tank is $\frac{3}{4}$ full, how long will it take to empty it when both pipes are open?
5. *A* can do a piece of work in two-thirds the time that it takes *B*, and they can do it together in 6 hours. How long will it take each alone to do the work?
6. A large pipe takes $\frac{2}{3}$ as much time to fill a tank as a smaller one. Together they can fill the tank in 48 minutes. How long will it take the smaller pipe alone to fill the tank?
7. *A* can do a piece of work in three-fourths the time required by *B*, and they can do it together in 24 minutes. How long will it take each alone to do the work?

8. A drain takes three times as long to empty a tank as a pipe can fill it. When both the pipe and the drain are open, it takes 30 minutes to fill the tank. How long will it take the pipe to fill the tank if the drain is closed?

9. One pipe can fill a tank in 2 hours, a second pipe takes 3 hours to fill it, and a third pipe takes 6 hours. If all three of the pipes are open, how long will it take to fill the tank?

10. A large pipe can fill a tank in 15 minutes and a small pipe can fill it in 20 minutes. If one large pipe and two small pipes are opened, how long will it take to fill the tank?

11. Each of two men can do a piece of work in 2 hours and a boy can do it in 2 hours and 20 minutes. If all three work together, how many minutes will it take to complete the work?

12. A man can do a piece of work in 1 hour and 30 minutes, his wife can do it in 1 hour and 45 minutes, and their son can do it in 2 hours and 6 minutes. If all three work together, how long will it take to complete the work?

13. George can do a piece of work in 45 minutes and with the help of Frank he can do it in 20 minutes. How long will it take Frank working alone to do the task?

14. It takes a pipe 10 minutes to fill a basin when the drain is closed, and 35 minutes to fill the basin when the drain is open. If the basin is half-filled, how long will it take the drain to empty it?

15. A workman started a job which usually took him 3 hours. After working 30 minutes, he was relieved by a second workman who finished the job in 2 hours and 5 minutes. How long would it have taken the second man to do the whole job?

16. One pipe can fill a tank in 1 hour. After this pipe has been running for 12 minutes, it is shut off and a second pipe is opened. The second pipe finishes filling the tank in 36 minutes. How long would it have taken the second pipe alone to fill the tank?

17. One pipe can fill a tank in 20 minutes and a second pipe can fill it in 30 minutes. These pipes together can fill the tank in 18 minutes when the tank is being emptied by a third pipe. How long will it take the third pipe alone to empty a full tank?

18. A can do a piece of work in 20 days and B can do it in 24 days. A, B, and C together can do it in 8 days. How long will it take C alone to do the work?

19. A man can do a piece of work in 6 days and his older son can do it in 10 days. The man and his older son work together for 2 days, and then

the man and his younger son finish the job in 2 days. How long will it take the younger son alone to do the work?

20. A can do a piece of work in 20 days and B can do it in 24 days. After A and B have been working together for 7 days, they are joined by C and the three finish the job in 3 more days. How long will it take C alone to do the job?

71. Problems Leading to Fractional Equations. In setting up the equations for many stated problems, we often obtain equations which are expressed in the form of fractions.

Example 1. The larger of two numbers is 6 more than the smaller, and $\frac{2}{3}$ of the smaller increased by $\frac{3}{4}$ of the larger equals 13. Find the numbers.

Analysis. Let $x =$ the smaller number; then $x + 6 =$ the larger number. Hence,

$$\tfrac{2}{3}x + \tfrac{3}{4}(x + 6) = 13.$$

Solution. Multiplying by 12 to clear the equation of fractions, we have

$$\begin{aligned} 8x + 9(x + 6) &= 156, \\ 8x + 9x + 54 &= 156, \\ 17x &= 102, \\ x &= 6. \end{aligned}$$

Answer. The numbers are 6 and 12.

Example 2. The denominator of a fraction is 4 more than the numerator. If both the numerator and the denominator of the fraction are increased by 1, the resulting fraction equals $\frac{1}{2}$. Find the fraction.

Analysis. Let $x =$ the numerator of the fraction; then $x + 4 =$ the denominator. Increasing both of these expressions by 1, we obtain the resulting equation

$$\frac{x + 1}{x + 5} = \frac{1}{2}.$$

Solution. Clearing of fractions, we have

$$\begin{aligned} 2x + 2 &= x + 5, \\ x &= 3. \end{aligned}$$

Answer. The fraction is $\frac{3}{7}$.

EXERCISE 53

1. Find a number such that the sum of its fourth and sixth parts is 25.

2. Find a number such that six-sevenths of it will exceed five-sixths of it by 4.

3. The fifth, fifteenth, and fiftieth parts of a number together make 43. Find the number.

4. Two consecutive numbers are such that one-fifth of the lesser exceeds one-seventh of the greater by 1. Find the numbers.

5. Find three consecutive numbers such that if they are divided by 2, 3, and 5 respectively, the sum of the quotients will be 40.

6. Find three consecutive even numbers such that if they be divided by 2, 3, and 4 respectively, the sum of the quotients will be 19.

7. The width of a room is five-sevenths of its length. If the width were 3 feet more and the length were 3 feet less, the room would be square. Find the dimensions of the room.

8. The length of a rectangle is 10 more than one-half the width, and the width is 5 more than one-third the length. Find its dimensions.

9. The smaller of two numbers is two-thirds of the larger, and the sum of their reciprocals is $\frac{1}{6}$. What are the numbers?

10. The smaller of two numbers is three-fifths of the larger, and the sum of their reciprocals is $\frac{4}{45}$. Find the numbers.

11. The tens' digit of a two-figured number is 5 more than the units' digit, and the number divided by the sum of its digits equals 8. Find the number.

12. The units' digit in a two-figured number is 3 more than the tens' digit, and the number divided by the number with its digits reversed equals $\frac{4}{7}$. Find the number.

13. The numerator of a certain fraction is 3 less than the denominator. If the numerator is decreased by 1 and the denominator is increased by 2, the resulting fraction will equal $\frac{1}{2}$. What is the fraction?

14. The denominator of a certain fraction exceeds the numerator by 7. If both the numerator and the denominator are increased by 3, the resulting fraction equals $\frac{1}{2}$. What is the fraction?

15. A boy's age is 1 year more than one-fifth of his father's age, and in four years his age will be 1 year less than one-third of his father's age. How old is the boy?

16. Paul's age is $\frac{2}{3}$ the age of John. In 5 years Paul's age will be $\frac{3}{4}$ John's age. Find their present ages.

17. Find the number which must be subtracted from both the numerator and the denominator of the fraction $\frac{65}{92}$ so that the resulting fraction will have a value of $\frac{4}{7}$.

18. The sum of two numbers is 400. When the larger is divided by the smaller, there is a quotient of 6 and a remainder of 8. Find the numbers.

19. When 176 is divided by a certain number, the quotient is 13 and the remainder is 7. What is the divisor?

20. The units' digit of a two-figured number is 2 more than the tens' digit. When the number is divided by the sum of its digits, the quotient is 4 and the remainder is 9. Find the number.

21. A and B together can do a piece of work in 12 hours. After working together for 8 hours, B quit and A finished in 10 hours more. How long would it take B alone to do the work?

22. The denominator of a fraction is 1 more than the numerator. If the numerator is increased by 9 and the denominator is increased by 12, the value of the fraction is unchanged. What is the fraction?

23. The denominator of a fraction is 3 less than twice the numerator. If the numerator is increased by 20 and the denominator is increased by 34, the value of the fraction is unchanged. What is the fraction?

24. A, B, and C have \$1470 among them. B's share is \$20 more than five-ninths of A's, and C's is less than four-sevenths of A's by \$24. Find the share of each.

25. The width of a rectangle is $\frac{4}{7}$ of the length. If both dimensions are increased by 2 feet, the area will be increased by 48 square feet. Find the dimensions of the rectangle.

26. A sum of money is divided between two persons. One receives a dollars more than one-third of the whole sum, and the other receives b dollars, the amount that is left. Find the original sum.

27. Find two consecutive numbers such that if they are divided by a and b respectively, the sum of the quotients will be c.

28. At a school 1 less than one-fifth of the students are seniors, 3 less than one-fourth are juniors, seven-twentieths are freshmen, and the remainder, 28 in number, are sophomores. Find the number attending the school.

29. A man traveled a distance of 60 miles at a constant rate of speed. If he had doubled his rate, he would have taken 1 hour and 12 minutes less time to travel the same distance. At what rate did he travel?

30. One person can do a piece of work in a minutes and a second person can do it in b minutes. How many minutes will it take them to do this job when they work together?

31. How much water must be added to n gallons of an $a\%$ solution of acid in order to reduce it to a $b\%$ solution of the acid?

32. A man has k coins in quarters and half dollars. If their value is D dollars, how many of each does he have?

33. A motorist travels n miles in t hours. How much faster must he travel if he wishes to cover the same distance in one hour less?

34. Out of a certain sum, a boy spent 5 cents for some candy. He spent half of the remainder for a show, and one-fourth of what he then had left he spent for some ice cream. He then had one-third of the original sum. How much did he have at first?

35. A man has a certain sum of money from which he gives his son \$4 and one-fourth of what remains; he then gives his daughter \$5 and one-fifth of what remains. He finds now that he has given away half of his money. How much did the son and daughter each receive?

36. In a bag of mixed nuts the walnuts composed 3 pounds more than one-third the total weight, the almonds 4 pounds more than one-fourth the total weight, and the peanuts 5 pounds more than one-sixth the total weight. What was the total weight of the nuts?

37. A bag contains white balls and black balls. Ten more than half the number of balls are black balls, and 6 more than half the number of black balls are white balls. How many white balls and black balls does the bag contain?

38. The denominator of a certain fraction is 2 more than the numerator. If the numerator is increased by 7 and the denominator is increased by 1, the resulting fraction equals the reciprocal of the original fraction. What is the fraction?

39. The denominator of a certain fraction exceeds the numerator by 1. If 8 is added to the numerator and 5 to the denominator, the resulting fraction is equal to the reciprocal of the given fraction. What is the fraction?

40. A cistern can be filled by any one of three pipes. The largest pipe can fill it alone in 6 hours. The largest and smallest together can fill it in 5 hours, and the two smaller pipes can fill it in 12 hours. How long would it take to fill the cistern if all three pipes were used?

REVIEW OF CHAPTER VI

A

Reduce the following fractions to lowest terms:

1. $\dfrac{52a^3b^2}{65ab^3}$ **2.** $\dfrac{42x^3yz^2}{49x^2y^2z^2}$ **3.** $\dfrac{-108a^3x^2y^4}{81a^2xy^2}$

4. $\dfrac{51m^6n^7}{68m^8n^3}$ **5.** $\dfrac{5a+15}{2a^2+6a}$ **6.** $\dfrac{5ab+5b^2}{a^2-b^2}$

7. $\dfrac{4a^2 - 12ab}{3ab - 9b^2}$ 8. $\dfrac{3x^3 - 15x}{4x^4 - 20x^2}$ 9. $\dfrac{ax^2 - ay^2}{bx + by}$

10. $\dfrac{4x^2y - 6xy^2}{6x^2y - 9xy^2}$ 11. $\dfrac{a^2 - 9}{a^2 - a - 6}$ 12. $\dfrac{2x^2 + x - 6}{2x^2 - 5x + 3}$

13. $\dfrac{3x^2 - 5x - 2}{6x^2 - x - 1}$ 14. $\dfrac{8a^2 - 18a + 9}{12a^2 - a - 6}$

15. $\dfrac{18x^3 - 57x^2 + 30x}{12x^3 - 60x^2 + 75x}$ 16. $\dfrac{12a^3x - 2a^2x^2 - 2ax^3}{45a^4x + 24a^3x^2 + 3a^2x^3}$

17. $\dfrac{a^2x^2 - ax}{a^2x^2 - 1}$ 18. $\dfrac{x^4 - 1}{x^3 - x^2 + x - 1}$

19. $\dfrac{4x^3 - 6x^2 - 2x + 3}{6x^3 + 4x^2 - 3x - 2}$ 20. $\dfrac{(a + b)^2 - c^2}{(a + c)^2 - b^2}$

B

Multiply:

1. $\dfrac{14a^3b^2}{9x^2y^2} \times \dfrac{6x^3y}{35a^2b^4}$ 2. $\dfrac{10r^2s^3}{3t^3} \times \dfrac{21rt^2}{4s^2}$

3. $\dfrac{6a^2b}{5a - 10b} \times \dfrac{2a - 4b}{9ab^2}$ 4. $\dfrac{x^2y^2 + 3xy}{4x^2 - 1} \times \dfrac{2x + 1}{xy + 3}$

5. $\dfrac{a^2 - b^2}{8c^3} \times \dfrac{16c^2}{(a - b)^2}$ 6. $\dfrac{x^2y + xy^2}{x^2y + y^3} \times \dfrac{x^4 - y^4}{3xy(x + y)^2}$

7. $\dfrac{x^2 + 4x + 3}{x^2 + x - 6} \times \dfrac{x^2 + 3x - 10}{x^2 + 6x + 5}$ 8. $\dfrac{x^2 + x - 12}{x^2 - x - 6} \times \dfrac{x^2 - 3x - 10}{x^2 + 9x + 20}$

9. $\dfrac{2x^4 - 9x^3y - 5x^2y^2}{3x^2y^2 - 16xy^3 + 5y^4} \times \dfrac{3x^2y^2 - 7xy^3 + 2y^4}{6x^4 - x^3y - 2x^2y^2}$

10. $\dfrac{a^3 - a^2 + 3a - 3}{2a^2 - 3a} \times \dfrac{2a^2 - a - 3}{a^2 + 3}$

Divide:

11. $\dfrac{15a^4x^3}{4b^2y} \div \dfrac{5a^3x^4}{8by^2}$ 12. $\dfrac{72m^3n^4}{25p^2q^3} \div \dfrac{18m^2n^5}{45p^4q}$

13. $\dfrac{4x^2 + 6xy}{3} \div \dfrac{4xy + 6y^2}{9}$ 14. $\dfrac{a^2b^2 - ab}{3a - 3b} \div \dfrac{2ab - 2}{9a - 9b}$

15. $\dfrac{m^4 - n^4}{m^4 - m^2n^2} \div \dfrac{m^2 + n^2}{m^3 + m^2n}$ 16. $\dfrac{a^2b + ab^2}{a^3 - a^2b} \div \dfrac{ab + b^2}{ab - b^2}$

17. $\dfrac{x^2 + xy - 2y^2}{x^2 - 4xy - 5y^2} \div \dfrac{x^2 - 3xy + 2y^2}{x^2 - xy - 2y^2}$

18. $\dfrac{2m^2 + 11mn + 5n^2}{3m^2 + mn - 2n^2} \div \dfrac{2m^2 - 3mn - 2n^2}{3m^2 + 7mn - 6n^2}$

19. $\dfrac{4ax^2 - 16a}{3x^3 - 21x^2 + 30x} \div \dfrac{2ax^2 + 6ax + 4a}{9x^3 - 36x^2 - 45x}$

20. $\dfrac{x^3 + x^2 + x + 1}{2x^3 + 2x^2 + x + 1} \div \dfrac{2x^3 - x^2 + 2x - 1}{4x^3 - 2x^2 + 2x - 1}$

C

Combine and simplify:

1. $\dfrac{2x - 3}{2} - \dfrac{3x - 2}{4}$

2. $\dfrac{5a + 2b}{2} + \dfrac{2a - 3b}{3}$

3. $\dfrac{2}{2x^3} - \dfrac{3}{5x^2} + \dfrac{7}{10x}$

4. $\dfrac{a}{b} + \dfrac{b}{c} + \dfrac{c}{a}$

5. $\dfrac{4a - 1}{2} - \dfrac{3a + 1}{3} + \dfrac{2a - 7}{6}$

6. $\dfrac{2x - 3y}{4} + \dfrac{x + y}{3} - \dfrac{3x - 5y}{6}$

7. $\dfrac{x + 2}{2x} - \dfrac{4x^2 - 3}{4x^2} + \dfrac{4x^3 - 5}{8x^3}$

8. $\dfrac{m + n}{m} - \dfrac{m - n}{n} + \dfrac{m^2 - n^2}{mn}$

9. $\dfrac{a + x}{ax} + \dfrac{a + y}{ay} - \dfrac{x + y}{xy}$

10. $\dfrac{x + 3a}{ax} + \dfrac{x + 4b}{bx} - \dfrac{7}{x}$

11. $\dfrac{3}{4a - 6} + \dfrac{1}{6a - 9} - \dfrac{5}{12a - 18}$

12. $\dfrac{5}{x - 5y} - \dfrac{3}{2x - 10y} + \dfrac{1}{3x - 15y}$

13. $\dfrac{4}{x + y} - \dfrac{3}{x + 2y}$

14. $\dfrac{4}{3 + 4x} + \dfrac{5}{2 - 5x}$

15. $\dfrac{x - 8}{x - 3} - \dfrac{x - 9}{x - 4}$

16. $\dfrac{3x + 5y}{2x - 3y} + \dfrac{5x - 4y}{3x - 2y}$

17. $\dfrac{x + 1}{3x - 3} - \dfrac{x + 2}{2x + 2} + \dfrac{x}{6(x + 1)}$

18. $\dfrac{1}{x - 2} + \dfrac{2}{x + 2} - \dfrac{3}{x - 3}$

19. $\dfrac{3}{x^2 - 5x + 4} - \dfrac{2}{x^2 - 6x + 8}$

20. $\dfrac{2}{x^2 + 3x + 2} + \dfrac{1}{x^2 + 4x + 3} + \dfrac{2}{x^2 + 5x + 6}$

D

Reduce the following to fractions:

1. $\dfrac{4a - 3}{a - 2} - 3$

2. $1 - \dfrac{m(m - n)}{1 - mn}$

3. $\dfrac{xy - 3x^2}{3x - y} + x - 2y$

4. $1 - \dfrac{3}{p - 1} + p$

Perform the indicated operations and simplify:

5. $\left(\dfrac{x^2}{4} - 4\right) \div \left(1 + \dfrac{x}{4}\right)$

6. $\left(\dfrac{9m^2}{n^2} - 1\right)\left(\dfrac{n}{3m + n} - 1\right)$

7. $\left(x - \frac{1}{x}\right) \div \left(1 - \frac{1}{x}\right)$

8. $\left(2 - \frac{1}{x}\right) \div \left(4 - \frac{1}{x^2}\right)$

9. $\left(\frac{a}{b} - \frac{b}{a}\right)\left(b + \frac{b^2}{a - b}\right)$

10. $\left(x - 2 - \frac{4}{x + 1}\right) \div \left(x - 1 - \frac{3}{x + 1}\right)$

11. $\left(a + 4 + \frac{14}{a - 5}\right)\left(a - 4 - \frac{7}{a + 2}\right)$

12. $\left(\frac{xy - y^2}{1 - xy} + 1\right)\left(x + \frac{x^2 + xy}{1 - y}\right)$

Simplify:

13. $\dfrac{3x + \frac{4y}{3}}{\frac{9x}{2} + 2y}$

14. $\dfrac{a - \frac{c^2}{b}}{\frac{ab}{c} - c}$

15. $\dfrac{\frac{x^2}{y^2} - 1}{\frac{x^2}{y^2} + \frac{2x}{y} + 1}$

16. $\dfrac{\frac{m}{n} - \frac{n}{m}}{\frac{1}{m} - \frac{1}{n}}$

17. $1 - \dfrac{1}{1 - \frac{1}{x}}$

18. $\dfrac{\frac{6a}{b} + 13 + \frac{6b}{a}}{\frac{3a}{b} - 10 - \frac{8b}{a}}$

19. $\dfrac{\frac{x}{x - y} + \frac{y}{x + y}}{\frac{x}{x - y} - \frac{y}{x + y}}$

20. $\dfrac{p - q - \frac{2p(p - q)}{p + q}}{\frac{p^2 + q^2}{pq + q^2} - 1}$

E

Solve:

1. $\frac{x + 1}{3} + \frac{x + 2}{4} - \frac{2x + 1}{5} = 1$

2. $\frac{2}{3}x - \frac{1}{5}(2x + 1) - \frac{1}{6}(x - 3) = 0$

3. $\frac{x + 2}{6} - \frac{2x + 1}{3} = \frac{3}{2} + x$

4. $\frac{2x - 3}{4} + \frac{x + 2}{6} = \frac{5x + 4}{12}$

5. $\frac{1}{5}(2x - 1) - \frac{1}{3}(x - 1) + \frac{1}{15}(3x - 4) = 0$

6. $\frac{1 - x}{2} + \frac{1 + x}{5} - \frac{3 - x}{3} = 0$

7. $\frac{3}{2x - 1} = \frac{7}{3x + 1}$

8. $\frac{3x + 4}{6x - 5} = \frac{2x + 5}{4x - 1}$

9. $\frac{2x - 3}{3x} = \frac{x + 1}{x - 1} - \frac{1}{3}$

10. $\frac{2x + 4}{x - 3} = \frac{1}{2}x - \frac{1}{6}(3x - 2)$

11. $\frac{3}{x + 2} + \frac{2x}{4 - x^2} = \frac{2}{x - 2}$

12. $\frac{3}{x + 1} + \frac{2}{x - 1} - \frac{2x + 1}{x^2 - 1} = 0$

13. $\frac{3}{x+1} - \frac{4}{x+2} + \frac{1}{x+3} = 0$

14. $\frac{x+1}{x+2} + \frac{x+3}{x+4} = 2$

15. $\frac{10}{1-3x} + \frac{3}{2x+3} = \frac{2x+6}{6x^2+7x-3}$

16. $\frac{3}{x+1} - \frac{3}{x-1} + \frac{8}{x^2+2x+1} = 0$

17. $\frac{1}{2x^2-7x-15} = \frac{3}{4x^2-9}$

18. $\frac{x+4}{2x^2+5x} = \frac{x+1}{2x^2+x-10}$

Solve for x:

19. $8a - 3x = 4(3x - a)$
20. $(x + 3c)(x - 2c) = (x + 4c)(x - c)$
21. $5ab - a^2 + 3(a - b)x = 2(a + b)x$

22. $\frac{x-a}{x+b} + \frac{b}{a} = 0$

23. $\frac{c+d}{x+1} = \frac{c-d}{x-1}$

24. $\frac{3x-m}{3x+m} = \frac{9x^2}{9x^2-m^2}$

25. $\frac{x+c}{x-d} - \frac{x-c}{x+d} = \frac{c^2-d^2}{x^2-d^2}$

26. $(x - p)^2 + (3p + q)(p - q) = (x - q)^2$

Solve each formula for the letter indicated:

27. $l = a + (n - 1)d$, for n

28. $\frac{P-P_1}{P-P_2} = k$, for P

29. $\frac{e}{E} = \frac{r}{R+r}$, for r

30. $F = M(a - a') + F'$, for a

F

1. Find three consecutive even numbers such that if they are divided by 5, 10, and 15 respectively, the sum of the quotients will be 10.
2. What number should be added to both the numerator and the denominator of $\frac{41}{57}$ so that the value of the resulting fraction will be $\frac{1}{2}$?
3. A baseball team has won 23 games and lost 12. How many games must they win in succession to bring their average up to 0.750?

4. The units' digit of a two-figured number is 3 more than the tens' digit. The number divided by the sum of its digits gives 4 as a quotient and 3 as the remainder. Find the number.

5. At a certain school 40 more than half of the students are boys, and the number of girls at the school is 4 less than half the number of boys. How many students attend the school?

6. A boy earns a certain sum of money. He gives half of it to his mother, puts two-thirds of the remainder in his bank, and then goes to a show with the remaining 20 cents. How much did he earn?

7. One pipe can fill a tank in 40 minutes and another can fill it in an hour. How long will it take the pipes together to fill the tank?

8. Two boys together can spade a garden in 2 days. If one of them could do the job alone in 3 days, how long would it take the other boy alone to do the job?

9. A drain takes four times as long to empty a tank as a pipe can fill it. When both the pipe and the drain are open, it takes 48 minutes to fill the tank. How long would it take the drain to empty half the tank?

10. To do a certain job B takes $1\frac{1}{6}$ times as long as A, and C takes $1\frac{2}{5}$ times as long as A. Working together they can finish the job in 35 hours. How long will it take each alone to do the job?

11. A alone can do a piece of work in 18 days and B alone can do it in 24 days. After A and B have been working together for 5 days, they are joined by C and they finish the job in 4 more days. How long would it take C alone to do the whole job?

12. A girl's age is 2 years less than one-fifth of her mother's age. In three years her age will be 1 year less than one-fourth of her mother's age. Find their present ages.

13. A man has enough fodder to last C cows for d days. How long will the fodder last if he buys n more cows?

14. If a man walks x yards in m minutes, how long will it take him to walk mx miles?

15. If a man can do a certain job in m hours and a boy can do the job in b hours, how long will it take them when working together?

16. The sum of the numerator and the denominator of a proper fraction is 74. If each is increased by 5, the value of the fraction becomes $\frac{3}{4}$. Find the fraction.

17. A man has two-thirds of his money invested at 5%, one-fourth at 4%, and the remainder at 3%. How much money does he have, if his annual income is $302.50?

18. A sum of money is divided among three persons. The first receives a dollars more than half of the whole sum; the second receives b dollars more than half of what remains; and the third receives c dollars, the amount that is left. What was the original sum?

19. The sum of two numbers is s, and the larger divided by the smaller gives a quotient q and a remainder r. Find the two numbers.

20. If each side of a square is decreased by 4 inches, the decrease in area will be the same as the increase in area would have been, had the sides been increased by 2 inches. Find the area of the square.

Are Fractions Necessary?

Show that the value of the product

$$\left(x^4 + x^3 + x^2 + x + 1 + \frac{2}{x-1}\right)\left(x^4 - x^3 + x^2 - x + 1 - \frac{2}{x+1}\right)$$

is not changed if we throw out the two fractions.

Brass and Bronze

Brass is an alloy of copper and zinc. Bronze is an alloy containing 80% copper, 4% zinc, and 16% tin. A fused mass of brass and bronze is found to contain 74% copper, 16% zinc, and 10% tin. Find the ratio of copper to zinc in the composition of brass.

More Work

To complete a certain job, A alone would take m times as many days as B and C working together; B alone would take n times as many days as A and C together; C alone would take p times as many days as A and B together. Prove that

$$\frac{m}{m+1} + \frac{n}{n+1} + \frac{p}{p+1} = 2.$$

Wieviel Uhr ist es?

At what time between 4 and 5 o'clock will the minute hand of a watch be 13 minutes in advance of the hour hand?

CHAPTER VII

EQUATIONS OF THE FIRST DEGREE IN THE UNKNOWNS

72. Indeterminate and Simultaneous Equations. In many algebraic problems there are two or more unknown quantities. It is sometimes convenient to represent these unknowns by different letters, rather than to attempt to represent each of them in terms of one letter.

Consider the problem: "The sum of two numbers is 20 and the difference of the two numbers is 4. Find the numbers."

If we let x be the larger number and y be the smaller number, then the conditions of the problem state that

$$(1)\quad x + y = 20,$$
$$(2)\quad x - y = 4.$$

It is clear that such an equation as (1) by itself does not define any definite values for x and y. It merely limits the choice of pairs of numbers to those which in this case add to 20, such as 20, 0; 17, 3; $11\frac{1}{2}$, $8\frac{1}{2}$; -5, 25; and so on. Such an equation is called *indeterminate.*

In the above problem, however, *both* equations are to be satisfied by the same values. Thus, if we solve (1) and (2) for x,

$$x = 20 - y,$$
$$x = 4 + y;$$

and since x represents identically the same number in both cases,

$$4 + y = 20 - y;$$

hence

$$2y = 16,$$
$$y = 8.$$

Thus, y must have the value 8 and if we substitute this value of y in (1) we find that x must have the value 12. In this manner we see that there is only one set of values which satisfies *both* the equations (1) and (2).

Definition. *When two or more equations are satisfied by the same values of the unknown quantities, they are called* ***simultaneous equations.***

73. Solution of a System of Equations. If a system of equations is satisfied by the same values of x and y, *any* equation formed by combining them will also be satisfied by these values. Our object will be to obtain an equivalent equation that contains *only one* of the unknowns.

The process by which we cause either of the unknown quantities to be eliminated is called *elimination.* It may be done in various ways, but two methods are particularly useful: (1) by **addition** or **subtraction;** (2) by **substitution.**

74. Elimination by Addition or Subtraction. The following examples illustrate this method of solution.

Example 1. Solve the following equations and check your result.

$$(1)\ 3x + 2y = 18,$$
$$(2)\ 5x - 2y = 14.$$

Solution. In solving a system of equations, it is immaterial which of the unknowns we choose to eliminate. In this example the unknown y can be eliminated more readily than x, so we proceed to eliminate y. We do this by adding the corresponding members of the two equations. Thus, we obtain the equivalent equation

$$8x = 32;$$

hence,

$$x = 4.$$

To find the corresponding value of y, we substitute the value just obtained for x in *either* of the two given equations. Thus, substituting 4 for x in (1), we have

$$12 + 2y = 18,$$
$$2y = 6,$$
$$y = 3.$$

Hence, the solution is $x = 4$ and $y = 3$.

Check. Substituting $x = 4$ and $y = 3$ in the given equations, we have

$$3(4) + 2(3) = 18, \quad \text{and} \quad 5(4) - 2(3) = 14,$$
$$12 + 6 = 18, \qquad 20 - 6 = 14,$$
$$18 = 18, \qquad 14 = 14.$$

Rule for Elimination by Addition or Subtraction

Multiply each equation, when necessary, by such positive numbers as will make the coefficients of the unknown that is to be eliminated the same in absolute value. Add the resulting equations if these coefficients are unlike in sign; subtract them if they are like in sign.

Example 2. Solve

$$(1)\ 3x + 5y = 5,$$
$$(2)\ 2x - 3y = 16.$$

Solution. To eliminate x we multiply (1) by 2 and (2) by 3, so as to make the coefficients of x numerically the same.

$$6x + 10y = 10,$$
$$6x - 9y = 48.$$

Subtracting the corresponding members of these equations, we have

$$19y = -38,$$
$$y = -2.$$

Substituting $y = -2$ in (2), we have

$$2x + 6 = 16,$$
$$2x = 10,$$
$$x = 5 \quad \text{and} \quad y = -2.$$

EXERCISE 54

Solve the following equations and check your results:

1. $3x - y = 7$, $2x + y = 8$
2. $2x + y = 9$, $2x - 3y = 5$
3. $x + 2y = 10$, $x - 2y = -6$

4. $5x - y = 2$, $4x + y = 7$
5. $2x - 3y = 1$, $2x - 4y = 2$
6. $x + 3y = -7$, $2x - 3y = 13$

7. $2x - 5y = 4$, $3x + 5y = 6$
8. $2x + 5y = 2$, $x + 3y = 2$
9. $x + 2y = 14$, $2x + y = 13$

10. $5x - 2y = 4$, $x + y = 5$
11. $3x - 5y = 13$, $6x + 7y = -8$
12. $5x - 2y = 6$, $3x - 4y = 12$

13. $7x - 4y = 12$, $5x - 2y = 12$
14. $4x - y = 6$, $5x - 4y = 13$
15. $2x + 3y = 18$, $3x + 2y = 17$

16. $7x - 2y = 4$, $9x - 3y = 3$
17. $4x + 7y = 2$, $3x + 5y = 1$
18. $5x + 3y = 3$, $3x - 2y = 17$

19. $4x + 3y = 3$
$10x - 6y = 3$

20. $8x - 6y = 9$
$20x + 4y = 13$

21. $2x + 7y = 5$
$3x + 8y = 6$

22. $3x + 5y = 3$
$9x + y = 2$

23. $15x + 7y = 37$
$9x - 10y = 8$

24. $6x - 5y = 23$
$9x + 7y = 20$

25. $20x + 23y = 11$
$14x - 13y = 95$

26. $18x - 25y = -13$
$26x - 35y = -11$

27. $17x - 13y = 7$
$19x - 11y = 29$

28. $15x - 7y = 103$
$39x + 23y = 103$

29. $29x + 17y = 22$
$58x + 31y = 35$

30. $27x + 16y = 17$
$60x - 38y = 1$

75. Elimination by Substitution. The following example illustrates this method of elimination.

Example 1. Solve

$$(1)\ 5x - 2y = 8,$$
$$(2)\ 2x + 3y = 26.$$

Solution. Transposing $-2y$ in (1) and dividing by 5, we have

$$x = \frac{2y + 8}{5}.$$

Substituting this value of x in (2) yields

$$2\left(\frac{2y + 8}{5}\right) + 3y = 26.$$

Solving, we have

$$\begin{aligned} 2(2y + 8) + 15y &= 130, \\ 4y + 16 + 15y &= 130, \\ 19y &= 114, \\ y &= 6. \end{aligned}$$

This value substituted in either (1) or (2) gives $x = 4$.

Check. Substituting $x = 4$ and $y = 6$ in the given equations gives:

$$\begin{aligned} 5(4) - 2(6) &= 8, \quad \text{and} & 2(4) + 3(6) &= 26, \\ 20 - 12 &= 8, & 8 + 18 &= 26, \\ 8 &= 8. & 26 &= 26. \end{aligned}$$

Rule for Elimination by Substitution

In one of the equations, find the value of one of the unknowns in terms of the other; then substitute this value in the other equation and solve.

NOTE. If any one of the four coefficients equals 1 in absolute value, we may avoid fractions by solving initially for that unknown.

Example 2. Solve

$$(1)\ 5x + 2y = 8,$$
$$(2)\ 3x - y = 7.$$

Solution. Since the coefficient of y in (2) is -1, we solve for y in (2). Thus,

$$y = 3x - 7.$$

Substituting in (1) gives

$$5x + 2(3x - 7) = 8,$$
$$11x = 22,$$
$$x = 2 \quad \text{and hence} \quad y = -1.$$

EXERCISE 55

Solve the following equations by substitution and check your results:

1. $2x - y = 7$, $x + y = 5$

2. $x - y = 1$, $x + y = 9$

3. $2x + 3y = 7$, $3x + y = 7$

4. $x - 3y = 5$, $3x - y = 7$

5. $y = 2x - 1$, $2x - 3y = 7$

6. $y = 3x - 5$, $x + 4y = 6$

7. $x - 2y = 0$, $3x - 5y = 2$

8. $y = 2x + 3$, $x = 2y - 3$

9. $y = 2x + 9$, $y = 3 - x$

10. $3x - 4y = 0$, $y = \dfrac{2x + 1}{3}$

11. $3x + 2y = 5$, $x = \dfrac{3y + 12}{2}$

12. $x = 3y + 4$, $x = 5y + 10$

13. $3x + 2y = 3$, $5x + 4y = 7$

14. $3x - 2y = 2$, $6x - 5y = -1$

15. $7x - 8y = 9$, $3x - 4y = 5$

16. $4x + 3y = 7$, $2x + 9y = 11$

17. $2x + 3y = 9$, $3x + 2y = 11$

18. $5x - 3y = 7$, $3x + 4y = 10$

19. $3x - 5y = 3$, $4x - 3y = 15$

20. $7x + 5y = -4$, $5x + 3y = 0$

21. $5x + 3y = 7$, $9x - y = 3$

22. $2x - 7y = 5$, $5x + 2y = 6$

23. $5x + 5y = 1$, $3x - 8y = 5$

24. $7x + 3y = 4$, $9x + 5y = 8$

25. $17x - 9y = 57$, $23x - 27y = 3$

26. $11x + 5y = 53$, $31x - 14y = 37$

27. $29x + 17y = 55$, $47x + 34y = 44$

28. $21x + 13y = 55$, $42x - 17y = 67$

29. $17x + 19y = 0$, $23x + 20y = 194$

30. $49x - 43y = 71$, $23x - 21y = 17$

76. Simplification of Equations. It is sometimes necessary to simplify the equations before proceeding to solve.

Example 1. Solve

$$(1)\ \frac{x}{4} + \frac{y}{6} = 1,$$

$$(2)\ x + 2(x - y) = 7.$$

Solution. Clearing of fractions in (1), we have

$$(3)\ 3x + 2y = 12.$$

Removing parentheses in (2) gives

$$(4)\ 3x - 2y = 7.$$

Adding (3) and (4), we find $x = \frac{19}{6}$.
Subtracting (4) from (3) gives $y = \frac{5}{4}$.

NOTE. Occasionally, as in the above example, the value of the second unknown is found more easily by elimination than by substituting the value of the unknown already found in one of the given equations.

Example 2. Solve

$$(1)\quad (x + 1)(y + 1) = (x + 4)(y - 1),$$

$$(2)\quad \frac{x + y}{2x - y} = 5.$$

Solution. Removing parentheses in (1) gives

$$xy + x + y + 1 = xy - x + 4y - 4$$

$$(3)\quad 2x - 3y = -5.$$

Clearing of fractions in (2), we have

$$x + y = 10x - 5y$$

$$(4)\quad 3x - 2y = 0.$$

From (3) and (4) we find that $x = 2$ and $y = 3$.

EXERCISE 56

Solve the equations:

1. $\frac{2x}{3} + y = 13$, $\quad x + \frac{y}{3} = 9$

2. $\frac{x}{4} = y$, $\quad \frac{x}{2} + \frac{y}{3} = 7$

3. $\frac{x}{3} + \frac{y}{5} = 12$, $\quad x - y = 4$

4. $x + y = 5$, $\quad \frac{x}{3} + \frac{y}{5} = 5$

5. $\frac{1}{2}x + \frac{1}{3}y = 0$, $\quad \frac{3}{4}x - \frac{1}{6}y = 8$

6. $\frac{1}{5}x + \frac{2}{3}y = 20$, $\quad \frac{3}{4}x - \frac{1}{6}y = 11$

7. $\dfrac{3x-1}{4} - y = 2$
$3x - 5y = 3$

8. $\dfrac{x+1}{2} + \dfrac{y-1}{3} = 14$
$9x - 5y = 7$

9. $\dfrac{5x-1}{3} - 2y - 1 = 0$
$x - \dfrac{3y+2}{2} + 2 = 0$

10. $3x + \frac{1}{3}y = 10$
$7x - \dfrac{y-1}{4} = 2$

11. $3(x+1) - 2(y-1) = 16$
$5x - 4y = 17$

12. $6(2x-3) - 5y = 21$
$3x - 5(2y+3) = 21$

13. $4(x+3) - 3(y-4) = 5$
$2(2x+7) + 5(y-5) = 10$

14. $5(2x-1) - 3(y-3) = 11$
$4(3x+5) + 3(2y+1) = 41$

15. $\dfrac{x+3}{y-1} = 3$
$\frac{1}{3}x + 3y = 10$

16. $\dfrac{x}{x+y} = \dfrac{2}{3}$
$3x - 4y = 4$

17. $\dfrac{1}{x+1} + \dfrac{1}{y+1} = 0$
$\dfrac{1}{2x+1} + \dfrac{1}{y} = 0$

18. $\dfrac{2x+3}{3y+2} = 1$
$\dfrac{3x}{y} - 7 = \dfrac{1}{y}$

19. $(x+1)(y+1) = (x+4)(y-1)$
$7x - 3y = 5$

20. $(x-1)(y+1) = xy$
$\frac{1}{2}x + \frac{1}{3}y = 3$

21. $x(y+2) = y(x-2)$
$5x - 7y = 12$

22. $(x+3)(y-1) = (x-1)(y+1)$
$(x-3)(y+5) = (x-9)(y-7)$

23. $\dfrac{x-1}{x} = \dfrac{y}{y+2}$
$\dfrac{x+9}{x} = \dfrac{y}{y-3}$

24. $\dfrac{x-3}{y+3} = \dfrac{x+1}{y-1}$
$\dfrac{x-3}{y-3} = \dfrac{x-2}{y-1}$

25. $\dfrac{x+1}{10} = \dfrac{4y-1}{3} = \dfrac{x-y}{8}$

26. $\dfrac{2x-1}{9} = \dfrac{2y-7}{5} = \dfrac{3x-2y}{3}$

77. Equations in Terms of Reciprocals of the Unknown. The reciprocals of x and y are $\dfrac{1}{x}$ and $\dfrac{1}{y}$ respectively; and in solving the following equations we consider $\dfrac{1}{x}$ and $\dfrac{1}{y}$ as the unknown quantities.

Example 1. Solve

$$(1) \qquad \frac{2}{x} + \frac{6}{y} = 3,$$

$$(2) \qquad \frac{4}{x} - \frac{3}{y} = 1.$$

Solution. Copying (1) and multiplying (2) by 2, we have

$$\frac{2}{x} + \frac{6}{y} = 3,$$

$$\frac{8}{x} - \frac{6}{y} = 2;$$

adding, $$\frac{10}{x} = 5,$$

clearing fractions, $$10 = 5x;$$

hence, $$x = 2,$$

and by substitution, $$y = 3.$$

Example 2. Solve

$$(1) \qquad \frac{1}{4x} + \frac{1}{3y} = 2,$$

$$(2) \qquad \frac{1}{y} - \frac{1}{2x} = 1.$$

Solution. To clear these equations of fractional coefficients, we first multiply (1) by 12 and (2) by 2.

$$(3) \qquad \frac{3}{x} + \frac{4}{y} = 24,$$

$$(4) \qquad \frac{2}{y} - \frac{1}{x} = 2.$$

We then multiply (4) by 3 and add to (3):

$$\frac{10}{y} = 30,$$

$$y = \tfrac{1}{3};$$

and by substitution, $$x = \tfrac{1}{4}.$$

EXERCISE 57

Solve the following equations and check your results:

1. $\frac{2}{x} + \frac{3}{y} = 2$, $\frac{4}{x} - \frac{3}{y} = 1$

2. $\frac{6}{x} + \frac{2}{y} = 1$, $\frac{6}{x} - \frac{6}{y} = 5$

3. $\frac{3}{x} + \frac{4}{y} = 1$, $\frac{4}{x} + \frac{5}{y} = 1$

4. $\frac{7}{x} - \frac{4}{y} = 5$, $\frac{5}{x} - \frac{2}{y} = 4$

5. $\frac{2}{x} - \frac{3}{y} = 2$, $\frac{3}{x} + \frac{2}{y} = 16$

6. $\frac{3}{x} - \frac{1}{y} = 10$, $\frac{1}{x} + \frac{2}{y} = -13$

7. $\frac{2}{x} - \frac{1}{y} = 3$, $\frac{3}{x} - \frac{2}{y} = 1$

8. $\frac{2}{x} + \frac{1}{y} = 8$, $\frac{3}{x} - \frac{2}{y} = 19$

9. $\frac{1}{x} + \frac{2}{y} = 3$, $\frac{2}{x} - \frac{1}{y} = 11$

10. $\frac{2}{x} - \frac{5}{y} = 1$, $\frac{1}{x} + \frac{1}{y} = 11$

11. $\frac{1}{2x} + \frac{1}{3y} = 2$, $\frac{2}{x} - \frac{1}{y} = 1$

12. $\frac{2}{3x} - \frac{3}{4y} = \frac{1}{12}$, $\frac{1}{2x} - \frac{2}{3y} = \frac{1}{36}$

13. $\frac{1}{2x} - \frac{1}{5y} = 4$, $\frac{1}{7x} + \frac{1}{15y} = 3$

14. $\frac{1}{5x} - \frac{1}{4y} = 0$, $\frac{2}{x} + \frac{1}{2y} = 12$

15. $\frac{1}{7x} + \frac{1}{4y} = \frac{4}{7}$, $\frac{1}{x} + \frac{1}{3y} = \frac{7}{6}$

16. $\frac{9}{5x} - \frac{4}{3y} = \frac{1}{12}$, $\frac{4}{x} + \frac{5}{6y} = \frac{3}{8}$

17. $\frac{2}{3x} - \frac{5}{6y} = \frac{3}{4}$, $\frac{3}{2x} - \frac{7}{6y} = \frac{4}{3}$

18. $\frac{5}{3x} + \frac{3}{5y} = \frac{8}{15}$, $\frac{7}{4x} + \frac{6}{5y} = \frac{3}{4}$

19. $\frac{4}{3x} - \frac{5}{6y} = \frac{1}{6}$, $\frac{7}{4x} + \frac{5}{8y} = \frac{9}{16}$

20. $\frac{4}{3x} - \frac{3}{4y} = \frac{23}{24}$, $\frac{5}{4x} - \frac{4}{5y} = \frac{17}{20}$

78. Literal Equations. If the system of equations has literal coefficients, the equations are solved in the same manner.

Example 1. Solve

$$(1)\quad 3ax + 2by = 7,$$
$$(2)\quad 4ax + 3by = 10.$$

Solution. Multiplying (1) by 3 and (2) by 2, we have

$$9ax + 6by = 21,$$
$$8ax + 6by = 20.$$

Subtracting,

$$ax = 1,$$
$$x = \frac{1}{a}.$$

Substituting $\frac{1}{a}$ for x in (1), we have

$$3 + 2by = 7,$$
$$2by = 4,$$
$$y = \frac{2}{b}.$$

Example 2. Solve and check,

$$(1) \qquad ax - by = a^2 + b^2,$$
$$(2) \qquad bx + ay = a^2 + b^2.$$

Solution. Multiplying (1) by a and (2) by b, we have

$$a^2x - aby = a(a^2 + b^2),$$
$$b^2x + aby = b(a^2 + b^2).$$

Adding, $$(a^2 + b^2)x = (a + b)(a^2 + b^2).$$

Hence, dividing both sides of the equation by $a^2 + b^2$ gives

$$x = a + b.$$

Substituting $a + b$ for x in (2), we have

$$ab + b^2 + ay = a^2 + b^2,$$
$$ay = a^2 - ab,$$
$$y = a - b.$$

Check. Substituting $x = a + b$ and $y = a - b$ in the given equations:

$$a(a + b) - b(a - b) = a^2 + b^2, \quad \text{and} \quad b(a + b) + a(a - b) = a^2 + b^2,$$
$$a^2 + ab - ab + b^2 = a^2 + b^2, \qquad ab + b^2 + a^2 - ab = a^2 + b^2,$$
$$a^2 + b^2 = a^2 + b^2. \qquad a^2 + b^2 = a^2 + b^2.$$

EXERCISE 58

Solve for x and y, and check your results:

1. $2x + y = 3c$
$x - y = 3d$

2. $x + y = 5a$
$3x - 2y = 5b$

3. $2x - y = 5m$
$x + 2y = 5n$

4. $2x - y = a$
$5x - 3y = 0$

5. $3x + 2y = p$
$4x - y = 5p$

6. $7x - 2y = 8s$
$2x + y = 7s$

7. $2x - 3y = a - 4b$
$x + y = 3a + 3b$

8. $2x + 3y = 5a - b$
$3x - 2y = a + 5b$

9. $4x - y = c - 9d$
$3x + 2y = 9c - 4d$

10. $3x - 4y = -c - 21d$
$x + y = 9c$

11. $ax + by = 4$
$ax - 2by = 1$

12. $4ax + 5by = 6$
$3ax - by = 14$

13. $2cx + dy = 7$
$cx + 2dy = -1$

14. $2cx - 3dy = 2$
$3cx + 2dy = 16$

15. $\dfrac{2x}{a} - \dfrac{y}{b} = 4$

16. $\dfrac{4x}{a} - \dfrac{3y}{b} = 1$

$\frac{3x}{a} - \frac{2y}{b} = 2$ | $\frac{3x}{a} - \frac{4y}{b} = 6$

17. $\frac{x}{c} - \frac{3y}{d} = 11$
$\frac{2x}{c} - \frac{y}{d} = 7$

18. $\frac{2x}{c} - \frac{3y}{d} = 7$
$\frac{x}{c} + \frac{y}{d} = 11$

19. $ax + 2by = 3c$
$2ax - by = 11c$

20. $3ax + by = 4ab$
$ax - 3by = 8ab$

21. $2ax - by = 2a^2 + b^2$
$x + y = 3a$

22. $ax + by = a^2 + b^2$
$y - x = 2b$

23. $ax + by = 2a^2 - 2b^2$
$x - y = a + b$

24. $ax - by = 2a^2 + 2b^2$
$2x + 3y = 13a$

25. $ax + by = a^2$
$bx - ay = ab$

26. $ax - by = 2a^2 - 2b^2$
$ay - bx = a^2 - b^2$

27. $x + y = a(a + b)$
$\frac{x}{y} = \frac{a}{b}$

28. $\frac{x + y}{a + b} = 4$
$\frac{ay - bx}{a - b} = a + b$

29. $ax - by = a^2$
$\frac{x + y}{x - y} = \frac{2a + b}{b}$

30. $\frac{x - y}{a - b} = 1$
$\frac{x - c}{a} + \frac{y - c}{b} = 2$

79. Systems of Equations Involving Three Unknowns. (Optional.) In order to solve a system of equations involving two unknown quantities we have seen that we must have two equations. Similarly, in order to solve a system of equations involving three unknowns we must have three *independent* equations.

The problem of solving three linear equations containing three unknowns may be reduced to a problem of solving two equations in two unknowns by either of the two processes of elimination just studied, that is,

1. elimination by addition or subtraction
2. elimination by substitution.

Example 1. Solve

$$(1)\quad 3x - y + 2z = 9,$$
$$(2)\quad 2x + y - z = 7,$$
$$(3)\quad x + 2y - 3z = 4.$$

First solution. Suppose that we choose to eliminate the unknown y. Adding (1) and (2) we get

$$(4) \qquad 5x + z = 16.$$

Multiplying (1) by 2 and adding to (3) gives

$$(5) \qquad 7x + z = 22.$$

The problem of solving the original equations is thus reduced to the problem of solving equations (4) and (5).

Subtracting (4) from (5) gives $\qquad 2x = 6,$

$$x = 3;$$

from (4), $\qquad z = 1,$

and from (2), $\qquad y = 2.$

Second solution. Suppose that we choose to eliminate the unknown y. Solving (2) for y gives

$$y = 7 - 2x + z.$$

Substituting this value of y in (1) and (3), we have

$$3x - (7 - 2x + z) + 2z = 9,$$
$$x + 2(7 - 2x + z) - 3z = 4.$$

Simplifying these equations, we have

$$5x + z = 16,$$
$$-3x - z = -10.$$

By adding, we obtain $x = 3$. Hence, as before, $z = 1$ and $y = 2$.

Check. Substituting $x = 3$, $y = 2$, and $z = 1$ in the given equations, we obtain

$3(3) - (2) + 2(1) = 9$, and $2(3) + (2) - (1) = 7$, and $(3) + 2(2) - 3(1) = 4$,

$9 - 2 + 2 = 9, \qquad 6 + 2 - 1 = 7, \qquad 3 + 4 - 3 = 4,$

$9 = 9. \qquad 7 = 7. \qquad 4 = 4.$

EXERCISE 59

Solve each of the following sets of equations and check your results:

1. $x + 2y + z = 9$
$x + y - z = 5$
$3x - y + 2z = 12$

2. $2x + y - 4z = 5$
$3x - 2y + 3z = 9$
$z = x + y$

3. $2x + y - 3z = 1$
$x - y + 2z = 1$
$x + 3y - z = 6$

4. $x - 2y + 3z = -1$
$2x - 3y + z = 11$
$3x - y + 2z = 8$

5. $x - 2y + z = 5$
$2x + 3z = 4$
$y + 2z = 2$

6. $3x + 4y - z = -3$
$4x + 5y + 7z = 14$
$x - 2y - 2z = 9$

7. $2x - 3y + 4z = 8$
$3x + 4y - 5z = -4$
$4x - 5y + 6z = 12$

8. $x - y - z = 1$
$2x + 3y + 2z = 8$
$4x + 3y - 2z = -2$

9. $2x - y + 3z = 4$
$x + 3y + 3z = -2$
$3x + 2y - 6z = 6$

10. $3x + 2y - z = 2$
$x - y - 9z = 1$
$4x + 3y + 3z = 0$

11. $\frac{1}{2}x + \frac{1}{3}y + \frac{1}{4}z = 3$
$x - \frac{5}{6}y + \frac{1}{2}z = \frac{3}{2}$
$\frac{3}{4}x + \frac{2}{3}y - \frac{5}{8}z = 1$

12. $x + \frac{1}{2}y + \frac{1}{3}z = \frac{1}{6}$
$y + \frac{1}{2}z + \frac{1}{3}x = \frac{4}{3}$
$z + \frac{1}{2}x + \frac{1}{3}y = -\frac{7}{12}$

13. $\frac{1}{x} + \frac{1}{y} + \frac{2}{z} = 1$
$\frac{2}{x} + \frac{1}{y} - \frac{2}{z} = 1$
$\frac{3}{x} + \frac{4}{y} - \frac{4}{z} = 2$

14. $\frac{4}{x} - \frac{1}{y} - \frac{1}{z} = 1$
$\frac{3}{x} + \frac{1}{y} + \frac{3}{z} = 1$
$\frac{2}{x} + \frac{1}{y} + \frac{1}{z} = 1$

15. $y - x = z - y = 4x - z = 2$

16. $\frac{x + y}{6} = \frac{x + z}{7} = \frac{y + z}{8}$
$x + y + z = 21$

17. $\frac{x}{y + z} = \frac{2}{3}, \quad \frac{y + 1}{x + z} = \frac{3}{8}$
$\frac{z - 2}{x + y} = 6$

18. $(x - 1)(z - 2) = xz - y$
$(y - 2)(z + 1) = z(y - 1)$
$(x + 2)(y - 1) = xy + 2$

19. $2x + y + z = 7a$
$x + 2y + z = 3a$
$x + y + 2z = 6a$

20. $x + y - z = 6a - 5b$
$x - y + z = 2a - b$
$x + y + z = 8a - 3b$

80. Motion Problems Involving Opposite Rates. We know that a boat which is allowed to drift in a body of moving water will travel with the current. The speed at which it moves we call the *rate of the stream* or the *rate of the current.* The rate at which a boat can be propelled with respect to the water itself is called the *rate of the boat in still water*, or simply the *rate in still water.*

In figuring the absolute rate of travel of a boat which is moving directly with or against a current, we have

Rate of boat downstream = Rate in still water + Rate of the stream

Rate of boat upstream = Rate in still water − Rate of the stream

The above analysis is also applicable to airplanes traveling in air currents. Thus, an airplane which can travel 100 miles per hour in still air will travel 140 miles per hour when traveling with a 40-mile-per-hour wind, and only 60 miles per hour when traveling directly against a 40-mile-per-hour wind.

Example 1. A man can row downstream 3 miles in 20 minutes, but it takes him 1 hour to return. How fast can he row in still water and what is the rate of the current?

Analysis. Let x = the rate of rowing in still water
and y = the rate of the stream.

	Rate ×	Time =	Distance
Downstream	$x + y$	$\frac{1}{3}$	3
Upstream	$x - y$	1	3

Solution.

$$\begin{cases} \frac{1}{3}(x + y) = 3, \\ x - y = 3, \end{cases}$$

$$\begin{cases} x + y = 9, \\ x - y = 3, \end{cases}$$

$$2x = 12,$$

$$x = 6,$$

$$y = 3.$$

Answer. He can row 6 miles per hour in still water and the rate of the current is 3 miles per hour.

Example 2. In a motorboat which can travel 18 miles per hour in still water, a boy travels down a stream for 30 minutes and then returns in $37\frac{1}{2}$ minutes. What is the rate of the current?

Analysis. Let x = the rate of the current
and y = the distance traveled down the stream.

	Rate ×	Time =	Distance
Downstream	$18 + x$	$\frac{1}{2}$	y
Upstream	$18 - x$	$\frac{5}{8}$	y

Solution.

$$\begin{cases} \frac{1}{2}(18 + x) = y, \\ \frac{5}{8}(18 - x) = y, \end{cases}$$

$$\frac{1}{2}(18 + x) = \frac{5}{8}(18 - x),$$

$$72 + 4x = 90 - 5x,$$

$$9x = 18,$$

$$x = 2.$$

Answer. The rate of the current is 2 miles per hour.

EXERCISE 60

Opposite Rates

1. A boy can row downstream at the rate of 5 miles per hour and upstream at a rate of 3 miles per hour. What is the rate of the current and at what rate can he row in still water?
2. A canoeist can go downstream 12 miles in 2 hours. He can return only 2 miles in 30 minutes. Find the rate of the current and that of the canoeist.
3. An airplane traveled 450 miles in 3 hours with the wind. It would have taken 5 hours to make the same trip against the wind. Find the speed of the plane and the velocity of the wind.
4. A crew can row 2 miles down a stream in 8 minutes, but requires 12 minutes to row the same distance up the stream. How fast can the men row in still water and what is the rate of the current?
5. A launch traveling with the tide takes 15 minutes to reach a buoy 5 miles away. It makes the return trip in 20 minutes. What is the rate of the launch in still water and what is the rate of the tide?
6. A motorboat travels 6 miles down a stream in 20 minutes and requires 30 minutes for the return trip upstream. What is the rate of the boat in still water and what is the rate of the current?
7. A man can row one mile down a stream in 15 minutes and one mile up the same stream in 20 minutes. What is the rate of the stream?
8. An oarsman can travel a two-mile course downstream in 15 minutes, and upstream in 20 minutes. What is his rate of rowing in still water?
9. An airplane cruising at half speed against a wind travels 60 miles in 40 minutes, and at full speed it covers 60 miles in 18 minutes. What is the velocity of the wind?
10. A motorboat travels 4 miles down a stream in 15 minutes. If the current had been twice as fast, the motorboat would have made the trip in 12 minutes. What is the rate of the current?

11. An airplane made a trip of 1260 miles against a wind in 7 hours. If the wind had been only half as strong, the plane would have taken one hour less time for the trip. Find the velocity of the wind.

12. A motorboat traveling at half speed up a stream covers 4 miles in an hour, and at full speed it covers 9 miles in an hour. What is its maximum speed in still water?

13. A motorboat traveling at full speed against a current goes 12 miles in an hour, and at half speed with the current it also goes 12 miles in an hour. What is the rate of the current?

14. An airplane cruising at half speed against a wind travels 400 miles in 4 hours. Traveling at full speed against a wind three times as strong, the plane covers 780 miles in 4 hours. What is the plane's maximum velocity in still air?

15. A boy can row 5 miles per hour in still water. He rows up a stream a certain distance in 45 minutes and returns in 30 minutes. What is the rate of the stream and how far up the stream did he row?

16. A motorboat travels for 2 hours against a tide which moves at the rate of 2 miles per hour. It then returns over the same route, making the trip in 1 hour and 40 minutes. What is the total distance traveled by the boat?

17. An airplane traveling from Los Angeles to San Francisco against a wind of 15 miles per hour makes the trip in $2\frac{1}{2}$ hours. On the return trip, with the aid of a tail wind blowing 25 miles per hour, the plane makes the trip in 2 hours. What is the distance between the two cities?

18. A motorboat whose speed in still water is 15 miles per hour travels up a stream a certain distance and returns in one hour less time than it took going up the stream. If the rate of the stream is 3 miles per hour, how far up the stream did the boat go?

19. An oarsman travels 12 miles in an hour and a half, rowing with the tide, and requires 4 hours to return rowing against a tide one-half as strong. Find the velocity of the stronger tide.

20. A man travels n miles in a hours rowing with the tide, and requires b hours to return against the tide. Find the velocity of the tide.

81. Problems Leading to Systems of Equations. In our previous work we saw that it is essential to have two equations when there are two unknowns to be determined. Consequently in the solution of stated problems using two unknowns the statement should contain two *independent conditions* from which we may derive these equations.

Example 1. Find two numbers such that twice the larger exceeds three times the smaller by 1, and three times the larger exceeds their sum by 7.

Analysis. Let x represent the larger number and y the smaller. The two conditions of the problem written algebraically then have the form

$$(1) \qquad 2x - 3y = 1,$$
$$3x - (x + y) = 7.$$

Solution. Collecting terms in the second equation, we have

$$(2) \qquad 2x - y = 7.$$

Subtracting (1) from (2) gives $2y = 6,$

$$y = 3.$$

Substituting 3 for y in (1) gives $x = 5.$

Answer. The numbers are 5 and 3.

Example 2. If 15 bats and 10 balls together cost \$24.25, and 25 bats and 13 balls together cost \$38.40, find the price of a bat and the price of a ball.

Analysis. Suppose that a bat costs x cents and a ball costs y cents. Then from the statement,

$$(1) \; 15x + 10y = 2425,$$
$$(2) \; 25x + 13y = 3840.$$

Solution. Multiplying (1) by 5 and (2) by 3 gives

$$75x + 50y = 12125,$$
$$75x + 39y = 11520.$$

Subtracting, we have $11y = 605,$

$$y = 55.$$

Substituting in (1): $15x + 550 = 2425,$

$$15x = 1875,$$
$$x = 125.$$

Answer. Each bat costs \$1.25 and each ball costs 55 cents.

Example 3. If the width of a rectangle is increased 1 foot and the length is decreased 2 feet, the area remains unaltered. Also if the width is decreased 1 foot and the length is increased 3 feet, the area remains unaltered. Find the dimensions of the rectangle.

Analysis. Let $x =$ the width and $y =$ the length of the given rectangle. The dimensions and areas of the figures under consideration are as follows:

	y			$y - 2$			$y + 3$
x	xy		$x + 1$	$(x + 1)(y - 2)$		$x - 1$	$(x - 1)(y + 3)$

Since the area of each of these rectangles is the same, we have

$$(x + 1)(y - 2) = xy,$$
$$(x - 1)(y + 3) = xy.$$

Solution. Multiplying the left members and simplifying, we have

$$-2x + y - 2 = 0,$$
$$3x - y - 3 = 0.$$

These equations yield the solution $x = 5$ and $y = 12$.

Answer. The width of the rectangle is 5 feet and the length is 12 feet.

Example 4. A certain job is to be finished in two days. The first day A worked 9 hours, B worked 8 hours, and they finished half the job. The second day A worked only 6 hours and B had to work 12 hours in order to finish the job. How long would it take each alone to do the whole job?

Analysis. Let $x =$ the time required by A to do the whole job alone and let $y =$ the time required by B to do the whole job alone.

		Rate $\times$	Time	= Fraction of Work
First day	A	$\frac{1}{x}$	9	$\frac{9}{x}$
	B	$\frac{1}{y}$	8	$\frac{8}{y}$
Second day	A	$\frac{1}{x}$	6	$\frac{6}{x}$
	B	$\frac{1}{y}$	12	$\frac{12}{y}$

Since half of the work is completed each day,

$$(1)\quad \frac{9}{x} + \frac{8}{y} = \frac{1}{2},$$
$$(2)\quad \frac{6}{x} + \frac{12}{y} = \frac{1}{2}.$$

Solution. Multiplying (1) by 3 and (2) by 2, we have

$$\frac{27}{x} + \frac{24}{y} = \frac{3}{2},$$

$$\frac{12}{x} + \frac{24}{y} = 1.$$

Subtracting gives

$$\frac{15}{x} = \frac{1}{2},$$

$$x = 30.$$

By substitution, we find $y = 40$.

Answer. A requires 30 hours and B requires 40 hours to do the whole job alone.

EXERCISE 61

1. The sum of two numbers is 91 and their difference is 15. Find the numbers.

2. Half the sum of two numbers is 40 and three times their difference is 18. Find the numbers.

3. Find a fraction which reduces to $\frac{1}{2}$ if 4 is added to its denominator, and which reduces to $\frac{2}{3}$ if 3 is added to its numerator.

4. If the numerator of a fraction is decreased by 1 and the denominator is increased by 3, the fraction reduces to $\frac{1}{2}$. If the numerator is increased by 3 and the denominator is decreased by 1, it reduces to $\frac{4}{5}$. What is the fraction?

5. Five horses and eight cows can be bought for $1435, and eight horses and five cows can be bought for $1165. What is the price of each animal?

6. Three pounds of tea and six pounds of coffee cost $6.33, and five pounds of tea and four pounds of coffee cost $6.23. What is the price per pound of the tea and the coffee?

7. In eight hours A walks 3 miles farther than B walks in nine hours; and in six hours B walks 2 miles farther than A walks in four hours. How many miles does each walk per hour?

8. A man walks 35 miles partly at the rate of 3 miles per hour and partly at the rate of 5 miles per hour. If he had walked at 5 miles per hour when he walked at 3 miles per hour, and vice versa, he would have covered 2 miles more in the same time. Find the time he took to walk the 35 miles.

9. A man has two investments, one paying 4% interest and the other 5%. His annual return on these investments is \$420. If the 4% investment paid 5%, and vice versa, the man would have an annual income of \$435. How much money does he have invested altogether?

10. A man owns two stores. In 1951 one store earned 10% of the investment in it while the other lost 5%, netting a gain of \$6250. In 1952 the first store earned 8% while the second earned 6%, netting a gain of \$9500. What is the value of each store?

11. If the length of a rectangle is increased by 2 feet and the width by 1 foot, the area is increased by 18 square feet. If the length is increased by 1 foot and the width by 2 feet, the area is increased by 22 square feet. Find the dimensions of the rectangle.

12. The sum of the perimeters of two squares of different sizes is 60 inches. If two sides of the squares are placed together, the outside perimeter of the figure formed is 48 inches. What is the size of each square?

13. Four times the larger of two numbers exceeds their sum by 25, and four times the smaller exceeds their difference by 1. Find the numbers.

14. Three times the larger of two numbers decreased by four times the smaller equals 5, and three times the smaller decreased by twice the larger equals 2. Find the numbers.

15. A man has a pile of coins, consisting of dimes and quarters, worth \$7.20. He observes that if the dimes were quarters and the quarters were dimes, he would have \$1.35 more money. How many of each coin has he?

16. A newsboy sells a certain number of *Posts* for 3 cents and *Times* for 5 cents, taking in \$2.70 all together. If he had sold twice as many *Posts* and only half as many *Times*, he would have taken in the same amount of money. How many of each did he sell?

17. A certain number with two digits is four times the sum of its digits, and if 27 is added to the number the digits will be reversed. Find the number.

18. A number between 10 and 100 is four times the sum of its digits, and if the number is doubled and then decreased by 12 the result will be the number with its digits reversed. What is the number?

19. Dan's age next year will be twice what Marvin's age was 5 years ago; and four times Marvin's age is 8 years more than three times what Dan's age was 3 years ago. Find their present ages.

20. A man has two sons, one five years older than the other. Two years ago his age was three times the sum of his boys' ages, and next year his age will be twice the sum of his boys' ages. Find their present ages.

21. The wages of 8 men and 7 boys amount to $94.10. If 3 men together receive $1.50 more than 4 boys, what are the wages of each man and each boy?

22. Eight shillings and seven francs are worth $1.14, and nine shillings and seventy francs are worth $1.46. What is the value of a shilling and the value of a franc?

23. Two persons 15 miles apart who start out at the same time will be together in 6 hours if they walk in the same direction, but in 2 hours if they walk toward each other. Find their rates of walking.

24. A and B starting at the same time from two points 22 miles apart walk toward each other and meet in 2 hours and 45 minutes. If A had walked twice as fast, they would have met in 2 hours. At what rate did B walk?

25. Four pounds of Grade A tea mixed with five pounds of Grade B tea is worth 60 cents a pound, and seven pounds of Grade A mixed with two pounds of Grade B is worth 66 cents a pound. What is the price of each per pound?

26. A chemist had a solution of 60% acid. He added some distilled water, reducing the concentration to 40%. He then added one quart more of water, reducing the concentration to 30%. How much of the 30% solution did he then have?

27. A, B, and C have $320 among them. A has twice as much as C, and B and C together have $20 less than A. How much does each have?

28. A, B, C, and D have $185 among them. A has twice as much as C, and B has three times as much as D; B and C together have $5 more than A and D. Find how much each has.

29. A boy can row 4 miles upstream and 2 miles downstream in 2 hours and 15 minutes, and he can row 2 miles upstream and 4 miles downstream in 1 hour and 30 minutes. Find the rate of the current.

30. A motorboat travels up a stream a distance of 12 miles in 2 hours. If the current had been twice as strong, the trip would have taken 3 hours. How long should it take for the return trip down the stream?

31. A man has a sum of money invested at a certain rate of interest. If he had $1000 more invested at a rate 1% lower, he would have the same annual return; and if he had $600 less invested at a rate 1% higher, his interest would be the same. How much does he have invested, and at what rate?

32. A man has $10,000 invested; part is at 4%, part is at 5%, and the remainder is in a business enterprise. One year the business lost 2% and the man's net income was $130. The next year the business paid 10%

and the man received altogether \$730. Find how much he has invested in the business enterprise.

33. If A works for 8 days and B for 15 days they can complete a certain job; and if A works for 12 days and B for 10 days they can complete the same job. How long will it take each working alone to do the job?

34. A tank is equipped with a pipe which drains it continuously. If a pipe that fills the tank is turned on full force, the tank is full in 3 hours. If the pipe is turned on only half speed, it takes 12 hours to fill the tank. How long does it take the drain to empty a full tank?

35. The width of a second rectangle is 1 foot more than half the length of the first rectangle. The length of the second rectangle is 1 foot more than twice the width of the first rectangle. If the perimeters of the rectangles are equal and the area of the second rectangle is 12 square feet more than the area of the first rectangle, find the dimensions of the two rectangles.

36. The perimeter of a rectangle divided by its length gives a quotient of 3 and a remainder of 9. The perimeter divided by the width gives a quotient of 4 and a remainder of 8. Find the dimensions of the rectangle.

37. It takes A 4 hours longer to walk 30 miles than it does B; but if he doubles his pace he takes 1 hour less than B. Find their rates of walking.

38. A hiker walks a certain distance. If he had walked one mile an hour faster, he would have taken four-fifths as much time; and if he had walked one mile an hour slower, he would have taken 30 minutes longer. Find the distance he walked.

39. A number consists of three digits, the middle one being zero. If the hundreds' digit is halved and the other two digits are interchanged, the number is decreased by 137; and if the digits are reversed, the number obtained is 297 larger than the original number. Find the number.

40. The middle digit of a number between 100 and 1000 is zero. If the tens' digit and the units' digit are interchanged, the number is increased by 45; and if the hundreds' digit and the units' digit are interchanged, the number is increased by 198. Find the number.

41. The circumference of the fore-wheel of a carriage is five-sixths the circumference of the hind-wheel. In traveling a mile the fore-wheel makes 88 more revolutions than the hind-wheel. What are the circumferences of the two wheels?

42. A man and a woman enter a streetcar on which there are $\frac{7}{10}$ as many men as women. After they have boarded the car there are $\frac{5}{7}$ as many men as women on the car. How many people were on the car before they entered it?

REVIEW OF CHAPTER VII

A

Solve and check:

1. $4x + y = 5$
   $3x - 2y = 12$

2. $5x - 2y = 7$
   $x + 3y = 15$

3. $x - 2y = -4$
   $2x + 5y = 1$

4. $3x + 2y = 7$
   $7x - y = 5$

5. $\frac{1}{2}x + \frac{1}{3}y = 3$
   $\frac{1}{4}x + y = 4$

6. $\frac{1}{2}x - \frac{1}{4}y = 5$
   $\frac{1}{3}x + \frac{1}{2}y = -2$

7. $0.4x - 0.2y = 2$
   $0.7x + 0.3y = 23$

8. $0.05x - 0.3y = -2.4$
   $0.4x + 0.17y = 6.5$

9. $2x + 3y = 2$
   $8x - 9y = 1$

10. $5x - 2y = 4$
    $15x + 8y = 5$

11. $7x + 6y = 1$
    $3x + 10y = 6$

12. $3x - 2y = 8$
    $13x + 10y = 2$

13. $\frac{1}{2}x - \frac{1}{3}y = \frac{4}{15}$
    $\frac{3}{4}x + \frac{5}{6}y = 2$

14. $\frac{3}{4}x + \frac{2}{3}y = -1\frac{1}{5}$
    $5x + 2y = 3$

15. $3.7x - 5.3y = 9.95$
    $2.9x + 4.7y = -0.17$

16. $2.5x - 3.3y = 1.21$
    $3.2x + 1.7y = 9.25$

B

Solve:

1. $\frac{2}{x} + \frac{9}{y} = 2$
   $\frac{6}{x} - \frac{3}{y} = 1$

2. $\frac{2}{x} - \frac{7}{y} = 8$
   $\frac{10}{x} + \frac{3}{y} = 2$

3. $\frac{3}{x} - \frac{2}{y} = 4$
   $\frac{5}{x} - \frac{3}{y} = 5$

4. $\frac{3}{x} + \frac{5}{y} = 10$
   $\frac{5}{x} - \frac{3}{y} = -23$

5. $\frac{7}{3x} + \frac{1}{2y} = 1$
   $\frac{8}{5x} - \frac{3}{5y} = 1$

6. $\frac{1}{2x} + \frac{3}{4y} = 4$
   $\frac{5}{3x} - \frac{2}{3y} = 7$

7. $\frac{7}{2x} - \frac{5}{4y} = \frac{37}{24}$
$\frac{1}{3x} - \frac{2}{5y} = -\frac{5}{9}$

8. $\frac{3}{2x} + \frac{2}{3y} = 3\frac{1}{2}$
$\frac{4}{3x} + \frac{3}{4y} = 2\frac{7}{8}$

9. $x + 2y = 4xy$
$2x - y = 11xy$

10. $2x - 3y = 5xy$
$6x - 5y = 19xy$

C

Solve for x and y:

1. $2x + 3y = 13a$
$3x - 2y = 13b$

2. $5x - 3y = 7a + 11b$
$4x - y = 7a + 6b$

3. $9mx - 4ny = 16$
$3mx + 5ny = 37$

4. $\frac{5x}{a} - \frac{y}{b} = 14$
$\frac{3x}{a} + \frac{2y}{b} = 24$

5. $x + y = (a + b)^2$
$bx - ay = 0$

6. $px - qy = 5b$
$4px + qy = 5a$

7. $x + 3y = 7ab$
$2x - y = 7$

8. $x + y = \frac{a}{b} + \frac{b}{a}$
$ax - by = 2b$

9. $2ax + by = 2a^2 + b^2$
$ay - bx = 2a^2 + b^2$

10. $ax - by = (2a - b)(a + b)$
$bx - 2ay = b^2$

D

Solve:

1. $\frac{x - 1}{3} + y = 6$
$5x - 3y = 5$

2. $3x + \frac{y - 3}{5} = 11$
$2x - 3y = 14$

3. $\frac{1}{3}x + \frac{1}{2}y = 3$
$\frac{x + 2}{2} + \frac{y + 2}{3} = 2$

4. $\frac{3x}{4} - \frac{2y}{3} = 3$
$\frac{2x + 5}{9} - \frac{y + 6}{3} = -3$

5. $5(x + 3) + 9(y + 1) = 5$
$3(2x + 1) - 7(y - 3) = 19$

6. $6(3x - 1) + 3(y + 5) = 16$
$3(x + 2) - 4(3y - 1) = 7$

7. $\frac{x + 1}{y - 3} = 2$
$\frac{1}{3}x + y = 6$

8. $\frac{4x - 3y}{3x - 4y} = \frac{13}{15}$
$15x + y = 4$

9. $\dfrac{1}{x-1} + \dfrac{1}{y-1} = 0$
$\dfrac{1}{2x+3} + \dfrac{1}{y} = 0$

10. $(x+3)(y+1) = (x+2)(y+2)$
$\frac{1}{2}x + \frac{1}{3}y = 2$

11. $\dfrac{x-2}{5} = \dfrac{y-3}{6} = \dfrac{5y-6x}{3}$

12. $\dfrac{x+5}{y-4} = \dfrac{x+8}{y-2}$
$5(x+3) = y - 1$

E (Optional)

Solve for x, y, and z:

1. $x - 2y + z = 3$
$2x + y - z = 7$
$3x - y + 2z = 6$

2. $4x + y - 2z = 1$
$x - 2y + z = 7$
$2x - y - 3z = 5$

3. $3x - y - 3z = 1$
$x - 3y + z = 1$
$5y - 3z = 1$

4. $x + y + z = 2$
$2x + z = 11$
$3x - 2y = 23$

5. $x + y + \dfrac{1}{z} = 3$
$2x - y + \dfrac{1}{z} = 4$
$x - y - \dfrac{2}{z} = 4$

6. $\frac{1}{2}(x+1) + \frac{1}{3}(y-1) + \frac{1}{4}(z+1) = 1$
$\frac{1}{4}(x-1) + \frac{1}{2}(y+2) + \frac{1}{3}z = 1$
$\frac{1}{3}(x-4) - \frac{1}{4}(y-2) + \frac{1}{2}(z-1) = 1$

7. $2x - y - z = 3$
$2x - 2y + z = n$
$x + 3y - 3z = n + 3$

8. $x + ay + z = a + 2$
$x - a^2y + z = 1$
$ax - a^3y + cz = c$

F

1. Three times the larger of two numbers exceeds four times the smaller by 3, and the sum of the numbers exceeds five times their difference by 4. What are the numbers?
2. A motorboat travels 4 miles downstream in 20 minutes, but requires 30 minutes for the return trip upstream. What is the rate of the boat in still water and what is the rate of the current?
3. Seven algebra books and six geometry books cost \$31.40, and five algebra books and ten geometry books cost \$37. What is the price for each algebra book and each geometry book?

4. A man's age next year will be five times as great as his son's present age, and his son's age next year will be one-fourth as great as the man's present age. Find their present ages.

5. A boy walked for 30 minutes and ran for 30 minutes traveling a distance of 6 miles. If he had walked for only 15 minutes and run for 45 minutes, he would have traveled one mile farther. Find his rate of walking and his rate of running.

6. A man has two investments, one paying 4% interest and the other 6% interest. His annual return on these investments is \$610. If all of his money were invested at 5% interest, he would receive \$55 more each year. How much does he have invested at each rate?

7. A alone can do a certain job in 15 days and B alone can do it in 20 days. If A works x days and B works y days, they complete the job; but if A works y days and B works x days, they complete only $\frac{13}{15}$ of the job. Find x and y.

8. The perimeter of a rectangle is 60 feet. If the length is decreased 1 foot and the width is increased 1 foot, the area of the rectangle will be increased by 5 square feet. Find the dimensions of the rectangle.

9. Fourteen cubic centimeters of one acid solution added to 6 cubic centimeters of another gives an acid mixture which is 37% acid. Again, if 2 cubic centimeters of the first solution is added to 8 cubic centimeters of the second, the mixture is 32% acid. Find the acid concentration of the two original acid solutions.

10. A boy has \$7 in nickels and dimes. If he had twice as many nickels and only half as many dimes, he would have 40 cents more money. How many nickels and how many dimes does he have?

11. A certain number with two digits is equal to five times the sum of its digits. If the number is doubled and then decreased by 36, the result will be the number with its digits reversed. Find the number.

12. A Boy Scout Troop purchased a boat, each scout paying an equal share. If there were 3 more scouts, each would have paid a dollar apiece less; but if there were 2 scouts less, each would have paid a dollar apiece more. How many scouts were there, and what did the boat cost?

13. Three pounds of tea and seven pounds of coffee cost \$6.30. If the price of tea should increase 10% and that of coffee should decrease 10%, the cost of the above quantities would be \$6.09. Find the price per pound of each.

14. A man rows upstream a distance of 1 mile in 2 hours. If he had rowed twice as hard and if the current had been half as strong, he would have made the trip in 30 minutes. What is the rate of the current?

15. If the numerator and the denominator of a fraction are both increased by 3, the value of the fraction becomes $\frac{4}{5}$; but if they are both decreased by 3, the value of the fraction becomes $\frac{5}{7}$. What is the fraction?

16. A man has two sons of different ages; his age is three times the sum of his sons' ages. In three years his age will be five times the age of his younger son and also twice the sum of his sons' ages. Find their present ages.

17. If the length of a rectangle is decreased 1 foot and the width is increased 2 feet, the area is increased by 17 square feet; but if the width is decreased 1 foot and the length is increased 2 feet, the area is decreased by 1 square foot. What are the dimensions of the rectangle?

18. A, B, and C together have \$3000 invested. A receives annually 6% on his investment, B 7%, and C 8%; and the sum of their incomes is \$219. If A received 7%, B 8%, and C 6%, the sum of their incomes would be \$204. How much has each invested?

19. A tank can be filled by two pipes, one of which runs 6 hours and the other 12 hours; or by the same two pipes, if the first runs 9 hours and the second 8 hours. How long would it take each pipe alone to fill the tank?

20. After going a certain distance in an automobile, a driver found that if he had gone 6 miles per hour slower he would have traveled the distance in 1 hour more time; and if he had gone 9 miles per hour faster, he would have traveled the distance in 1 hour less time. What was the distance traveled?

Vitamins

A druggist has three different mixtures containing vitamins A, B, and C. If the first mixture contains 1 part A to 3 parts B to 4 parts C, the second 3 parts A to 2 parts B to 3 parts C, and the third 3 parts A to 3 parts B to 2 parts C, how much of each mixture should be taken in order to obtain 30 grams of another mixture consisting of equal parts of A, B, and C?

Hup, Two, Three, Four

A column of soldiers is marching at a constant rate toward a town 80 miles away. A messenger is sent ahead to the town and he reports back to the column $6\frac{2}{3}$ hours later. He is immediately sent back to the town with another message and he returns to the column $4\frac{4}{9}$ hours later. Find the rate at which the soldiers are marching.

CHAPTER VIII

EXPONENTS, ROOTS, AND RADICALS

82. Laws of Exponents. If n is any positive integer and a is any number, we define the expression a^n to mean the product of n factors each of which is equal to a.

Illustrations. (1) $a^1 = a$,
(2) $a^4 = a \cdot a \cdot a \cdot a$,
(3) $a^n = a \cdot a \cdot \cdots \cdot a$ (n factors).

We call a^n the *nth power of a*; n is the *exponent*, and a is the *base*. In the following statements of the **laws of exponents,** we shall limit our proofs to those cases in which the exponents are positive integers. The laws are true, however, without this restriction.

I. $\boldsymbol{a^m \cdot a^n = a^{m+n}}$.

Proof.

$$\begin{aligned} a^m \cdot a^n &= (a \cdot a \cdot a \cdot \cdots \text{ to } m \text{ factors})(a \cdot a \cdot a \cdot \cdots \text{ to } n \text{ factors}), \\ &= a \cdot a \cdot a \cdot \cdots \text{ to } (m+n) \text{ factors}, \\ &= a^{m+n} \end{aligned}$$

Illustrations. $x^5 \cdot x^3 = x^8$; $(a+b)^2 \cdot (a+b)^4 = (a+b)^6$.

II. (1) $\boldsymbol{\dfrac{a^m}{a^n} = a^{m-n}}$, if m is greater than n.

(2) $\boldsymbol{\dfrac{a^m}{a^n} = 1}$, if m is equal to n.

(3) $\boldsymbol{\dfrac{a^m}{a^n} = \dfrac{1}{a^{n-m}}}$, if m is less than n.

Proof. (1)

$$\begin{aligned} \frac{a^m}{a^n} &= \frac{a \cdot a \cdot a \cdot \cdots \text{ to } m \text{ factors}}{a \cdot a \cdot a \cdot \cdots \text{ to } n \text{ factors}}, \\ &= \frac{a \cdot a \cdot a \cdot \cdots \text{ to } (m-n) \text{ factors}}{1}, \\ &= a^{m-n}. \end{aligned}$$

The proof of (2) and (3) is left to the student.

Illustrations. (1) $\dfrac{x^7}{x^3} = x^{7-3} = x^4.$

(2) $\dfrac{y^5}{y^5} = 1.$

(3) $\dfrac{x^2}{x^6} = \dfrac{1}{x^{6-2}} = \dfrac{1}{x^4}.$

III. $(a^m)^n = a^{mn}.$

Proof.

$$\begin{aligned}(a^m)^n &= a^m \cdot a^m \cdot a^m \cdot \cdots \text{ to } n \text{ factors,}\\ &= a^{m+m+m+\cdots \text{ to } n \text{ terms}},\\ &= a^{mn}.\end{aligned}$$

Illustrations. $(x^4)^3 = x^{12}$; $(2^3)^2 = 2^6 = 64.$

IV. $(ab)^n = a^n b^n.$

Proof.

$$\begin{aligned}(ab)^n &= (ab) \cdot (ab) \cdot (ab) \cdot \cdots \text{ to } n \text{ factors,}\\ &= (a \cdot a \cdot a \cdot \cdots \text{ to } n \text{ factors})(b \cdot b \cdot b \cdot \cdots \text{ to } n \text{ factors}),\\ &= a^n b^n.\end{aligned}$$

Illustrations.

$$(2x)^3 = 2^3x^3 = 8x^3; \quad (3x^2y)^4 = (3)^4(x^2)^4(y)^4 = 81x^8y^4.$$

V. $\left(\dfrac{a}{b}\right)^n = \dfrac{a^n}{b^n}.$

Proof.

$$\begin{aligned}\left(\frac{a}{b}\right)^n &= \left(\frac{a}{b}\right) \cdot \left(\frac{a}{b}\right) \cdot \left(\frac{a}{b}\right) \cdot \cdots \text{ to } n \text{ factors,}\\ &= \frac{a \cdot a \cdot a \cdot \cdots \text{ to } n \text{ factors}}{b \cdot b \cdot b \cdot \cdots \text{ to } n \text{ factors}},\\ &= \frac{a^n}{b^n}.\end{aligned}$$

Illustrations. $\left(\dfrac{x}{3}\right)^2 = \dfrac{x^2}{3^2} = \dfrac{x^2}{9}$; $\left(\dfrac{2x^2}{y^3}\right)^3 = \dfrac{(2)^3(x^2)^3}{(y^3)^3} = \dfrac{8x^6}{y^9}.$

NOTE. Observe, especially, that *an exponent indicates the power to be taken of that quantity and only that quantity to which it is attached.* Thus,

(1) $5x^2$ means $5 \cdot x \cdot x$, whereas $(5x)^2$ means $5 \cdot 5 \cdot x \cdot x$.

(2) $-(x^3)^2$ means $-(x^3)(x^3) = -x^6$, whereas $(-x^3)^2$ means $(-x^3)(-x^3) = x^6$.

EXERCISE 62

Perform the indicated operations, using the laws of exponents:

1. $x^3 \cdot x^4$
2. $m^2 \cdot m^6$
3. $a \cdot a^2 \cdot a^3$
4. $3^2 \cdot 3 \cdot 3^4$
5. $y^8 \cdot y^{10} \cdot y^5$
6. $(-2) \cdot (-2)^4 \cdot (-2)^2$
7. $a^2x^3 \cdot a^3x^4$
8. $ab^2c \cdot a^2bc^3$
9. $x^2y \cdot x^3y \cdot xy^2$
10. $mn \cdot m^2n^3 \cdot m^2n$
11. $\dfrac{x^5}{x^2}$
12. $\dfrac{y^4}{y^6}$
13. $\dfrac{z^7}{z^7}$
14. $\dfrac{x^3y^4}{xy^2}$
15. $\dfrac{a^3b}{a^2b^2}$
16. $\dfrac{xy^2}{x^3y^3}$
17. $\dfrac{a^2b^2}{ab^2}$
18. $\dfrac{a^3x^2}{a^3x^2}$
19. $\dfrac{x^3yz^2}{xyz^4}$
20. $\dfrac{abc^5}{ab^2c^3}$
21. $(x^3)^5$
22. $(-x)^4$
23. $-(b^8)^2$
24. $(a^5)^5$
25. $(-m^2)^3$
26. $(y^6)^4$
27. $(x^2)^3 \cdot (y^3)^2$
28. $(a^3)^5 \cdot (x^4)^4$
29. $(a^2)^2 \cdot (b^3)^3 \cdot (c^4)^4$
30. $(a^2)^4 \cdot (x^3)^3 \cdot (y^4)^2$
31. $(2x^3)^3$
32. $(-3a)^2$
33. $-(a^2b^3)^4$
34. $(-x^5y^4)^3$
35. $(2ab^2)^3$
36. $(ax^3)^2$
37. $(a^2b^3c)^4$
38. $(x^3y^2z^4)^7$
39. $(2ax^2y^3)^4$
40. $(-3m^2nx)^3$
41. $(\frac{3}{4})^3$
42. $\left(-\dfrac{x}{4}\right)^2$
43. $\left(\dfrac{a}{x^2}\right)^4$
44. $\left(\dfrac{5}{p}\right)^3$
45. $\left(\dfrac{a^3}{b^5}\right)^5$
46. $\left(\dfrac{2x}{3y^2}\right)^3$
47. $\left(\dfrac{5a^3}{6b^2}\right)^2$
48. $\left(\dfrac{ab^2}{2x^3}\right)^5$
49. $\left(\dfrac{ab^2c}{x^2y^3z}\right)^{10}$
50. $\left(-\dfrac{x^2y^3}{2a^2b}\right)^8$

Perform the indicated operations and simplify:

51. $(ax^2)^2 \cdot (ax^3)^3$
52. $(mn)^2 \cdot (2m^2n)^3$
53. $\dfrac{(x^2y^3)^3}{(x^4y)^4}$
54. $\dfrac{(b^3c^5)^3}{(b^2c)^2}$
55. $(x^3y)^2 \cdot (x^2y^2)^3$
56. $(ab^2)^4 \cdot (a^2b)^3$
57. $(2ax^2)^3 \cdot (3bx^3)^2$
58. $(5a^2x^3)^2 \cdot (b^3y^4)^3$
59. $\dfrac{(ab^2)^3}{(ab^3)^2}$
60. $\dfrac{(3p^2q)^3}{(6pq^2)^2}$
61. $\dfrac{(xy^2)^3 \cdot (x^2y^3)^2}{(x^5y)^2}$
62. $\dfrac{(6a^5b^3)^4}{(3ab^2)^3(2a^2b)^5}$
63. $\left(\dfrac{m^2}{n^3}\right)^3\left(\dfrac{n}{m^3}\right)^2$
64. $\left(\dfrac{2a^2}{x}\right)^3\left(\dfrac{x^2}{2a}\right)^4$
65. $\left(\dfrac{ax^2}{b^2y}\right)^3\left(\dfrac{by^2}{a^2x}\right)^2$

66. $\left(\dfrac{x^2y}{z^3}\right)^2\left(\dfrac{x^3z^2}{y^2}\right)^3$ **67.** $(a^2b^3)^4\left(\dfrac{x^2}{a^2}\right)^2\left(\dfrac{y}{b^3}\right)^3$ **68.** $\left(\dfrac{ax^2}{y}\right)^2\left(\dfrac{x^2y^3}{a}\right)^3\left(\dfrac{ay}{x^3}\right)^2$

69. $\left(\dfrac{2a}{3c}\right)^2\left(\dfrac{6c}{b^2}\right)^3\left(\dfrac{b^3}{4a}\right)^2$ **70.** $\left(\dfrac{a^2b}{x}\right)^4\left(\dfrac{a}{b^2x^3}\right)^2\left(\dfrac{x^2}{a^2}\right)^5$

83. Roots. We know from arithmetic that

$$5 \times 5 = 25,$$
$$5 \times 5 \times 5 = 125,$$
$$5 \times 5 \times 5 \times 5 = 625; \text{ and so on.}$$

In mathematics, we often need to know the equal factors which when multiplied yield a given number. We refer to these factors as the roots of the given number. That is,

5 is the *square root* of 25,
5 is the *cube root* of 125,
5 is the *fourth root* of 625; and so on.

In general, *the nth root of a number is one of the n equal factors of the given number.*

The process of extracting roots is indicated by the sign $\sqrt{\ }$, called the *radical sign.* To indicate which type of root is meant, a number is placed in the crook of the radical sign. This number is called the *index* or *order* of the root. The number or expression appearing under the radical sign is called the *radicand.*

Thus, in the expression $\sqrt[3]{8} = 2$, 8 is the radicand, 2 is the root, and 3 is the index. The radical is said to be of the third order.

NOTE. If the index of a root is not explicitly written, it is implied that it is 2. Thus, $\sqrt{9}$ means $\sqrt[2]{9}$, which equals 3.

Since $(+2)^2 = 4$ and $(-2)^2 = 4$, we see that the number 4 has *two* square roots, namely, $+2$ and -2. In using radical signs for expressions that have both a positive and a negative root, it is agreed that only the positive root will be indicated. Thus, although the square roots of 4 are ± 2, the radical expression $\sqrt{4}$ represents only the root 2. This single value is called the *principal root* of the radicand.

Example 1. What are the square roots of $49a^2$?
Solution. Since $7a \cdot 7a = 49a^2$, the square roots of $49a^2$ are $\pm 7a$.

Example 2. What is the cube root of $-27a^3b^6$?
Solution. Since $(-3ab^2)^3 = -27a^3b^6$, the cube root of $-27a^3b^6$ is $-3ab^2$.

Example 3. Evaluate $\sqrt[4]{16a^4} + \sqrt[7]{a^7}$.
Solution. Since $(2a)^4 = 16a^4$ and $(a)^7 = a^7$,

$$\sqrt[4]{16a^4} + \sqrt[7]{a^7} = 2a + a = 3a.$$

NOTE. In evaluating an expression such as $\sqrt{4a^2}$, observe that if a is negative the principal root is $-2a$.

EXERCISE 63

Give the square roots of the following:

1. 16	**2.** 36	**3.** 64	**4.** 81	**5.** a^2
6. b^4	**7.** x^4y^2	**8.** $49m^2$	**9.** $25a^4$	**10.** $400c^{10}$

Give the cube root of each of the following:

11. 8	**12.** 27	**13.** -64	**14.** 125	**15.** a^3
16. $-c^9$	**17.** $64x^3y^3$	**18.** $1000m^6$	**19.** $-a^6b^3$	**20.** $\frac{8}{27}a^3$

Give the roots as indicated in the following:

21. $\sqrt{121}$	**22.** $-\sqrt{144}$	**23.** $\sqrt[3]{27}$	**24.** $\sqrt{225}$
25. $\sqrt[4]{16}$	**26.** $\sqrt[3]{729}$	**27.** $\sqrt[3]{-8}$	**28.** $-\sqrt{625}$
29. $\sqrt[3]{\frac{1}{64}}$	**30.** $\sqrt[5]{32}$	**31.** $\sqrt[4]{b^8}$	**32.** $\sqrt[3]{64m^3n^3}$

Simplify and combine the following:

33. $\sqrt{4} + \sqrt[3]{8}$	**34.** $\sqrt[4]{16} + \sqrt[3]{-27}$	**35.** $\sqrt[3]{64a^3} - \sqrt{a^2}$
36. $\sqrt[5]{a^5} + \sqrt{9a^2}$	**37.** $\sqrt[3]{-8} + \sqrt{144}$	**38.** $\sqrt[4]{81} - \sqrt[5]{1}$
39. $\sqrt{81a^4} - \sqrt[3]{-8a^6}$	**40.** $\sqrt[4]{16a^4b^4} - \sqrt[7]{a^7b^7}$	**41.** $\sqrt[4]{\frac{16}{81}} - \sqrt{\frac{1}{4}}$

84. Fractional Exponents. The laws of exponents were derived under the supposition that the exponents considered were positive whole numbers. Thus, a^n has a meaning if n is a positive integer. The question arises, however, as to the possible significance of such an expression as $a^{\frac{1}{2}}$. We should expect this quantity to obey the laws of exponents and hence by the first law of exponents, we would have

$$a^{\frac{1}{2}} \cdot a^{\frac{1}{2}} = a^{\frac{1}{2}+\frac{1}{2}} = a^1 = a.$$

However, from the definition of roots,

$$\sqrt{a} \cdot \sqrt{a} = a.$$

Hence, we see that $a^{\frac{1}{2}}$ *has the same meaning as* $\sqrt{a}$.

In general for the expression $a^{\frac{1}{n}}$,

$$a^{\frac{1}{n}} \cdot a^{\frac{1}{n}} \cdot \cdots \text{to } n \text{ factors} = a^{\frac{1}{n}+\frac{1}{n}+\cdots \text{to } n \text{ terms}} = a^1 = a.$$

Again, we see that $a^{\frac{1}{n}}$ has the same meaning as $\sqrt[n]{a}$. For this reason, we define the expression $a^{\frac{1}{n}}$ to mean the principal nth root of a; thus

$$a^{\frac{1}{n}} = \sqrt[n]{a}.$$

From the third law of exponents, we know that $a^{\frac{m}{n}} = (a^{\frac{1}{n}})^m = (a^m)^{\frac{1}{n}}$. Hence it follows that

$$a^{\frac{m}{n}} = (\sqrt[n]{a})^m = \sqrt[n]{a^m}.$$

Rule for Fractional Exponents

For any quantity with a fractional exponent, (1) the numerator of the exponent denotes the power to which the quantity is to be raised, and (2) the denominator denotes the root which is to be taken.

Example 1. Evaluate $8^{\frac{2}{3}}$.

Solution. From the definition,

$$8^{\frac{2}{3}} = \sqrt[3]{8^2} = \sqrt[3]{64} = 4,$$

or

$$8^{\frac{2}{3}} = (\sqrt[3]{8})^2 = (2)^2 = 4.$$

The laws of exponents may be applied to expressions containing fractional exponents, as is illustrated in the following example.

Example 2. Using the laws of exponents, find (1) $x^{\frac{1}{2}} \cdot x^{\frac{1}{4}}$; (2) $\dfrac{x^{\frac{2}{3}}}{x^{\frac{1}{3}}}$; (3) $(x^{\frac{2}{3}})^6$; (4) $(x^6y^3)^{\frac{1}{3}}$; (5) $\left(\dfrac{x^{\frac{1}{3}}}{y^2}\right)^{\frac{1}{2}}$.

Solution.

$$(1)\quad x^{\frac{1}{2}}x^{\frac{1}{4}} = x^{\frac{1}{2}+\frac{1}{4}} = x^{\frac{3}{4}},$$

$$(2)\quad \frac{x^{\frac{2}{3}}}{x^{\frac{1}{3}}} = x^{\frac{2}{3}-\frac{1}{3}} = x^{\frac{1}{3}},$$

$$(3)\quad (x^{\frac{2}{3}})^6 = x^{\frac{12}{3}} = x^4,$$

$$(4)\quad (x^6y^3)^{\frac{1}{3}} = (x^6)^{\frac{1}{3}}(y^3)^{\frac{1}{3}} = x^2y,$$

$$(5)\quad \left(\frac{x^{\frac{1}{3}}}{y^2}\right)^{\frac{1}{2}} = \frac{(x^{\frac{1}{3}})^{\frac{1}{2}}}{(y^2)^{\frac{1}{2}}} = \frac{x^{\frac{1}{6}}}{y}.$$

EXERCISE 64

Express the following in their equivalent radical form:

1. $x^{\frac{1}{2}}, x^{\frac{1}{3}}, y^{\frac{2}{3}}, y^{\frac{3}{4}}, a^{\frac{1}{7}}, a^{\frac{3}{5}}, b^{0.5}, b^{0.4}$
2. $x^{\frac{3}{2}}, x^{1\frac{1}{3}}, y^{\frac{5}{6}}, y^{\frac{6}{5}}, a^{2\frac{1}{2}}, a^{\frac{2}{5}}, (a+b)^{\frac{1}{2}}, (a-b)^{\frac{1}{3}}$
3. $3x^{\frac{1}{2}}, (3x)^{\frac{1}{2}}, 8y^{\frac{1}{3}}, 5y^{\frac{3}{5}}, 2a^{\frac{4}{3}}, 2a^{\frac{3}{4}}, (a^2+b^2)^{\frac{1}{2}}, (a^3-b^3)^{\frac{1}{3}}$
4. $4x^{\frac{4}{3}}, 2x^{3\frac{1}{2}}, (4y)^{\frac{2}{3}}, 4y^{\frac{2}{3}}, 3a^{1\frac{1}{2}}, (2a)^{\frac{4}{3}}, 7(a+b)^{\frac{1}{2}}, [7(a+b)]^{\frac{1}{2}}$

Express the following in their equivalent exponent form:

5. $\sqrt{a}, \sqrt[3]{a}, \sqrt[3]{b^2}, \sqrt{b^3}, \sqrt[5]{x^2}, \sqrt{x^5}, \sqrt[4]{y^3}, \sqrt[3]{y^5}$
6. $\sqrt[3]{a^4}, \sqrt[4]{a}, \sqrt[6]{b^2}, \sqrt[3]{b^7}, \sqrt{x^9}, \sqrt[4]{x^7}, \sqrt{x+y}, \sqrt{x^2+y^2}$
7. $2\sqrt{a}, \sqrt{2a}, 7\sqrt[3]{b^2}, 3\sqrt[5]{b^3}, \sqrt[4]{16x^3}, 5\sqrt{x^5}, 2\sqrt{x-y}, 4\sqrt{x^2-y^2}$
8. $a\sqrt{b}, \sqrt{ab}, b\sqrt{a}, a\sqrt[3]{b^2}, \sqrt[3]{ab^2}, b^2\sqrt[3]{a}, x\sqrt{x+y}, x\sqrt[3]{x-y}$

Evaluate each of the following:

9. $25^{\frac{1}{2}}$ 10. $8^{\frac{1}{3}}$ 11. $16^{\frac{1}{4}}$ 12. $27^{\frac{2}{3}}$
13. $4^{\frac{3}{2}}$ 14. $(-8)^{\frac{2}{3}}$ 15. $144^{\frac{1}{2}}$ 16. $32^{\frac{1}{5}}$
17. $(-1)^{\frac{3}{5}}$ 18. $1^{\frac{4}{7}}$ 19. $9^{\frac{3}{2}}$ 20. $81^{\frac{3}{4}}$
21. $(\frac{1}{8})^{\frac{1}{3}}$ 22. $(\frac{25}{36})^{\frac{1}{2}}$ 23. $(\frac{8}{27})^{\frac{2}{3}}$ 24. $(\frac{4}{9})^{\frac{3}{2}}$
25. $(0.25)^{\frac{1}{2}}$ 26. $(1.21)^{\frac{1}{2}}$ 27. $(0.027)^{\frac{1}{3}}$ 28. $(0.125)^{\frac{2}{3}}$

Multiply the following in accordance with the laws of exponents:

29. $a^{\frac{1}{2}} \cdot a^{\frac{1}{4}}$ 30. $x^{\frac{1}{3}} \cdot x^{\frac{1}{3}}$ 31. $b \cdot b^{\frac{1}{2}}$ 32. $y^{\frac{1}{3}} \cdot y^{\frac{2}{3}}$
33. $2^{\frac{1}{2}} \cdot 2^{\frac{1}{3}}$ 34. $x^{\frac{3}{2}} \cdot x^{\frac{2}{3}}$ 35. $n^{\frac{1}{2}} \cdot n^{\frac{1}{4}} \cdot n^{\frac{1}{4}}$ 36. $a^{\frac{1}{3}} \cdot a^{\frac{2}{3}} \cdot a$

Divide the following in accordance with the laws of exponents:

37. $a^{\frac{3}{4}} \div a^{\frac{1}{4}}$ 38. $x^{\frac{5}{6}} \div x^{\frac{1}{3}}$ 39. $y^{\frac{2}{3}} \div y^{\frac{1}{4}}$ 40. $2^{\frac{5}{4}} \div 2^{\frac{1}{4}}$
41. $\dfrac{n^{\frac{4}{3}}}{n^{\frac{3}{4}}}$ 42. $\dfrac{a^{0.5}}{a^{0.25}}$ 43. $\dfrac{x^{\frac{1}{2}}}{x^{\frac{1}{3}}}$ 44. $\dfrac{m^{2\frac{1}{2}}}{m^{1\frac{1}{3}}}$

Perform the indicated operations:

45. $(a^{\frac{2}{3}})^6$ 46. $(a^6)^{\frac{2}{3}}$ 47. $(x^{\frac{2}{3}})^{\frac{1}{2}}$ 48. $(y^{\frac{2}{3}})^{\frac{3}{2}}$
49. $(4^{\frac{3}{4}})^{\frac{2}{3}}$ 50. $(z^{10})^{\frac{2}{5}}$ 51. $(x^{\frac{1}{2}}y^{\frac{1}{3}})^6$ 52. $(a^2b^3)^{\frac{1}{6}}$

53. $\left(\frac{x^2}{y^3}\right)^{\frac{5}{6}}$ **54.** $\left(\frac{a^{\frac{3}{2}}}{b}\right)^4$ **55.** $\left(\frac{b^{\frac{3}{4}}}{c^{\frac{2}{3}}}\right)^6$ **56.** $\left(\frac{x^{\frac{1}{2}}}{y^2}\right)^3$

Simplify:

57. $\left(\frac{x^{\frac{1}{2}} \cdot x^{\frac{1}{3}}}{x^{\frac{5}{12}}}\right)^6$ **58.** $\frac{(ax^2)^{\frac{1}{2}} \cdot (a^2x)^{\frac{1}{4}}}{x^{\frac{1}{4}}}$ **59.** $\frac{(n^{\frac{2}{3}})^2 \cdot (n^{\frac{3}{5}})^{10}}{(n^{\frac{4}{3}})^3}$

60. $\frac{(y^{\frac{3}{2}})^{\frac{1}{2}} \cdot (y^3)^{\frac{1}{6}}}{(y^{\frac{5}{2}})^{\frac{1}{2}}}$ **61.** $\left(\frac{a^{\frac{1}{2}}}{b^2}\right)^2 \cdot \left(\frac{b^{10}}{a}\right)^{\frac{1}{2}}$ **62.** $\frac{a^{\frac{1}{2}}x^{\frac{2}{3}} \cdot a^{\frac{1}{3}}x^{\frac{1}{4}}}{a^{\frac{2}{3}}x^{\frac{5}{12}}}$

85. Zero and Negative Exponents. In order for the law of exponents $a^m \cdot a^n = a^{m+n}$ to hold for a zero exponent, it is necessary that we define the quantity a^0 so that

$$a^0 \cdot a^n = a^{0+n} = a^n.$$

Hence, we define

$$\mathbf{a^0 = 1, \text{ if } a \neq 0.}$$

Illustrations. By definition $(-5)^0 = 1$, $(\frac{2}{3})^0 = 1$, $(3x^2)^0 = 1$.

Also if the law of exponents for multiplication is to hold for negative exponents, we must have

$$a^m \cdot a^{-m} = a^{m-m} = a^0 = 1.$$

Hence, we define

$$\mathbf{a^{-m} = \frac{1}{a^m}, \text{ if } a \neq 0.}$$

Illustrations. $2^{-3} = \frac{1}{2^3} = \frac{1}{8}$, $9^{-\frac{1}{2}} = \frac{1}{9^{\frac{1}{2}}} = \frac{1}{3}$, $x^{-1} = \frac{1}{x}$.

From the definitions of zero and negative exponents, it follows that

Rule for Negative Exponents

Any factor of the numerator of a fraction may be changed to the denominator, or any factor of the denominator may be changed to the numerator, if the sign of its exponent is changed at the same time.

For example, either

$$\frac{ax^{-n}}{b} = \frac{ax^{-n}x^n}{bx^n} = \frac{ax^0}{bx^n} = \frac{a}{bx^n},$$

or

$$\frac{a}{bx^{-n}} = \frac{ax^n}{bx^{-n}x^n} = \frac{ax^n}{bx^0} = \frac{ax^n}{b}.$$

Illustration. $\dfrac{x^{-2}y^3}{a^{-1}b^{-4}} = \dfrac{ax^{-2}}{b^{-4}y^{-3}} = \dfrac{ab^4}{x^2y^{-3}} = \dfrac{ab^4y^3}{x^2}.$

NOTE. In applying the above rule, it is necessary to distinguish carefully between *factors* and *terms.*

Thus, $\dfrac{1}{a^{-1}b^{-1}}$ becomes ab.

But $\dfrac{1}{a^{-1}+b^{-1}}$ becomes $\dfrac{1}{\dfrac{1}{a}+\dfrac{1}{b}}$ or $\dfrac{ab}{a+b}.$

Example 1. Express $\left(\dfrac{ax^{-2}}{y^3}\right)^{-2}$ in the simplest form with positive exponents.

Solution. Applying the law of exponents before eliminating negative exponents, we have

$$\left(\frac{ax^{-2}}{y^3}\right)^{-2} = \frac{a^{-2}x^4}{y^{-6}} = \frac{x^4y^6}{a^2}.$$

EXERCISE 65

Find the numerical values of the following:

1. 4^{-1} **2.** 3^0 **3.** 2^{-3} **4.** 10^{-2}

5. $(-1)^0$ **6.** $(\frac{2}{3})^{-2}$ **7.** $\dfrac{1}{5^{-1}}$ **8.** $(7^{-2})^0$

9. $(\frac{1}{2})^{-2} \cdot 4^{-3}$ **10.** $(\frac{2}{5})^{-3} \cdot (\frac{3}{4})^0$ **11.** $8^{-\frac{1}{3}}$ **12.** 3×2^0

13. $\dfrac{3^{-2}}{2^{-3}}$ **14.** $4^{-\frac{1}{2}} \cdot (\frac{1}{2})^{-4}$ **15.** 3×10^{-4} **16.** $4^{-\frac{3}{2}} \div 8^{-\frac{2}{3}}$

17. $(0.25)^{-\frac{1}{2}} \cdot (-\frac{1}{2})^0$ **18.** $(\frac{2}{3})^{-5} \cdot (\frac{3}{4})^{-3}$

19. $\sqrt[3]{-8^{-2}}$ **20.** $\sqrt[6]{64^{-1}} \cdot \sqrt[5]{10^0}$

Perform the indicated operations, using the laws of exponents:

21. $x^{-2} \cdot x^3$ **22.** $a^2 \cdot a^{-\frac{1}{2}}$ **23.** $m^{-1} \cdot m^{-3}$ **24.** $b^2 \div b^{-2}$

25. $x^0 \div x^{-4}$ **26.** $y^3 \div y^{-4}$ **27.** $(x^{-2})^{-3}$ **28.** $(a^{-2})^4$

29. $(y^{-\frac{1}{2}})^{-6}$ **30.** $(x^{-2}y^{-4})^{-1}$ **31.** $(a^{-\frac{1}{2}}b^2)^{-2}$ **32.** $(a^3x^{-1})^{-4}$

Express the following in the simplest form with positive exponents:

33. a^{-4} **34.** $a^{-2}b^3$ **35.** $\dfrac{x^{-2}}{5}$ **36.** $3y^{-3}$

37. $\dfrac{a^{-3}}{x^{-2}}$ **38.** $\dfrac{a^{-1}b^{-2}}{c^{-3}}$ **39.** $\dfrac{2x^{-4}}{3y^{-1}}$ **40.** $\dfrac{a^2b^{-3}}{x^2y^{-3}}$

41. $x^{-1} - y^{-1}$ **42.** $\dfrac{a}{b^{-1}} + \dfrac{b}{a^{-1}}$ **43.** $(x - y)^{-1}$ **44.** $\dfrac{4^{-1}m^{-2}n^{-3}}{(2mn)^{-5}}$

45. $\dfrac{x}{y} + \dfrac{y^{-1}}{x^{-1}}$ **46.** $\dfrac{x^{-1} + y^{-1}}{z^{-1}}$ **47.** $ab^{-1} + a^{-1}b$ **48.** $\sqrt{(x + y)^{-2}}$

49. $(m^{-1} + n^{-1})(m^{-1} - n^{-1})$ **50.** $a(a + b)^{-1} + b(a - b)^{-1}$

51. $\dfrac{x^2 - y^2}{x^{-1} + y^{-1}}$ **52.** $\dfrac{(a + b)^{-1}}{a^{-1} + b^{-1}}$

Perform the indicated operations:

53. $a^n \cdot a^{3-n}$ **54.** $x^m \cdot x^{m-2}$ **55.** $\dfrac{b^{n+1}}{b^{2n-1}}$ **56.** $\dfrac{y^{-n-2}}{y^{-5}}$

57. $\left(\dfrac{x^n}{y^3}\right)^{n-1}$ **58.** $\left(\dfrac{a^{3r}}{b^r}\right)^{\frac{1}{r}}$ **59.** $(a^2b^n)^n$ **60.** $(x^{-a}y^{1-a})^{-2}$

61. $(a^kb^2)^3 \cdot (a^{k-1}b^3)^{-2}$ **62.** $x^{n-1}y^{1-n} \cdot x^{2-n}y^{\frac{n}{2}}$

63. $\dfrac{(p^rq^{r-1})^r}{(p^{r-1}q^r)^r}$ **64.** $\dfrac{m^{x+2}n^{2x-1}}{(m^xn^{1-x})^2}$

86. Rational and Irrational Numbers. **A rational number** is defined as a number which can be expressed as the quotient of two (positive or negative) integers. Any real number which is not rational is called an **irrational number.** (See Article 93, p. 201.)

Illustration. The numbers $2\frac{2}{3}$, -5, 2.17 can be written as $\frac{8}{3}$, $\dfrac{-5}{1}$, $\dfrac{217}{100}$ and hence are rational numbers. The numbers represented by π, $\sqrt{2}$, $\sqrt[3]{5}$ are examples of irrational numbers. Proof of the irrationality of these and other numbers is given in more advanced courses.

An irrational number which is an indicated root of a rational number is called a *surd.*

Thus, $\sqrt{2}$, $\sqrt[3]{5}$ are surds. $\sqrt[3]{8}$ is not a surd because $\sqrt[3]{8} = 2$ is a rational number. $\sqrt{1 + \sqrt{2}}$ is not a surd because $1 + \sqrt{2}$ is not a rational number.

A surd of the second order is called a **quadratic surd.** A binomial which contains at least one quadratic surd is called a **binomial quadratic surd.**

Thus, $2 - \sqrt{5}$ and $2\sqrt{3} + \sqrt{7}$ are binomial quadratic surds.

87. Multiplication of Radicals. From the definition of radicals,

$$\sqrt{p^2q^2} = \sqrt{(pq)^2} = (pq) = p \cdot q = \sqrt{p^2} \cdot \sqrt{q^2}.$$

If we let $a = p^2$ and $b = q^2$, this result reduces to

$$\sqrt{ab} = \sqrt{a} \cdot \sqrt{b}.$$

In general, for roots of any order,

$$\sqrt[n]{p^nq^n} = \sqrt[n]{(pq)^n} = p \cdot q = \sqrt[n]{p^n} \cdot \sqrt[n]{q^n}.$$

Again, letting $a = p^n$ and $b = q^n$, we have

$$\sqrt[n]{ab} = \sqrt[n]{a} \cdot \sqrt[n]{b}.$$

The converse of this relationship is also evidently true; that is,

$$\boldsymbol{\sqrt[n]{a} \cdot \sqrt[n]{b} = \sqrt[n]{ab}.}$$

Thus,

Rule for Multiplication of Radicals

1. ***The product of two radicals of the SAME order is equal to the like root of the product of their radicands.***
2. ***The root of a product is equal to the product of the like roots of its factors.***

The second relation above may be used to *simplify* radical expressions when the radicand contains a factor whose root can be found exactly.

Example 1. Simplify the radicals $\sqrt{18}$, $\sqrt[3]{-16}$, $\sqrt{8a^2b^5}$.

Solution. By the above rule,

$$\sqrt{18} = \sqrt{9 \cdot 2} = \sqrt{9} \cdot \sqrt{2} = 3\sqrt{2},$$

$$\sqrt[3]{16} = \sqrt[3]{-8 \cdot 2} = \sqrt[3]{-8} \cdot \sqrt[3]{2} = -2\sqrt[3]{2},$$

$$\sqrt{8a^2b^5} = \sqrt{4a^2b^4 \cdot 2b} = \sqrt{4a^2b^4} \cdot \sqrt{2b} = 2ab^2\sqrt{2b}.$$

Example 2. Multiply and simplify $2\sqrt{6} \cdot 3\sqrt{10}$.

Solution. Multiplying, we have

$$2\sqrt{6} \cdot 3\sqrt{10} = 2 \cdot 3 \cdot \sqrt{6} \cdot \sqrt{10} = 6 \cdot \sqrt{60}.$$

Simplifying, we have

$$6 \cdot \sqrt{60} = 6 \cdot \sqrt{4 \cdot 15} = 6 \cdot \sqrt{4} \cdot \sqrt{15} = 6 \cdot 2 \cdot \sqrt{15} = 12\sqrt{15}.$$

Example 3. Multiply and simplify $\sqrt{65} \cdot \sqrt{91}$.

Solution. When two radicands of a product are seen to contain a common factor, the simplification can be made easier, as follows:

$$\sqrt{65} \cdot \sqrt{91} = \sqrt{13 \cdot 5} \cdot \sqrt{13 \cdot 7} = \sqrt{13} \cdot \sqrt{5} \cdot \sqrt{13} \cdot \sqrt{7}$$
$$= \sqrt{13} \cdot \sqrt{13} \cdot \sqrt{5} \cdot \sqrt{7}, = 13\sqrt{35}.$$

EXERCISE 66

Multiply and evaluate:

1. $\sqrt{2} \cdot \sqrt{8}$ **2.** $\sqrt{3} \cdot \sqrt{12}$ **3.** $\sqrt{5} \cdot \sqrt{5}$
4. $\sqrt[3]{9} \cdot \sqrt[3]{3}$ **5.** $\sqrt[3]{4} \cdot \sqrt[3]{-2}$ **6.** $\sqrt[4]{8} \cdot \sqrt[4]{2}$
7. $\sqrt{a} \cdot \sqrt{4a}$ **8.** $\sqrt{\frac{1}{2}x^3} \cdot \sqrt{18x}$ **9.** $\sqrt[3]{9b^4} \cdot \sqrt[3]{-3b^2}$

Simplify each of the following radicals:

10. $\sqrt{12}$ **11.** $\sqrt{50}$ **12.** $\sqrt{72}$ **13.** $\sqrt{242}$
14. $\sqrt{75}$ **15.** $\sqrt{162}$ **16.** $\sqrt{147}$ **17.** $\sqrt{800}$
18. $\sqrt[3]{24}$ **19.** $\sqrt[3]{54}$ **20.** $\sqrt[3]{-40}$ **21.** $\sqrt[3]{-250}$
22. $\sqrt[4]{32}$ **23.** $\sqrt[4]{48}$ **24.** $\sqrt{20x^3}$ **25.** $\sqrt{45a^3b^2}$
26. $\sqrt{175xy^2}$ **27.** $\sqrt{63a^2}$ **28.** $\sqrt[3]{48x^5}$ **29.** $\sqrt[3]{81a^5b^9}$

Multiply and simplify:

30. $\sqrt{2} \cdot \sqrt{10}$ **31.** $\sqrt{7} \cdot \sqrt{21}$ **32.** $\sqrt{6} \cdot \sqrt{15}$
33. $\sqrt{35} \cdot \sqrt{45}$ **34.** $\sqrt{42} \cdot \sqrt{77}$ **35.** $\sqrt{30} \cdot \sqrt{35}$
36. $\sqrt[3]{3} \cdot \sqrt[3]{18}$ **37.** $\sqrt[3]{4} \cdot \sqrt[3]{14}$ **38.** $3\sqrt{10} \cdot 2\sqrt{15}$
39. $(2a\sqrt{5a})^2$ **40.** $4\sqrt[3]{6} \cdot 7\sqrt[3]{12}$

88. Division of Radicals. The division of radicals can be expressed in the same manner as the multiplication of radicals.

$$\sqrt{\frac{p^2}{q^2}} = \sqrt{\left(\frac{p}{q}\right)^2} = \frac{p}{q} = \frac{\sqrt{p^2}}{\sqrt{q^2}}.$$

Hence, letting $a = p^2$ and $b = q^2$, we have

$$\sqrt{\frac{a}{b}} = \frac{\sqrt{a}}{\sqrt{b}}, \text{ and } \frac{\sqrt{a}}{\sqrt{b}} = \sqrt{\frac{a}{b}}.$$

As before by a similar reduction, we see that this relationship is true for roots of any order; hence,

Rule for Division of Radicals

1. ***The quotient of two radicals of the SAME order is equal to the like root of the quotient of their radicands.***
2. ***The root of a quotient is equal to the quotient of the like roots of the dividend and divisor respectively.***

The second relation above may be used to *simplify* radicals in which the radicands are fractional expressions.

Example 1. Simplify $\sqrt{\frac{2}{3}}$.

Solution. Multiply the numerator and denominator of the fraction by the smallest number which will make the denominator a "square number." In this case we multiply by 3; thus,

$$\sqrt{\frac{2}{3}} = \sqrt{\frac{6}{9}} = \frac{\sqrt{6}}{\sqrt{9}} = \frac{\sqrt{6}}{3} = \frac{1}{3}\sqrt{6}.$$

Example 2. Simplify $\frac{2}{3}\sqrt[3]{6\frac{3}{4}}$.

Solution. For cube roots, we wish to have the denominator of the radical a perfect cube; hence, we proceed as follows:

$$\frac{2}{3}\sqrt[3]{\frac{27}{4}} = \frac{2}{3}\sqrt[3]{\frac{27 \cdot 2}{8}} = \frac{2}{3} \cdot \frac{\sqrt[3]{27}\sqrt[3]{2}}{\sqrt[3]{8}} = \frac{2}{3} \cdot \frac{3\sqrt[3]{2}}{2} = \sqrt[3]{2}.$$

Example 3. Simplify $\dfrac{6ab^2}{\sqrt{12ab^3}}$.

Solution. Multiplying numerator and denominator by $\sqrt{3ab}$ gives

$$\frac{6ab^2 \cdot \sqrt{3ab}}{\sqrt{12ab^3} \cdot \sqrt{3ab}} = \frac{6ab^2\sqrt{3ab}}{\sqrt{36a^2b^4}} = \frac{6ab^2\sqrt{3ab}}{6ab^2} = \sqrt{3ab}.$$

EXERCISE 67

Divide and evaluate:

1. $\dfrac{\sqrt{12}}{\sqrt{3}}$ 2. $\dfrac{\sqrt{72}}{\sqrt{2}}$ 3. $\dfrac{\sqrt{294}}{\sqrt{6}}$ 4. $\dfrac{\sqrt{180}}{\sqrt{5}}$

5. $\dfrac{\sqrt[3]{40}}{\sqrt[3]{5}}$ 6. $\dfrac{\sqrt[3]{108}}{\sqrt[3]{4}}$ 7. $\dfrac{\sqrt[3]{-56}}{\sqrt[3]{7}}$ 8. $\dfrac{\sqrt[4]{48}}{\sqrt[4]{3}}$

9. $\dfrac{\sqrt{8a^3b}}{\sqrt{2ab}}$ 10. $\dfrac{\sqrt{75x^7y^8}}{\sqrt{3x^3y^2}}$ 11. $\dfrac{\sqrt[3]{x^8}}{\sqrt[3]{x^2}}$ 12. $\dfrac{\sqrt[4]{a^9b^6}}{\sqrt[4]{ab^2}}$

Simplify:

13. $\sqrt{\frac{3}{5}}$ 14. $\sqrt{\frac{5}{7}}$ 15. $\sqrt{\frac{5}{12}}$ 16. $\sqrt{\frac{7}{18}}$

17. $\sqrt{2\frac{1}{2}}$ 18. $\sqrt{3\frac{5}{9}}$ 19. $\sqrt{5\frac{2}{5}}$ 20. $\sqrt{8\frac{1}{6}}$

21. $\sqrt[3]{\frac{3}{4}}$ 22. $\sqrt[3]{7\frac{1}{9}}$ 23. $\sqrt{\dfrac{3a}{2b^3}}$ 24. $\sqrt{\dfrac{28}{45n}}$

25. $\sqrt[3]{\dfrac{x}{y}}$ 26. $\sqrt[3]{\dfrac{2a^2}{3b^2}}$ 27. $\dfrac{2a}{b}\sqrt{\dfrac{b^3}{2a}}$ 28. $\dfrac{12}{x}\sqrt[3]{\dfrac{x^3}{12}}$

Divide and simplify:

29. $\dfrac{\sqrt{7}}{\sqrt{10}}$ 30. $\dfrac{\sqrt{75}}{\sqrt{8}}$ 31. $\dfrac{\sqrt{28}}{\sqrt{5}}$ 32. $\dfrac{3}{\sqrt{12}}$

33. $\dfrac{2\sqrt{3}}{5\sqrt{20}}$ 34. $\dfrac{3\sqrt{5}}{2\sqrt{18}}$ 35. $\dfrac{3\sqrt{8}}{4\sqrt{27}}$ 36. $\dfrac{5\sqrt{32}}{4\sqrt{45}}$

37. $\dfrac{3\sqrt[3]{5}}{2\sqrt[3]{9}}$ 38. $\dfrac{5\sqrt[3]{15}}{3\sqrt[3]{5}}$ 39. $\dfrac{2\sqrt[3]{54}}{3\sqrt[3]{4}}$ 40. $\dfrac{6y\sqrt{2x^3y}}{5x\sqrt{8xy^3}}$

41. $\dfrac{3ab\sqrt{20c^2}}{2c\sqrt{9a^2b}}$ 42. $\dfrac{a\sqrt[3]{5a}}{3\sqrt[3]{4a^2}}$ 43. $\dfrac{2x\sqrt{9xy}}{3\sqrt{8x}}$ 44. $\dfrac{\sqrt{ax-ay}}{\sqrt{bx-by}}$

89. Addition and Subtraction of Radicals. Radicals of the same order which have the same radicand are called *similar* or *like* radicals.

Thus, $5\sqrt{2}$, $\frac{1}{2}\sqrt{2}$, and $x\sqrt{2}$ are *similar* radicals.

But $\sqrt{2}$, $\sqrt{3}$, and $\sqrt[3]{2}$ are *dissimilar* radicals.

Similar radicals may be regarded as similar terms, and hence can be combined by addition or subtraction in accordance with the rule for combining similar terms.

Thus, $3\sqrt{5} + 4\sqrt{5} = 7\sqrt{5}$, and $8\sqrt[3]{3} - 7\sqrt[3]{3} = \sqrt[3]{3}$.

Unlike radicals *cannot* be combined by addition or subtraction unless it is possible to reduce them to like radicals.

Example 1. Combine $\sqrt{18} + \sqrt{72} - \sqrt{32}$.

Solution. Simplifying each radical before combining, we have

$$\sqrt{18} + \sqrt{72} - \sqrt{32} = 3\sqrt{2} + 6\sqrt{2} - 4\sqrt{2} = 5\sqrt{2}.$$

Example 2. Combine $\sqrt[3]{16} + \sqrt[3]{64} - \sqrt[3]{54}$.

Solution.

$$\sqrt[3]{16} + \sqrt[3]{64} - \sqrt[3]{54} = 2\sqrt[3]{2} + 4 - 3\sqrt[3]{2} = 4 - \sqrt[3]{2}.$$

Example 3. Combine $\sqrt{12} - \dfrac{1}{\sqrt{2}} - \sqrt{\tfrac{4}{3}} + \sqrt{3\tfrac{1}{8}}$.

Solution.

$$\begin{aligned}\sqrt{12} - \frac{1}{\sqrt{2}} - \sqrt{\tfrac{4}{3}} + \sqrt{3\tfrac{1}{8}} &= \sqrt{4 \cdot 3} - \frac{\sqrt{2}}{\sqrt{2}\sqrt{2}} - \sqrt{\tfrac{4}{9} \cdot 3} + \sqrt{\tfrac{25}{16} \cdot 2},\\ &= 2\sqrt{3} - \tfrac{1}{2}\sqrt{2} - \tfrac{2}{3}\sqrt{3} + \tfrac{5}{4}\sqrt{2},\\ &= \tfrac{4}{3}\sqrt{3} + \tfrac{3}{4}\sqrt{2}.\end{aligned}$$

EXERCISE 68

Combine:

1. $\sqrt{50} + \sqrt{8}$
2. $\sqrt{48} + \sqrt{27}$
3. $\sqrt{45} - \sqrt{20}$
4. $\sqrt{98} - \sqrt{72}$
5. $\sqrt{75} - \sqrt{12}$
6. $\sqrt{80} + \sqrt{45}$
7. $\sqrt{128} + \sqrt{18}$
8. $\sqrt{108} - \sqrt{48}$
9. $\sqrt{162} - \sqrt{72}$
10. $\sqrt{180} + \sqrt{125}$
11. $2\sqrt{125} - 3\sqrt{20}$
12. $5\sqrt{12} - 2\sqrt{27}$
13. $6\sqrt{28} + 2\sqrt{63}$
14. $2\sqrt{40} - 3\sqrt{90}$
15. $9\sqrt{8} - \sqrt{72}$
16. $2\sqrt{98} - 5\sqrt{18}$
17. $\sqrt{80} - \sqrt{320}$
18. $\frac{1}{3}\sqrt{288} + \frac{2}{5}\sqrt{450}$
19. $\sqrt[3]{3} + \sqrt[3]{24}$
20. $\sqrt[3]{54} - \sqrt[3]{16}$
21. $\sqrt[3]{40} + \sqrt[3]{5}$
22. $\sqrt[3]{56} + \sqrt[3]{-7}$
23. $3\sqrt[3]{162} + 2\sqrt[3]{48}$
24. $4\sqrt[3]{250} - 3\sqrt[3]{432}$
25. $3\sqrt[3]{192} - 5\sqrt[3]{24}$
26. $3\sqrt{45} - \sqrt{25} + 2\sqrt{5}$
27. $4\sqrt{63} - 5\sqrt{28} + 3\sqrt{20}$
28. $3\sqrt{98} + \frac{1}{2}\sqrt{48} + \sqrt{50}$
29. $5\sqrt{75} + 7\sqrt{108} - 6\sqrt{245}$
30. $\sqrt{\frac{1}{2}} + \sqrt{\frac{2}{3}} + \sqrt{\frac{1}{6}}$
31. $8\sqrt{\dfrac{1}{12}} - \dfrac{5}{\sqrt{75}} + 10\sqrt{\dfrac{3}{5}}$
32. $3\sqrt{8\frac{1}{3}} - 2\sqrt{5\frac{1}{3}} - 2\sqrt{1\frac{1}{3}}$
33. $\sqrt{\frac{4}{3}} - \sqrt{\frac{3}{4}} - \sqrt{\frac{1}{12}}$
34. $5\sqrt[3]{2} + 4\sqrt[3]{16} + 3\sqrt[3]{24}$
35. $2\sqrt[3]{64} - \sqrt[3]{48} - 6\sqrt[3]{\frac{1}{36}}$
36. $\sqrt[3]{8} - \sqrt[3]{-40} + \sqrt[3]{125}$

90. Multiplication of Multinomials Involving Radicals. The rules observed in the multiplication of multinomials likewise apply if the terms of the multinomials contain radicals.

Example 1. Multiply $3\sqrt{2} + 2\sqrt{6} - \sqrt{10}$ by $5\sqrt{2}$.

Solution. Multiplying each term of the trinomial by $5\sqrt{2}$, we have

$$5\sqrt{2}(3\sqrt{2} + 2\sqrt{6} - \sqrt{10}) = 15\sqrt{4} + 10\sqrt{12} - 5\sqrt{20},$$
$$= 30 + 20\sqrt{3} - 10\sqrt{5}.$$

Example 2. Multiply $\sqrt{45} - \sqrt{32}$ by $\sqrt{20} + \sqrt{18}$.

Solution. Working with radicals is often made easier by simplifying the radicals before carrying through the indicated operation. Thus, we first simplify the expressions in this example:

$$(\sqrt{45} - \sqrt{32})\ \sqrt{20} + \sqrt{18}) = (3\sqrt{5} - 4\sqrt{2})(2\sqrt{5} + 3\sqrt{2}),$$
$$= 6\sqrt{25} + 9\sqrt{10} - 8\sqrt{10} - 12\sqrt{4},$$
$$= 30 + \sqrt{10} - 24,$$
$$= 6 + \sqrt{10}.$$

Example 3. Evaluate $3x^2 - 6x - 17$ when $x = 1 + 2\sqrt{2}$.

Solution. Since $x = 1 + 2\sqrt{2}$,

$$x^2 = (1 + 2\sqrt{2})^2 = 1 + 4\sqrt{2} + 8 = 9 + 4\sqrt{2}.$$

By substitution:

$$3x^2 - 6x - 17 = 3(9 + 4\sqrt{2}) - 6(1 + 2\sqrt{2}) - 17,$$
$$= 27 + 12\sqrt{2} - 6 - 12\sqrt{2} - 17,$$
$$= 4.$$

EXERCISE 69

Multiply and simplify:

1. $\sqrt{3}(\sqrt{6} + \sqrt{12})$
2. $\sqrt{5}(\sqrt{5} + \sqrt{15})$
3. $\sqrt{2}(\sqrt{2} + \sqrt{3} - \sqrt{6})$
4. $\sqrt{6}(\sqrt{10} - \sqrt{21} - \sqrt{8})$
5. $\sqrt{6}(3\sqrt{2} - 2\sqrt{3} - \sqrt{6})$
6. $\sqrt{10}(2\sqrt{5} - 5\sqrt{2} + 3\sqrt{15})$
7. $4\sqrt{2}(5\sqrt{2} - 6\sqrt{6} - 2\sqrt{10})$
8. $3\sqrt{6}(4\sqrt{2} - 5\sqrt{3} + 3\sqrt{6})$
9. $2\sqrt{12}(3\sqrt{27} - 5\sqrt{8} - 2\sqrt{15})$
10. $5\sqrt{18}(2\sqrt{12} + 3\sqrt{10} + \sqrt{98})$
11. $(\sqrt{3} - \sqrt{2})(\sqrt{3} + \sqrt{2})$
12. $(\sqrt{5} + \sqrt{2})^2$
13. $(3\sqrt{7} + 2\sqrt{3})(2\sqrt{7} - 5\sqrt{3})$
14. $(5\sqrt{5} + \sqrt{3})(2\sqrt{5} - 9\sqrt{3})$
15. $(2\sqrt{2} + 3\sqrt{3})^2$
16. $(3\sqrt{2} - 2\sqrt{3})(3\sqrt{2} + 2\sqrt{3})$
17. $(\sqrt{12} + 2\sqrt{8})(2\sqrt{27} - \sqrt{50})$
18. $(3\sqrt{18} - \sqrt{75})(2\sqrt{8} - \sqrt{48})$

19. $(3\sqrt{72} - 2\sqrt{27})^2$

20. $(4\sqrt{50} + 5\sqrt{12})^2$

21. $(\sqrt[3]{16} + \sqrt[3]{9})(\sqrt[3]{4} - \sqrt[3]{3})$

22. $(1 + \sqrt[3]{2})(1 - \sqrt[3]{2} + \sqrt[3]{4})$

23. $(2\sqrt{3} - 3\sqrt{2})^3$

24. $(\sqrt{6} + \sqrt{10} + \sqrt{15})^2$

25. $\sqrt{x}(\sqrt{ax} - \sqrt{bx} + \sqrt{cx})$

26. $(\sqrt{x} + \sqrt{y})(\sqrt{x} - \sqrt{y})$

27. $(\sqrt{a} - \sqrt{b})^2$

28. $\sqrt{ab}(\sqrt{a} + \sqrt{b})$

29. $(\sqrt{n} - \sqrt{1 - n})^2$

30. $(\sqrt{x + y} + \sqrt{x - y})^2$

31. $(1 + \sqrt{a - 3})^3$

32. $(x\sqrt{2} - \sqrt{1 - 2x^2})^2$

Find the value of each of the following:

33. $x^2 + x + 1$, when $x = 2 + \sqrt{2}$

34. $x^2 - 3x - 7$, when $x = 1 - \sqrt{3}$

35. $x^2 - 2x - 17$, when $x = 1 - 3\sqrt{2}$

36. $x^2 + 4x + 1$, when $x = \sqrt{3} - 2$

37. $2x^2 - x + 7$, when $x = \sqrt{3} - \sqrt{2}$

38. $3x^2 - 2x - 5$, when $x = 3 + 2\sqrt{5}$

91. Division of Multinomials Involving Radicals. To divide a multinomial involving radicals by a single radical expression, we may divide term by term.

Example 1. Divide $4\sqrt{3} - 12\sqrt{2} + 6\sqrt{6}$ by $2\sqrt{6}$.
Solution. Dividing term by term gives

$$\frac{4\sqrt{3} - 12\sqrt{2} + 6\sqrt{6}}{2\sqrt{6}} = 2\sqrt{\tfrac{1}{2}} - 6\sqrt{\tfrac{1}{3}} + 3,$$

$$= \sqrt{2} - 2\sqrt{3} + 3.$$

In general, however, it is usually more convenient to multiply the dividend and the divisor by such a radical as will eliminate the radical in the divisor. This process is known as *rationalizing* the divisor. Thus, in the above example, we would multiply the dividend and divisor by $\sqrt{6}$.

$$\frac{4\sqrt{3} - 12\sqrt{2} + 6\sqrt{6}}{2\sqrt{6}} = \frac{(4\sqrt{3} - 12\sqrt{2} + 6\sqrt{6}) \cdot \sqrt{6}}{2\sqrt{6} \cdot \sqrt{6}},$$

$$= \frac{12\sqrt{2} - 24\sqrt{3} + 36}{12},$$

$$= \sqrt{2} - 2\sqrt{3} + 3.$$

If we multiply the sum and the difference of two quadratic radicals, we obtain a *rational product*. Thus,

$$(\sqrt{a} + \sqrt{b})(\sqrt{a} - \sqrt{b}) = (\sqrt{a})^2 - (\sqrt{b})^2 = a - b,$$

$$(4\sqrt{3} + 3\sqrt{5})(4\sqrt{3} - 3\sqrt{5}) = (4\sqrt{3})^2 - (3\sqrt{5})^2 = 48 - 45 = 3.$$

Hence, in the division of radicals, if the divisor is a binomial quadratic surd of the form $a\sqrt{b} + c\sqrt{d}$, we may *rationalize* the divisor by multiplying both the dividend and the divisor by $a\sqrt{b} - c\sqrt{d}$.

Example 2. Divide $\sqrt{14}$ by $2\sqrt{2} - \sqrt{7}$.

Solution. Multiplying the numerator and the denominator by $2\sqrt{2} + \sqrt{7}$, we have

$$\frac{\sqrt{14}}{2\sqrt{2} - \sqrt{7}} = \frac{\sqrt{14}(2\sqrt{2} + \sqrt{7})}{(2\sqrt{2} - \sqrt{7})(2\sqrt{2} + \sqrt{7})} = \frac{4\sqrt{7} + 7\sqrt{2}}{8 - 7} = 4\sqrt{7} + 7\sqrt{2}.$$

EXERCISE 70

Divide and simplify:

1. $\dfrac{3\sqrt{6} - 9\sqrt{15} + 12\sqrt{21}}{3\sqrt{3}}$

2. $\dfrac{15\sqrt{6} - 10\sqrt{10} - 25\sqrt{22}}{5\sqrt{2}}$

3. $\dfrac{4\sqrt{2} - 6\sqrt{3} + 2\sqrt{5}}{\sqrt{2}}$

4. $\dfrac{12\sqrt{5} + 18\sqrt{7} + 36\sqrt{11}}{2\sqrt{3}}$

5. $\dfrac{5\sqrt{5} + 10\sqrt{10} - 15\sqrt{15}}{\sqrt{30}}$

6. $\dfrac{7\sqrt{6} - 28\sqrt{10} - 21\sqrt{26}}{\sqrt{14}}$

7. $\dfrac{3\sqrt{5} + 4\sqrt{13} - 2\sqrt{19}}{2\sqrt{6}}$

8. $\dfrac{\sqrt{10} - 5\sqrt{7} - 3\sqrt{15}}{4\sqrt{5}}$

9. $\dfrac{2}{\sqrt{3} - 1}$

10. $\dfrac{4}{3 - \sqrt{5}}$

11. $\dfrac{\sqrt{3}}{\sqrt{3} + \sqrt{2}}$

12. $\dfrac{1 + \sqrt{6}}{\sqrt{2} + \sqrt{3}}$

13. $\dfrac{1 - \sqrt{10}}{\sqrt{5} + \sqrt{2}}$

14. $\dfrac{3\sqrt{7} + 12}{2\sqrt{7} - 1}$

15. $\dfrac{\sqrt{5} + 2}{\sqrt{5} - 2}$

16. $\dfrac{\sqrt{7} - \sqrt{5}}{\sqrt{7} + \sqrt{5}}$

17. $\dfrac{2\sqrt{3}}{3\sqrt{2} - 2\sqrt{3}}$

18. $\dfrac{16 - \sqrt{35}}{2\sqrt{5} + \sqrt{7}}$

19. $\dfrac{1}{\sqrt{a} + \sqrt{b}}$

20. $\dfrac{\sqrt{x} + \sqrt{y}}{\sqrt{x} - \sqrt{y}}$

Perform the indicated operation and simplify:

21. $\dfrac{4\sqrt{3} - 3\sqrt{2}}{\sqrt{6}} \div \dfrac{\sqrt{10}}{4\sqrt{3} + 3\sqrt{2}}$

22. $\dfrac{\sqrt{2}}{\sqrt{2} + 2} \cdot \dfrac{5 + 4\sqrt{2}}{\sqrt{2} + 3}$

23. $\dfrac{\sqrt{2}}{\sqrt{2}-1}+\dfrac{7\sqrt{2}}{2\sqrt{2}-1}$ **24.** $\dfrac{3\sqrt{2}+2\sqrt{3}}{\sqrt{3}-\sqrt{2}}-\dfrac{5\sqrt{2}-2\sqrt{3}}{\sqrt{3}+\sqrt{2}}$

25. $\dfrac{1}{\sqrt{3}-\sqrt{2}}\cdot\dfrac{4-\sqrt{6}}{1+\sqrt{6}}$ **26.** $\dfrac{8-2\sqrt{15}}{\sqrt{5}-\sqrt{3}}\div\dfrac{\sqrt{6}}{6+2\sqrt{15}}$

92. Radicals of Different Orders. (Optional.) In the discussion of fractional exponents we learned that $\sqrt[n]{a^m}=a^{\frac{m}{n}}$. Hence, it follows that

$$\sqrt[4]{a^2}=a^{\frac{2}{4}}=a^{\frac{1}{2}}=\sqrt{a},$$
$$\sqrt[9]{a^6}=a^{\frac{6}{9}}=a^{\frac{2}{3}}=\sqrt[3]{a^2};\text{ and so on.}$$

Thus, in general, if m, n, and r are any numbers,

$$\sqrt[nr]{a^{mr}}=a^{\frac{mr}{nr}}=a^{\frac{m}{n}}=\sqrt[n]{a^m}$$

and conversely,

$$\sqrt[n]{a^m}=\sqrt[nr]{a^{mr}}.$$

This property of radicals may sometimes be used to reduce radicals to simpler form.

Example 1. Simplify $\sqrt[4]{9}$, $\sqrt[12]{8}$, and $\sqrt[10]{10{,}000}$.

Solution. Writing each of the radicands in exponent form, we have

$$\sqrt[4]{9}=\sqrt[4]{3^2}=\sqrt{3},$$
$$\sqrt[12]{8}=\sqrt[12]{2^3}=\sqrt[4]{2},$$
$$\sqrt[10]{10{,}000}=\sqrt[10]{10^4}=\sqrt[5]{10^2}=\sqrt[5]{100}.$$

NOTE. It is sometimes convenient to think of higher-order radicals as though they were *compound* radicals; thus,

$$\sqrt[4]{a}=a^{\frac{1}{4}}=(a^{\frac{1}{2}})^{\frac{1}{2}}=\sqrt{\sqrt{a}},$$

$$\sqrt[6]{a}=a^{\frac{1}{6}}=(a^{\frac{1}{2}})^{\frac{1}{3}}\text{ or }(a^{\frac{1}{3}})^{\frac{1}{2}}=\sqrt[3]{\sqrt{a}}\text{ or }\sqrt{\sqrt[3]{a}},$$

$$\sqrt[12]{a}=\sqrt[4]{\sqrt[3]{a}},\text{ or }\sqrt[6]{\sqrt{a}},\text{ etc.}$$

Example 2. Simplify $\sqrt[6]{27}$ and $\sqrt[8]{25m^2}$.

Solution.

$$\sqrt[6]{27}=\sqrt{\sqrt[3]{27}}=\sqrt{3}\text{ and }\sqrt[8]{25m^2}=\sqrt[4]{\sqrt{25m^2}}=\sqrt[4]{5m}.$$

To multiply or divide radicals of different orders, we must first express the given radicals as radicals of some common index. This process is illustrated in the following example.

Example 3. Multiply $\sqrt{2}$ by $\sqrt[3]{4}$.

Solution. Reducing each of the given radicals to the same lowest order gives

$$\sqrt{2} = \sqrt[6]{2^3} \text{ and } \sqrt[3]{4} = \sqrt[6]{4^2}.$$

Since $4 = 2^2$, on multiplying and simplifying, we have

$$\sqrt{2} \cdot \sqrt[3]{4} = \sqrt[6]{2^3} \cdot \sqrt[6]{4^2} = \sqrt[6]{2^3 \cdot 4^2} = \sqrt[6]{2^7} = 2\sqrt[6]{2}.$$

EXERCISE 71

Reduce the order of the following radicals:

1. $\sqrt[4]{4}$ 2. $\sqrt[6]{25}$ 3. $\sqrt[6]{8}$ 4. $\sqrt[8]{49}$
5. $\sqrt[6]{81}$ 6. $\sqrt[9]{64}$ 7. $\sqrt[8]{16}$ 8. $\sqrt[12]{125}$
9. $\sqrt[10]{32}$ 10. $\sqrt[12]{144}$ 11. $\sqrt[10]{121}$ 12. $\sqrt[8]{81}$
13. $\sqrt[4]{9x^2y^2}$ 14. $\sqrt[6]{8a^3}$ 15. $\sqrt[8]{16m^4n^4}$ 16. $\sqrt[12]{64a^3b^6c^9}$

Express as radicals of the same lowest order:

17. $\sqrt{3}, \sqrt[3]{5}$ 18. $\sqrt[4]{7}, \sqrt{9}$ 19. $\sqrt[4]{2}, \sqrt[5]{3}$
20. $\sqrt[3]{4}, \sqrt[4]{3}$ 21. $\sqrt[3]{7}, \sqrt[6]{42}$ 22. $\sqrt[3]{3}, \sqrt[5]{5}$
23. $\sqrt{2}, \sqrt[4]{4}, \sqrt[8]{8}$ 24. $\sqrt[3]{2}, \sqrt[4]{3}, \sqrt[6]{5}$ 25. $\sqrt{3}, \sqrt[3]{5}, \sqrt[4]{7}$
26. $\sqrt{2a}, \sqrt[3]{3b^2}$ 27. $\sqrt[4]{x^3}, \sqrt[6]{3x^5}$ 28. $\sqrt[3]{3m}, \sqrt[4]{2n^3}$

Multiply:

29. $\sqrt{5} \cdot \sqrt[3]{3}$ 30. $\sqrt[3]{2} \cdot \sqrt[4]{3}$ 31. $\sqrt{7} \cdot \sqrt[4]{11}$
32. $\sqrt[3]{2} \cdot \sqrt[6]{6}$ 33. $\sqrt[4]{4} \cdot \sqrt[6]{5}$ 34. $\sqrt{2} \cdot \sqrt[10]{7}$
35. $\sqrt{ab} \cdot \sqrt[5]{a^4b}$ 36. $\sqrt[3]{x^2} \cdot \sqrt[7]{x^5}$ 37. $\sqrt[4]{2m^3n} \cdot \sqrt[12]{5m^5n^9}$
38. $\sqrt{3a} \cdot \sqrt[3]{9a^2}$ 39. $\sqrt[5]{a^4b^3} \cdot \sqrt[6]{a^3b^5}$ 40. $\sqrt[5]{3x^4y^2} \cdot \sqrt[10]{7x^2y^6}$

93. Imaginary Numbers. (Optional.) In the study of radicals we learned that $\sqrt{2}$ represents a number whose square is 2. Similarly, we should expect the radical $\sqrt{-2}$ to represent a number whose square is -2. However, by the rule of signs, we know that no number exists whose square is negative.

Since the square roots of negative quantities occur repeatedly in mathematics, it is of value to the applied sciences as well as to mathematics to define a new kind of number which we shall call an imaginary number.

Definition. *An indicated square root of a negative number is called an* ***imaginary number.***

Illustrations. $\sqrt{-4}$, $\sqrt{-\frac{3}{4}}$, $\sqrt{-9a^2}$ are imaginary numbers.

Further to distinguish imaginary numbers, the ordinary numbers with which we are familiar are referred to as **real numbers.**

By the laws of radicals, it is always possible to write any imaginary number, such as $\sqrt{-a}$, in the form $\sqrt{a}\sqrt{-1}$. Hence to aid in the expression of imaginary quantities, we give the following definition.

Definition. *The imaginary number* $\sqrt{-1}$ *is called the* ***imaginary unit*** *and is designated by the letter* $\boldsymbol{i}$.

Any imaginary number can be expressed in terms of the imaginary unit. Thus, for the examples given above, we have

$$\sqrt{-4} = \sqrt{4}\sqrt{-1} = 2i, \quad \sqrt{-\tfrac{3}{4}} = \sqrt{\tfrac{3}{4}}\sqrt{-1} = \frac{\sqrt{3}}{2}i,$$
$$\sqrt{-9a^2} = \sqrt{9a^2}\sqrt{-1} = 3ai.$$

From the above definition of the imaginary unit, i, it follows that $i^2 = -1$, $i^3 = i^2 \cdot i = -i$, $i^4 = i^2 \cdot i^2 = (-1)(-1) = +1$, etc.

In a similar manner we see that higher powers of the imaginary unit can be reduced to one of the four values i, -1, $-i$, $+1$ in accordance with the following table:

$$\begin{aligned} i &= i^5 = i^9 = \cdots \text{ etc.} = i, \\ i^2 &= i^6 = i^{10} = \cdots \text{ etc.} = -1, \\ i^3 &= i^7 = i^{11} = \cdots \text{ etc.} = -i, \\ i^4 &= i^8 = i^{12} = \cdots \text{ etc.} = +1. \end{aligned}$$

EXERCISE 72

Express each of the following in terms of the imaginary unit and simplify:

1. $\sqrt{-4}$ **2.** $\sqrt{-25}$ **3.** $\sqrt{-64}$ **4.** $\sqrt{-8}$
5. $\sqrt{-12}$ **6.** $2\sqrt{-9}$ **7.** $3\sqrt{-49}$ **8.** $2\sqrt{-50}$
9. $5\sqrt{-18}$ **10.** $4\sqrt{-45}$ **11.** $\sqrt{-4a^2}$ **12.** $\sqrt{-32a^2x^2}$
13. $\sqrt{-20x^4}$ **14.** $\sqrt{-9x^3}$ **15.** $\sqrt{-8a^5}$

Given the imaginary unit $i = \sqrt{-1}$, express each of the following in terms of i or a real number:

16. i^7 **17.** i^{12} **18.** i^{17} **19.** i^{25} **20.** i^{13}
21. i^{10} **22.** i^{19} **23.** i^{82} **24.** i^{57} **25.** i^{100}
26. $(-i)^3$ **27.** $(-i^3)^4$ **28.** $(-i)^6$ **29.** $(-i^7)^7$ **30.** $(-i^4)^3$

94. Equations Involving Radicals. An equation containing the unknown quantity under a radical is called a *radical equation.*

Illustrations. $\sqrt{x} = 7$, $2\sqrt[3]{x} = 5$, and $\sqrt{x^2 - 5} = x + 1$ are radical equations, whereas $x\sqrt{2} = 3$ is not.

Radicals may be eliminated from equations by means of the principle that if both members of an equation are raised to the same power, the resulting powers are equal.

Example 1. Solve $\sqrt{x-2}=7$.

Solution. Squaring both sides of this equation, we have

$$x-2=49,$$
$$x=51.$$

NOTE. Not all radical equations have solutions. For example, consider $\sqrt{x}=-5$. By squaring both sides, it appears that $x=25$ is a solution. But this value substituted for x does not satisfy the given radical equation, because $\sqrt{25}$ equals 5 and not -5.

Such apparent solutions are called *extraneous solutions* and should be rejected when solving radical equations. Hence, it is *necessary* to check all solutions of radical equations.

Example 2. Solve $\sqrt{4x-11}=2\sqrt{x}-1$.

Solution. Squaring both sides of this equation gives

$$4x-11=4x-4\sqrt{x}+1,$$
$$4\sqrt{x}=12,$$
$$\sqrt{x}=3.$$

Squaring this new equation, we have

$$x=9.$$

Hence, $x=9$ is the only *possible* solution.

Check.

$$\sqrt{4\cdot 9-11}=2\sqrt{9}-1,$$
$$\sqrt{25}=2\cdot 3-1,$$
$$5=5.$$

Thus, $x=9$ is a valid solution.

Example 3. Solve $2\sqrt{x+6}+3\sqrt{x+1}=0$.

Solution. Transposing and squaring, we have

$$2\sqrt{x+6}=-3\sqrt{x+1},$$
$$4(x+6)=9(x+1),$$
$$5x=15,$$
$$x=3.$$

Hence, $x=3$ is the only *possible* solution.

Check.

$$2\sqrt{3+6}+3\sqrt{3+1}=0,$$
$$2\cdot 3+3\cdot 2=0,$$
$$12=0.$$

Thus, $x = 3$ is extraneous and there are *no* solutions to the given radical equation.

EXERCISE 73

Solve:

1. $\sqrt{x-4} = 3$
2. $\sqrt{2x-1} = 5$
3. $4 + \sqrt{x+3} = 0$
4. $\sqrt{x^2-1} = x - 1$
5. $\sqrt{9x^2-5} = 3x - 5$
6. $\sqrt{5x+6} = -2$
7. $\sqrt{4x^2+9} = 2x + 1$
8. $\sqrt{4x^2+17} = 2x + 1$
9. $\sqrt[3]{x+2} = -3$
10. $\sqrt[3]{2x-1} = 4$
11. $x + \sqrt{x^2+1} = 1$
12. $3x + \sqrt{9x^2-8} = 2$
13. $3x + \sqrt{9x^2-5} = 1$
14. $x - \sqrt{x^2-9} = 1$
15. $\sqrt{5x-1} = 2\sqrt{x+1}$
16. $\sqrt[3]{9x-4} = 2\sqrt[3]{x-1}$
17. $\sqrt{x} + \sqrt{x+5} = 5$
18. $\sqrt{x-4} + 3 = \sqrt{x+11}$
19. $\sqrt{2x-1} + 2 = \sqrt{2x+11}$
20. $\sqrt{x-5} + \sqrt{x+10} = 3$
21. $\sqrt{x+4} + \sqrt{x+16} = 2$
22. $\sqrt{5x} + \sqrt{5x-9} = 1$
23. $\sqrt{x+31} - \sqrt{x+14} = 1$
24. $\sqrt{x+12} - \sqrt{x-12} = 2$

95. Evaluation of Quadratic Radicals: Square Root. Consider the following abbreviated table of squares:

$1^2 = 1$ $9^2 = 81$	$0.1^2 = 0.01$ $0.9^2 = 0.81$
$10^2 = 1'00$ $99^2 = 98'01$	$0.01^2 = 0.00'01$ $0.99^2 = 0.98'01$
$100^2 = 1'00'00$ $999^2 = 99'80'01$	$0.001^2 = 0.00'00'01$ $0.999^2 = 0.99'80'01$

An inspection of this table suggests the following:

Rule for Evaluation of Quadratic Radicals

If a number representing a square is separated into groups of two figures each, beginning at the decimal point, the square root of the number will have as many figures as there are groups.

Thus, the square root of 729 (= 7′29) will be a two-figured number.

Let x be the tens' digit and y the units' digit of this number. It follows that

$$729 = (10x + y)^2,$$

or

$$729 = 100x^2 + 20xy + y^2.$$

Since x and y are positive digits, it is evident in this equation that $100x^2$ cannot be greater than 700. Also, since x must be an integer, we see that $x^2 = 4$, or $x = 2$. Hence, $400 = 100x^2$, and if we subtract from the above equation,

$$329 = 20xy + y^2.$$

However, $x = 2$, so this equation becomes

$$329 = 40y + y^2,$$

or

$$329 = (40 + y)y.$$

Since y represents a digit, 40 may be taken as a trial divisor for $(40 + y)$. Hence dividing 329 by 40 we see that y might be 8. However, since $(40 + 8)8 = 384$, this value of y is too large; so we try a smaller value, $y = 7$. We now have $(40 + 7)7 = 329$ which does satisfy the above equation. Hence, since $x = 2$ and $y = 7$, the square root of 729 is 27.

In arithmetic we were taught a simplified method of computing square roots. This method was derived from the above algebraic analysis. Note the similarity between the algebraic process and the following arithmetic process.

First Step	Second Step
7′29 | 2	7′29 | 27
4	4
4 | 329	47 | 329
	329

When the number is a perfect square, this process of computation is continued until the remainder is zero.

Example 1. Find the square root of 424.36.

Solution.

$$
\begin{array}{r|l}
 & 4'24'.36 \mid \underline{20.6} \\
 & 4 \\
\underline{40} & 24 \\
 & 00 \\
\underline{406} & 2436 \\
 & \underline{2436}
\end{array}
$$

NOTE. The step involving the subtraction of zero may be and usually is omitted.

When the number is not a perfect square, this process of computation may be continued to any desired accuracy.

Example 2. Find the square root of 10.416 correct to the nearest hundredth.

Solution.

$$
\begin{array}{r|l}
 & 10'.41'60'00 \mid \underline{3.227} \\
 & 9 \\
\underline{62} & 141 \\
 & 124 \\
\underline{642} & 1760 \\
 & 1284 \\
\underline{6447} & 47600 \\
 & \underline{45129} \\
 & 2471
\end{array}
$$

Hence, the square root of 10.416 correct to the nearest hundredth is 3.23.

EXERCISE 74

Find the square roots of the following numbers:

1. 8.8804 **2.** 790321 **3.** 5012.64 **4.** 299.29
5. 0.881721 **6.** 76.0384 **7.** 11881 **8.** 4.9729
9. 0.247009 **10.** 0.029241 **11.** 0.000484 **12.** 3672.36

Find the square root of the following numbers correct to the nearest hundredth:

13. 17.06 **14.** 1.887 **15.** 23.95 **16.** 9.4061
17. 0.801 **18.** 67.85 **19.** 0.0515 **20.** 5.001
21. 20.0102 **22.** 12.3456 **23.** 16.6666 **24.** 253

96. Approximation of Square Roots to Many Places. (Optional.) If the square root of a number consists of $2n + 1$ figures, when the first $n + 1$ of these have been obtained by the ordinary method, the remaining n may be obtained by division.

Let N represent the given number; r the part of the square root already found, that is, the first $n + 1$ figures found by the arithmetic process followed by n ciphers; and x the remaining part of the root.

Hence,
$$\sqrt{N} = r + x,$$
$$N = r^2 + 2rx + x^2,$$
$$\frac{N - r^2}{2r} = x + \frac{x^2}{2r}.$$

The quantity $N - r^2$ is the remainder in the square root process after $n + 1$ figures of the root have been found, and $2r$ is the trial divisor for this remainder. From the above equation we see that $N - r^2$ divided by $2r$ gives x, the rest of the root required, increased by $\frac{x^2}{2r}$. However, $\frac{x^2}{2r}$ is a *proper fraction:* for x contains n figures, and therefore x^2 contains at most $2n$ figures; also r is a number of $2n + 1$ figures (the last n being ciphers) and hence $2r$ contains at least $2n + 1$ figures.

Therefore, neglecting the remainder arising from the division of $N - r^2$ by $2r$, we obtain x, the rest of the root.

Example 1. Find the square root of 410 to five decimal places.

Solution. Computing the square root, we have

$$\begin{array}{r|l}
 & 4'10 \,|\, \underline{20.24} \\
 & 4 \\ \hline
402 & 1000 \\
 & 804 \\ \hline
4044 & 19600 \\
 & 16176 \\ \hline
4048 & 3424
\end{array}$$

Having obtained four figures in the square root by the ordinary method, we may obtain three more by division alone. Dividing the remainder 342400 by its trial divisor 40480, we have

$$
\begin{array}{r|l}
 & 845 \\
\hline
4048 & 34240 \\
 & 32384 \\
\cline{2-2}
 & 18560 \\
 & 16192 \\
\cline{2-2}
 & 23680 \\
 & 20240 \\
\cline{2-2}
 & 3440
\end{array}
$$

Therefore, to five decimal places $\sqrt{410} = 20.24846$.

EXERCISE 75

Find the square roots of the following numbers to four decimal places:

1. 2 **2.** 3 **3.** 5 **4.** 7 **5.** 10

Find the square roots of the following numbers to five decimal places:

6. 397 **7.** 692 **8.** 131.6 **9.** 555.83 **10.** 3261
11. 65 **12.** 93.7 **13.** 9.009 **14.** 17692 **15.** 37.218

Find the square roots of the following numbers to six decimal places:

16. 29 **17.** 6 **18.** 83.67 **19.** 137 **20.** 768

97. Evaluation of Square Roots, Using Tables. The table on p. 209 lists the squares and square roots of the numbers from 1 to 100. For example, to find the *square* of 29, locate the number 29 in the column headed "Number." The value of 29^2, namely 841, appears to the right in the column headed "Square." Similarly, the *square root* of 29 is given in the second column to the right opposite 29 and is found to be 5.385, correct to the nearest thousandth.

Observe also that any number in the column headed "Square" has its square root in the column headed "Number." Thus, the square root of 6724 is 82.

In evaluating square roots it is often necessary to modify or simplify the radicand in order that it may be found in the table. The following examples indicate some of the devices employed.

Example 1. Evaluate $\sqrt{140}$, using tables.

Solution. Although the number 140 is not in the table, we may first simplify the radical. Thus,

$$\sqrt{140} = \sqrt{4 \cdot 35} = 2\sqrt{35} = 2(5.916) = 11.832.$$

TABLE OF SQUARES AND SQUARE ROOTS

Number	Square	Square Root	Number	Square	Square Root
1	1	1.000	51	2601	7.141
2	4	1.414	52	2704	7.211
3	9	1.732	53	2809	7.280
4	16	2.000	54	2916	7.348
5	25	2.236	55	3025	7.416
6	36	2.449	56	3136	7.483
7	49	2.646	57	3249	7.550
8	64	2.828	58	3364	7.616
9	81	3.000	59	3481	7.681
10	100	3.162	60	3600	7.746
11	121	3.317	61	3721	7.810
12	144	3.464	62	3844	7.874
13	169	3.606	63	3969	7.937
14	196	3.742	64	4096	8.000
15	225	3.873	65	4225	8.062
16	256	4.000	66	4356	8.124
17	289	4.123	67	4489	8.185
18	324	4.243	68	4624	8.246
19	361	4.359	69	4761	8.307
20	400	4.472	70	4900	8.367
21	441	4.583	71	5041	8.426
22	484	4.690	72	5184	8.485
23	529	4.796	73	5329	8.544
24	576	4.899	74	5476	8.602
25	625	5.000	75	5625	8.660
26	676	5.099	76	5776	8.718
27	729	5.196	77	5929	8.775
28	784	5.292	78	6084	8.832
29	841	5.385	79	6241	8.888
30	900	5.477	80	6400	8.944
31	961	5.568	81	6561	9.000
32	1024	5.657	82	6724	9.055
33	1089	5.745	83	6889	9.110
34	1156	5.831	84	7056	9.165
35	1225	5.916	85	7225	9.220
36	1296	6.000	86	7396	9.274
37	1369	6.083	87	7569	9.327
38	1444	6.164	88	7744	9.381
39	1521	6.245	89	7921	9.434
40	1600	6.325	90	8100	9.487
41	1681	6.403	91	8281	9.539
42	1764	6.481	92	8464	9.592
43	1849	6.557	93	8649	9.644
44	1936	6.633	94	8836	9.695
45	2025	6.708	95	9025	9.747
46	2116	6.782	96	9216	9.798
47	2209	6.856	97	9409	9.849
48	2304	6.928	98	9604	9.899
49	2401	7.000	99	9801	9.950
50	2500	7.071	100	10000	10.000

Example 2. Evaluate $\sqrt{2\frac{1}{2}}$, using tables.

Solution. We have $\sqrt{2\frac{1}{2}} = \sqrt{\frac{5}{2}}$; and although both $\sqrt{5}$ and $\sqrt{2}$ are in the table, it is advisable to simplify further. Thus,

$$\sqrt{2\tfrac{1}{2}} = \sqrt{\tfrac{5}{2}} = \sqrt{\tfrac{10}{4}} = \tfrac{1}{2}\sqrt{10} = \tfrac{1}{2}(3.162) = 1.581.$$

Example 3. Evaluate $\sqrt{1.71}$, using tables.

Solution. First removing decimals, we have

$$\sqrt{1.71} = \sqrt{\tfrac{171}{100}} = \tfrac{1}{10}\sqrt{171} = \tfrac{1}{10}\sqrt{9 \cdot 19} = \tfrac{3}{10}\sqrt{19} = 0.3(4.359) = 1.3077.$$

Example 4. Evaluate $\dfrac{1}{\sqrt{2} + 1}$, using tables.

Solution. Although the $\sqrt{2}$ can be found in the tables, it is advisable to simplify first. Thus,

$$\frac{2}{\sqrt{2} + 1} = \frac{2(\sqrt{2} - 1)}{(\sqrt{2} + 1)(\sqrt{2} - 1)} = \frac{2(\sqrt{2} - 1)}{2 - 1} = 2(1.414 - 1) = 0.828.$$

EXERCISE 76

Find the square roots of the following numbers in the table:

1. 19	2. 94	3. 66	4. 41	5. 87
6. 529	7. 9216	8. 3364	9. 1936	10. 5776

Using the table, find the square roots of the following numbers:

11. 228	12. 369	13. 175	14. 147	15. 605
16. $3\frac{2}{5}$	17. $2\frac{3}{4}$	18. $20\frac{4}{7}$	19. $34\frac{2}{3}$	20. $14\frac{2}{5}$
21. 1.8	22. 8.91	23. 3.25	24. 2.94	25. 0.608

Simplify and evaluate, using the table:

26. $\dfrac{\sqrt{5}}{\sqrt{3} - \sqrt{2}}$ 27. $\dfrac{3 + \sqrt{5}}{3 - \sqrt{5}}$ 28. $\dfrac{7}{2\sqrt{2} + 1}$

29. $\dfrac{5(2 + \sqrt{6})}{\sqrt{2} + 2\sqrt{3}}$ 30. $\dfrac{1 + \sqrt{\frac{1}{2}}}{1 + \sqrt{\frac{1}{3}}}$

98. Applications of Square Root. One of the most important relations of geometry was first proved by the Greek mathematician Pythagoras about 550 B.C. It is as follows:

Pythagorean Relation

In any right triangle the square on the hypotenuse is equal to the sum of the squares on the other two sides.

If a and b represent the length of the sides of a right triangle and c represents the length of the hypotenuse, this relation states that

$$c^2 = a^2 + b^2.$$

From this equation it is also evident that

$$a^2 = c^2 - b^2, \quad \text{or} \quad b^2 = c^2 - a^2.$$

Example 1. Find the hypotenuse of a right triangle whose sides are 24 feet and 45 feet respectively.

Solution. Let c represent the length of the hypotenuse; then by the Pythagorean relation, we have

$$c^2 = 24^2 + 45^2 = 576 + 2025 = 2601.$$

Extracting the square root, we find $c = \sqrt{2601} = 51$.
Hence, the hypotenuse is 51 feet long.

Example 2. A ladder 30 feet long just reaches the lower edge of a window when the foot of the ladder is 9 feet from the wall. Find, to the nearest tenth of a foot, how high the window is above the ground.

Solution. From the Pythagorean relation,

$$a = \sqrt{30^2 - 9^2} = \sqrt{(30+9)(30-9)} = \sqrt{39 \cdot 21} = \sqrt{3 \cdot 13 \cdot 3 \cdot 7} = 3\sqrt{91}.$$

From the table of square roots,

$$a = 3(9.539) = 28.617.$$

Hence, the window is 28.6 feet above the ground.

EXERCISE 77

In the following exercises a, b, and c represent the sides of a right triangle; c is the hypotenuse. Find the missing side:

1. $a = 6, b = 8$ **2.** $a = 30, c = 34$ **3.** $b = 10, c = 26$
4. $a = 20, b = 21$ **5.** $b = 36, c = 45$ **6.** $a = 40, c = 85$

7. Find, to the nearest tenth of an inch, the diagonal of a square whose area is 389 square inches.

8. Find, to the nearest tenth of an inch, the diameter of a circle whose area is 99 square inches. (Take $\pi = \frac{22}{7}$.)

9. A rope attached to the top of a flagpole is 112 feet long. When the rope is pulled taut, the loose end reaches the ground at a point 15 feet from the foot of the pole. How high is the flagpole?

10. Two vertical poles 20 feet and 30 feet high respectively are 60 feet apart. A rope is to be stretched from the top of one pole to a point on the ground halfway between the poles, and then to the top of the other pole. How long a rope is needed?

11. A telephone pole 32 feet high stands 100 feet away from a building 65 feet high. How long a wire is needed to stretch from the top of the pole to the top of the building?

12. If $m^2 + n^2$ represents the hypotenuse of a right triangle and $m^2 - n^2$ represents one of the sides, what expression in terms of m and n represents the third side of the triangle?

REVIEW OF CHAPTER VIII

A

Simplify the following radical expressions:

1. $\sqrt{242}$ **2.** $\sqrt{108}$ **3.** $\sqrt{12a^3}$

4. $6\sqrt{200}$ **5.** $\sqrt{98(x-y)^2}$ **6.** $3m^2n\sqrt{180m^5n^3}$

7. $\sqrt{(a^2-b^2)(a+b)}$ **8.** $5\sqrt[3]{-96}$ **9.** $\sqrt[3]{64a^5b^3c^2}$

10. $\sqrt{\dfrac{7a^3}{18d^5}}$ **11.** $6\sqrt{5\frac{1}{3}}$ **12.** $\sqrt{\dfrac{4ax}{27by}}$

13. $\dfrac{12}{7\sqrt{3}}$ **14.** $\sqrt[3]{13\frac{1}{2}}$ **15.** $\sqrt[3]{\dfrac{8a^4}{3x^2}}$

Combine:

16. $3\sqrt{50} - 2\sqrt{18}$ **17.** $x\sqrt{8} + 4\sqrt{2x^2}$ **18.** $\sqrt{150} - 6\sqrt{2\frac{2}{3}}$

19. $2\sqrt[3]{16} + \sqrt[3]{-54}$ **20.** $\sqrt[3]{\frac{1}{2}} - \sqrt[3]{\frac{4}{27}}$ **21.** $\sqrt{2a^2} - \sqrt{2b^2}$

22. $\sqrt{3\frac{1}{5}} + \sqrt{1\frac{1}{2}} - \sqrt{11\frac{1}{4}}$ **23.** $3\sqrt{245} - 3\sqrt{338} - 2\sqrt{125}$

24. $\sqrt{72} + \dfrac{\sqrt{243}}{3} - \dfrac{8}{\sqrt{2}} + 3\sqrt{\frac{1}{3}}$ **25.** $\sqrt[3]{1\frac{1}{3}} + \sqrt[3]{16} + \sqrt[3]{4\frac{1}{2}}$

26. $\sqrt{a^2b} - \sqrt{b^3} + \sqrt{bc^2}$ **27.** $\sqrt{a^3 - a^2b} - \sqrt{ab^2 - b^3}$

28. $\sqrt{xy^2} + \sqrt{x^3y^4} + \sqrt{x^5y^6}$ **29.** $\sqrt[3]{p^7} + \sqrt[3]{8p^4q^3} + \sqrt[3]{pq^6}$

30. $\sqrt{\dfrac{yz}{x}} + \sqrt{\dfrac{zx}{y}} + \sqrt{\dfrac{xy}{z}}$ **31.** $\sqrt{\dfrac{a}{b} + \dfrac{b}{a} + 2} - \sqrt{\dfrac{b}{a}}$

32. $\sqrt{ax^2 + 2ax + a} + \sqrt{ax^2 - 2ax + a}$

33. $\sqrt{(x+2)(x^2-4)} - \sqrt{x^3 - 2x^2}$

B

Multiply:

1. $4\sqrt{\frac{3}{2}} \cdot 3\sqrt{\frac{3}{8}}$ **2.** $3\sqrt{27} \cdot 2\sqrt{75}$

3. $4\sqrt[3]{12} \cdot 5\sqrt[3]{18}$ **4.** $\frac{2}{5}\sqrt[3]{\frac{9}{16}} \cdot \frac{2}{9}\sqrt[3]{\frac{3}{4}}$

5. $2\sqrt{27} \cdot 3\sqrt{21} \cdot 5\sqrt{28}$

6. $\frac{1}{2}\sqrt{\frac{4}{5}} \cdot \frac{2}{3}\sqrt{\frac{5}{8}} \cdot \frac{3}{5}\sqrt{\frac{1}{3}}$

7. $2\sqrt{3}(2\sqrt{15} - 3\sqrt{12} + 4\sqrt{21})$

8. $\sqrt{abc}(\sqrt{ab} + \sqrt{bc} + \sqrt{ac})$

9. $(\sqrt{2} + \sqrt{5})(\sqrt{2} + 3\sqrt{5})$

10. $(3\sqrt{5} - 2\sqrt{10})^2$

11. $(\sqrt[3]{2} + \sqrt[3]{4})(\sqrt[3]{2} - \sqrt[3]{4})$

12. $(\sqrt{7} - 1)^2 + 2(\sqrt{7} - 1) + 5$

Divide:

13. $4\sqrt{51} \div \sqrt{34}$

14. $5\sqrt{228} \div 2\sqrt{19}$

15. $18\sqrt[3]{13} \div 3\sqrt[3]{26}$

16. $10\sqrt[3]{15} \div 5\sqrt[3]{40}$

17. $(8\sqrt{35} - 2\sqrt{15} - 4\sqrt{20}) \div 2\sqrt{5}$

18. $(4\sqrt{18} + 3\sqrt{60} - 9\sqrt{2}) \div 3\sqrt{12}$

19. $\dfrac{6}{\sqrt{7} - 2}$

20. $\dfrac{5\sqrt{5}}{4\sqrt{2} - 3\sqrt{3}}$

21. $\dfrac{\sqrt{27} - \sqrt{18}}{\sqrt{75} - \sqrt{72}}$

22. $\dfrac{a - b}{\sqrt{a} - \sqrt{b}}$

23. $\dfrac{2x + 3y + 5\sqrt{xy}}{\sqrt{x} + \sqrt{y}}$

24. $\dfrac{a + \sqrt{ab}}{b + \sqrt{ab}}$

C

Find the numerical values of the following:

1. $9^{\frac{1}{2}} \cdot 3^{-1}$

2. $8^{\frac{2}{3}} \cdot (4x)^0$

3. $(\frac{2}{3})^{-2} + (\frac{1}{64})^{\frac{1}{3}}$

4. $(2x^3)^0 - (-2x^0)^3$

5. $(-\frac{1}{8})^{-\frac{2}{3}}$

6. $2^0 + 2^{-1} + 2^{-2}$

7. $27^{-\frac{2}{3}} \cdot 9^{\frac{3}{2}}$

8. $2^{-35} \cdot 2^{37}$

9. $(\frac{1}{4})^{\frac{1}{2}} + (4)^{-\frac{1}{2}}$

10. $16^{\frac{3}{4}}(8^{-\frac{2}{3}} + 4^{-\frac{3}{2}})$

11. $(1^0 + 2^0 + 3^0 + 4^0)^{-\frac{1}{2}}$

12. $(3^{-1} + 3^{-2})^{\frac{1}{2}}$

Perform the indicated operations, using the laws of exponents:

13. $a^4 \cdot a^{-1} \cdot a^{-2}$

14. $(x^{-2})^{-2} \cdot (x^2)^{-1}$

15. $5ax^{\frac{1}{2}} \cdot 7a^{\frac{1}{3}}x^{-\frac{1}{3}}$

16. $(xy^{\frac{1}{2}})^3 \cdot (x^{\frac{1}{2}}y)^4$

17. $\left(\dfrac{m^{-2}}{n}\right)^{-1} \cdot \left(\dfrac{m}{n^{-2}}\right)^2$

18. $\dfrac{9b^2x^{\frac{1}{2}}}{6bx^{-\frac{1}{2}}}$

19. $\dfrac{(ab^2)^{\frac{1}{2}}}{(a^2b^3)^{-1}}$

20. $\dfrac{3ax^{-1} \cdot 4a^2x^{\frac{1}{2}}}{6a^{-1}x^{-2}}$

21. $\dfrac{(2x^{\frac{1}{2}}y^{\frac{1}{3}})^6}{8x^{-1}y}$

Write with positive exponents, and simplify:

22. $(a^{\frac{1}{2}} - a^{-\frac{1}{2}})^2$

23. $(x^{-1} + y^{-1})(x + y)^{-1}$

24. $\left(\dfrac{m^{\frac{1}{2}}}{n^{\frac{1}{2}}} - \dfrac{n^{\frac{1}{2}}}{m^{\frac{1}{2}}}\right) \cdot \dfrac{1}{n^{-1} - m^{-1}}$

25. $(1 - x^{-1})^{-1}$

26. $\dfrac{ab^{-1} - a^{-1}b}{a^{-1} - b^{-1}}$

27. $\dfrac{x - y^{-1}z^2}{xyz^{-1} - z}$

Perform the indicated operations, and simplify:

28. $a^n \cdot a^{-\frac{n}{2}} \cdot a^{1-\frac{n}{2}}$

29. $x^{2n-1}y^n \cdot x^{3-n}y^{1-n}$

30. $(a^kb^2)^{-3} \cdot (a^{2-2k}b^{2k})^{\frac{1}{2}}$

D

(Optional)

Multiply:

1. $\sqrt[3]{9}\cdot\sqrt{6}$
2. $\sqrt{\frac{3}{2}}\cdot\sqrt[3]{\frac{4}{3}}$
3. $\sqrt[3]{9}\cdot\sqrt[4]{27}$
4. $\sqrt{18}\cdot\sqrt[4]{8}$
5. $\sqrt{\frac{1}{2}}\cdot\sqrt[3]{\frac{3}{4}}\cdot\sqrt[6]{\frac{8}{9}}$
6. $\sqrt{\frac{1}{2}}\cdot\sqrt[4]{\frac{2}{3}}\cdot\sqrt[8]{\frac{9}{64}}$
7. $(\sqrt{2}+\sqrt[3]{2})(\sqrt{2}-\sqrt[3]{2})$
8. $(\sqrt{2}+\sqrt[4]{6})(\sqrt{3}-\sqrt[4]{6})$
9. $\sqrt[4]{a^3b^2}\cdot\sqrt[5]{a^3b^4}$
10. $\sqrt[3]{4x^2y^2}\cdot\sqrt[4]{8x^3y^3}$
11. $\sqrt[3]{9m^2n^2}\div\sqrt{3mn}$
12. $2a^2b\sqrt[3]{3ab}\div 3b\sqrt[6]{9a^3b}$

E

Solve:

1. $x+\sqrt{x^2+7}=1$
2. $4+\sqrt{2x-1}=0$
3. $2+\sqrt[3]{x-4}=0$
4. $\sqrt[3]{x^3+3x^2}=x+1$
5. $1+\sqrt{x}=\sqrt{x+3}$
6. $2x-\sqrt{4x^2-11}=1$
7. $\sqrt{4x+1}=3-2\sqrt{x}$
8. $\sqrt{x-4}+\sqrt{x+4}=2$
9. $\sqrt{9x+40}=3\sqrt{x}+2$
10. $2+\sqrt{x+3}=3+\sqrt{x+2}$
11. $\sqrt{x+5}-1=\dfrac{x}{\sqrt{x+5}}$
12. $\sqrt{x-3}+\sqrt{x}=\dfrac{6}{\sqrt{x}}$

F

1. Find the hypotenuse of a right triangle whose legs are 28 and 45 respectively.
2. Find the smallest whole number which must be added to 4731 to make a square number.
3. The area of a sphere is given by the formula $S = 4\pi r^2$. Find the radius of a sphere whose area is 20 square inches.
4. A city park is a half mile long and 350 yards wide. How much distance is saved by walking diagonally through the park rather than along two sides?
5. A ladder 20 feet long just reaches a window when the base is 4 feet from the wall. How high above the ground is the window?
6. Two buildings are 22 feet apart. A ladder 25 feet long is placed with its base on the ground between the two buildings. If it reaches a window 20 feet from the ground on one building, how high will it reach on the other building if its base remains at the same point?
7. A gridiron is 100 yards long and 160 feet wide. How far is it from one corner to the opposite corner?

8. In the formula $V = \pi r^2 h$, what is the value of r when $h = 6$ and $V = 231$?

9. A boat leaves a dock and travels eastward for 38 miles, southward for 36 miles, and then eastward for 39 miles more. How far is the boat from the dock?

10. A ladder 26 feet long is placed against a wall and it reaches a point 24 feet from the ground. If its base were placed only half as far from the wall, how much higher would the ladder reach up the wall?

Candlelight

Although a certain three candles are of the same length, it is known that they will burn for different lengths of time. After being lighted, it is observed, at intervals of 2, $2\frac{1}{2}$, and 3 hours, that the remainder of one candle is twice as long as the remainder of one of the others. How many hours will each candle burn?

A Square Trick

Find the simplified numerical value of

$$\sqrt{7 + \sqrt{48}} + \sqrt{7 - \sqrt{48}}.$$

Archimedes' Principle

If 19 ounces of gold weigh but 18 ounces when submerged in water, and 10 ounces of silver then weigh 9 ounces, how many ounces of silver and gold are there in a mass of an alloy of the two metals which weighs 387 ounces in air and 351 ounces in water?

Squares Squared

Show that the product $(a^2 + b^2)(c^2 + d^2)$ can be written as the sum of two squares.

CHAPTER IX

GRAPHICAL METHODS

99. Constants and Variables. A general number is called a *constant*, if it is assumed that its value does not change in the course of a problem. Constants themselves are of two kinds: (1) *absolute constants*, such as 8, $\frac{1}{3}$, -4, π, each of which has the same value in any problem; and (2) *arbitrary constants*, each of which retains its value without change throughout a problem, but which may change from one problem to another.

A general number which may assume different values in the course of a problem is called a *variable*. It may happen, however, that a variable does not change its value. For example, during a certain period of time we should expect the population of a town to vary, but it is entirely possible that it may remain the same. Although the population in this case remains constant, we should consider it a variable since initially we had no way of knowing that it would remain constant.

Example 1. The formula,

$$V = \tfrac{4}{3}\pi r^3,$$

gives the volume of a sphere in terms of its radius. In it, $\frac{4}{3}$ and π are absolute constants. V and r would be considered variables, because the formula is valid for any value of r and there is no assumption that either r or V must maintain a fixed value.

Example 2. The formula,

$$V = \tfrac{1}{3}hs^2,$$

gives the volume of a pyramid having a height h and a square base, each side of which is represented by s. In this formula V, h, and s are variables. However, if we are interested only in pyramids having a fixed base, then s will be an arbitrary constant and only V and h will be considered variables.

100. Functions. By means of the formula,

$$A = \pi r^2,$$

we can find the area, A, of a circle when we know its radius, r. This dependence of A on r is expressed mathematically by saying that A is a *function* of r.

Definition. *If a variable* y *depends upon a variable* x *in such a way that when* x *assumes a value we can obtain at least one value for* y, *then* y *is said to be a function of* x.

Since the implication of this definition is that y depends on x, we call y the *dependent variable* and x the *independent variable.*

The definition of a function does not state explicitly in what way we find the value of the function when we know the value of the independent variable; it merely asserts that it can be found. In general, these functional values are usually determined in one of two ways: either (1) from an equation or formula, or (2) from a statistically determined table of values.

Example 1. Suppose that the variable y is a function of the variable x as expressed by the equation

$$y = x^2 + 2x - 3.$$

To find the value of the function y when x assumes any value whatsoever, say $x = -5$, we merely substitute the given value in the equation for x:

$$\begin{aligned} y &= (-5)^2 + 2(-5) - 3, \\ &= 25 - 10 - 3 = 12. \end{aligned}$$

Example 2. The following table based on data compiled by the United States government gives the average heights of boys according to their age.

Age, years	1	2	3	4	5	6	7	8	9	10	11	12	13	14	15	16
Height, inches	29.4	33.8	37.1	39.5	41.6	43.8	45.7	47.8	49.7	51.7	53.3	55.1	57.2	59.9	62.3	65.0

This table was made by making a great many measurements and averaging them. Since we can find the value of H, the average height, when we know A, the age, it follows that H is a function of A. Observe, however, that there is no formula for H in terms of A.

101. Statistical Graphs. When statistical data showing the functional relationship between two variables are given by means of a table of values, it is not always easy to understand the significance of the data as a whole. To aid us in this respect, we may picture the data by making a drawing that represents the numbers in the table. Such a drawing is called a **graph.**

Fig. 2 is a graphic portrayal of the number of men and women graduating from colleges and universities in the United States.

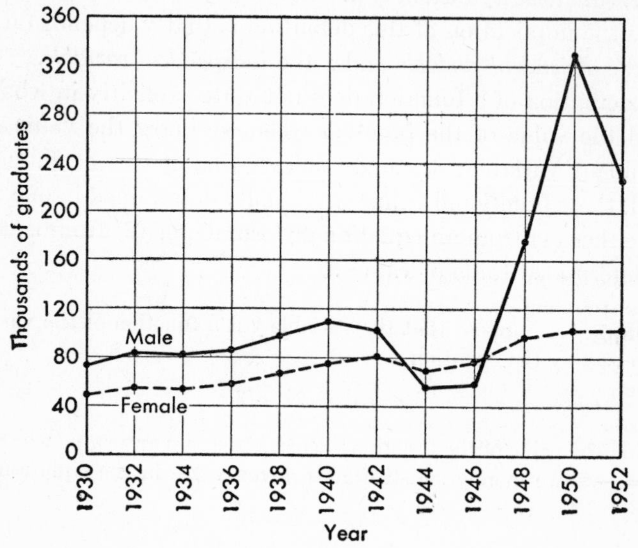

FIG. 2

College Graduates—Number, by Sex

Year Graduated	Male	Female	Year Graduated	Male	Female
1930	73,615	48,869	1942	103,889	81,457
1932	83,271	54,792	1944	55,876	69,999
1934	82,341	53,815	1946	58,664	77,510
1936	86,067	57,058	1948	175,456	95,563
1938	97,678	67,265	1950	329,819	103,915
1940	109,546	76,954	1952	227,029	104,895

Rules for Making Statistical Graphs

1. On a sheet of ruled paper draw two heavy lines, called **axes**, at right angles to each other.
2. Mark off equal divisions on each axis, and number them in succession. A suitable scale should be chosen on each axis so that the graph will fall on the paper.
3. Each pair of values in the table determines a point on the paper. Plot each pair of numbers in the table and connect the succession of points thus obtained by straight lines.

NOTE. There is no absolute rule as to which quantity should be plotted on the vertical axis and which quantity should be plotted on the horizontal axis. However, if one of the varying quantities increases or decreases by uniform amounts, it should usually be plotted on the horizontal axis.

EXERCISE 78

1. For the age group from 5 to 17 years inclusive, the percentages of persons attending school are given in the following table. Plot with respect to time.

Year	1870	'80	'90	1900	'10	'20	'30	'40	'44	'48
Percent	61.5	65.5	68.6	72.4	73.5	77.8	81.3	85.3	80.4	79.4

2. The annual fire losses in the United States are given in the following table. Plot with respect to time.

Year	1927	'29	'31	'33	'35	'37	'39	'41	'43	'45	'47	'49	'51
Millions of dollars	473	459	452	271	235	255	275	304	373	484	648	668	698

3. The wholesale price index for all commodities is indicated in the following table. The index number 100 is chosen to represent the average prices in 1926. Graph the numbers with respect to time and discuss the extremes.

Year	1860	'64	'70	'80	'90	1900	'10	'20	'30	'32	'40	'50
Index no.	61	134	86	65	56	56	70	154	80	65	79	160

4. The following table gives the average weights of boys according to their age. Graph weight as a function of age.

Age, years	1	2	3	4	5	6	7	8	9	10	11	12	13	14	15	16
Weight, pounds	21.9	27.1	32.2	35.9	41.1	45.2	49.1	53.9	59.2	65.3	70.2	76.9	84.4	94.9	107.1	121.0

5. Using the data in Example 2 on p. 217, graph the average height of boys as a function of their age.

6. The following table gives the automotive fatalities in the United States to the nearest thousand. Graph deaths as a function of the year.

Year	1925	'27	'29	'31	'33	'35	'37	'39	'41	'43	'45	'47	'49	'51
Deaths	22	26	31	33	31	34	37	30	38	22	26	30	32	35

7. The monthly precipitation in inches for Seattle and Denver is given in the following table. Plot each with respect to time on the same chart, and compare.

Month	Jan.	Feb.	Mar.	Apr.	May	June	July	Aug.	Sept.	Oct.	Nov.	Dec.
Seattle	4.9	3.9	3.0	2.4	1.9	1.3	0.6	0.7	1.8	2.8	5.0	5.6
Denver	0.4	0.5	1.0	2.1	2.2	1.4	1.7	1.4	1.0	1.0	0.6	0.7

8. The birth and death rates in the United States are given in the following table. Graph each with respect to time on the same chart. (The rates in the table are per 1000 population.)

Year	1931	'33	'35	'37	'39	'41	'43	'45	'47	'49	'51
Birth rate	18.0	16.5	16.9	17.1	17.3	18.9	21.5	19.6	25.8	24.0	23.1
Death rate	11.1	10.7	10.9	11.3	10.6	10.5	10.9	10.6	10.1	9.7	9.6

9. The monthly mean temperatures (Fahrenheit) for New York City, Chicago, and Los Angeles are given in the following table. Plot each with respect to time on the same chart, and compare.

Month	Jan.	Feb.	Mar.	Apr.	May	June	July	Aug.	Sept.	Oct.	Nov.	Dec.
N.Y.C.	31	31	38	49	61	69	74	73	67	56	44	35
Chicago	24	26	35	47	58	67	72	72	65	54	40	29
L.A.	55	56	58	59	62	66	70	71	69	65	61	57

10. The death rates in the United States per 100,000 population for various causes are given in the following table. Plot each with respect to time on the same chart, and discuss.

Year	1900	'05	'10	'15	'20	'25	'30	'35	'40	'45	'50
Cancer	64.0	72.1	76.2	81.1	83.4	92.6	97.4	107.9	120.0	134.5	139.1
Cirrhosis	12.5	14.8	13.3	12.6	7.1	7.3	7.2	7.9	8.6	9.5	9.2
Diabetes	11.0	13.0	15.3	17.5	16.1	16.9	19.1	22.2	26.5	26.6	16.9
Diphtheria	40.3	23.8	21.1	15.7	15.3	7.8	4.9	3.1	1.1	1.2	0.1

102. A Coordinate System. In somewhat the same manner as we have pictured statistical data, it is possible also to picture mathematical formulas. To attain this end, we form what is called a *system of coordinates.* First draw a horizontal straight line XX' and a vertical straight line YY'. These lines divide the plane into four quadrants numbered as shown in Fig. 3. Their point of intersection (O) is called the origin, XX' is called the x *axis*, and YY' is called the y *axis.* The position of any point in the plane may be described by giving its distances from these two axes.

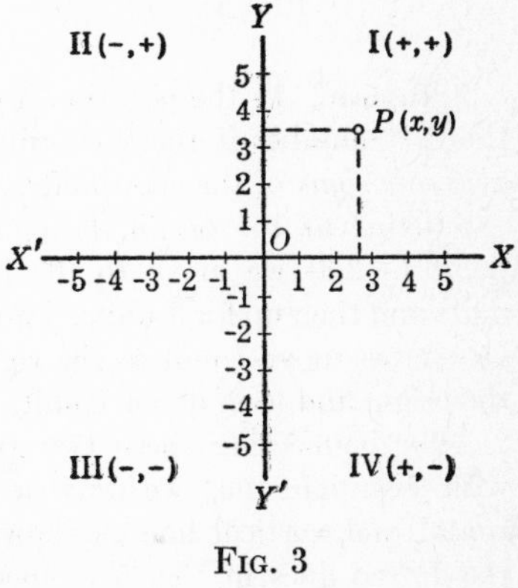

FIG. 3

To measure such distances, we arbitrarily choose some unit of length and agree that horizontal distances measured to the *right* of the y axis shall be *positive* and those to the *left* shall be *negative.* Similarly, vertical distances measured *upward* from the x axis, we agree, shall be *positive*, and those measured *downward* shall be *negative.*

Let P be any point on the plane, and let its horizontal distance from the y axis be x and its vertical distance from the x axis be y. Then x is said to be the *abscissa* or the x *coordinate* of the point P, and y is said to be the *ordinate* or the y *coordinate* of the point P. Together these values are called the *coordinates* of the point P and we write them in the form (x,y). The position of a point is determined if its coordinates are known. Conversely, if the position of a point is known, its coordinates can be found by measuring its distance from the two axes.

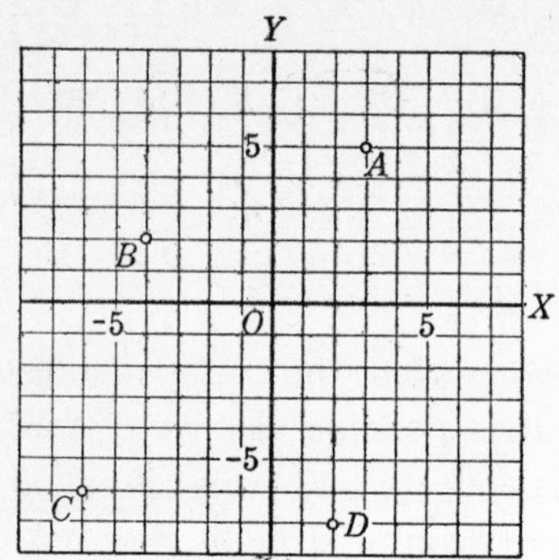

FIG. 4

The process of locating points by means of their coordinates is called *plotting* the points. Thus, in Fig. 4 we have *plotted* the points $A(3,5)$, $B(-4,2)$, $C(-6,-6)$, and $D(2,-7)$. Observe, as is indicated in Fig. 3, the character of the algebraic signs of the coordinates in each of the four quadrants.

Example 1. Draw a line connecting points $A(-2,3)$ and $B(1,-1)$, and find its length.

Solution. In the notation used for representing points, remember that the first number is the x coordinate, and pay particular attention to the *algebraic signs* of the coordinates.

Starting at the origin, to plot the point $(-2,3)$ we move to the left 2 units and then up for 3 units. To plot $(1,-1)$ we move 1 unit to the right of the origin and then down 1 unit.

After connecting these two points with a straight line, we draw a horizontal and vertical line as shown by the dotted lines in Fig. 5. Since the triangle thus formed is a right triangle, we can find the length of the hypotenuse by using the Pythagorean relation. Observe that the lengths of the dotted lines are easily found to be 3 and 4 respectively. Thus,

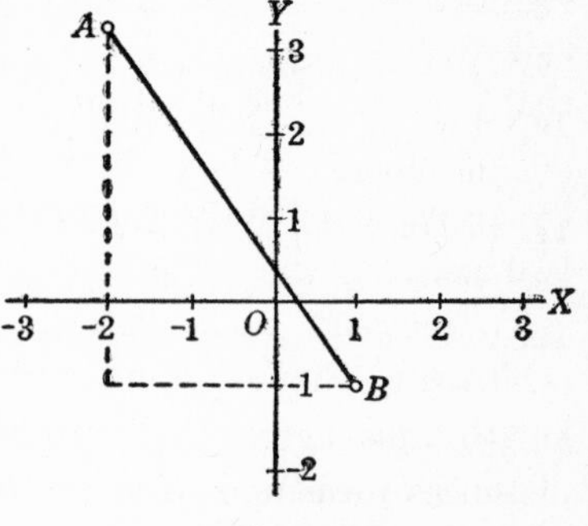

FIG. 5

$$AB = \sqrt{3^2 + 4^2} = 5.$$

EXERCISE 79

1. On squared paper draw a system of coordinates and locate the points whose coordinates are as follows: $A(-2,1)$, $B(3,4)$, $C(0,-3)$, $D(-5,-1)$, $E(5,0)$, $F(1\frac{1}{2},-2\frac{1}{2})$.
2. In Fig. 6, what are the coordinates of points A, B, C, D, E, and F?
3. Plot the points $A(3,6)$, $B(3,1)$, $C(-2,-2)$, $D(3,-2)$. How long are the line segments AB and CD?

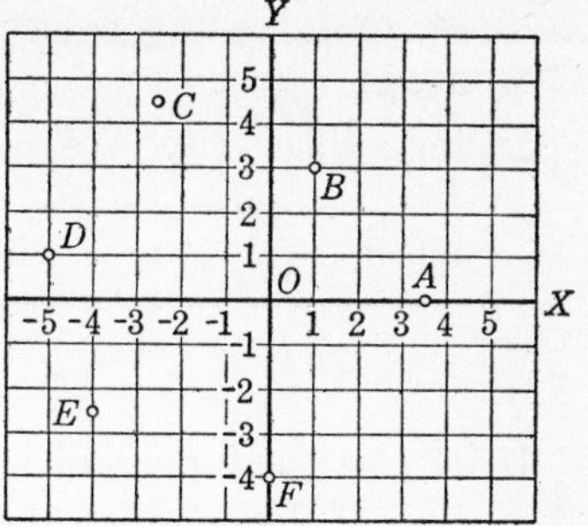

FIG. 6

4. (*a*) Plot the points $A(-1,-1)$, $B(5,-1)$, $C(5,3)$, $D(-1,3)$. (*b*) Connect these points in succession. What kind of figure is $ABCD$? (*c*) Draw the diagonals AC and BD. What are the coordinates of their point of intersection?

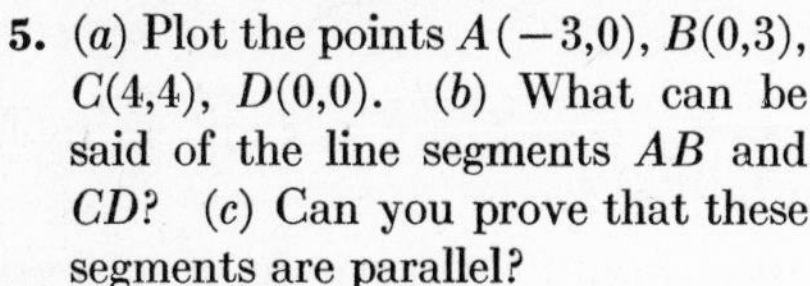

5. (*a*) Plot the points $A(-3,0)$, $B(0,3)$, $C(4,4)$, $D(0,0)$. (*b*) What can be said of the line segments AB and CD? (*c*) Can you prove that these segments are parallel?

6. Plot the points $A(-5,6)$, $B(3,-2)$, $C(-3,7)$, $D(-1,-1)$. What is the point of intersection of AB and CD?

7. Plot the points $A(-4,-3)$, $B(-1,2)$, $C(3,4)$, $D(6,-5)$. By extending the segments AB and DC, find their point of intersection.

8. Find the distance between points $(6,2)$ and $(-2,-4)$.

9. Find the distance between points $(-5,4)$ and $(7,-1)$.

10. How far are each of the points $(3,4)$, $(-4,3)$, $(-4,-3)$, $(3,-4)$ from the origin?

11. (*a*) Plot the points $A(-5,-2)$, $B(0,-2)$, $C(-2\frac{1}{2},4)$. What kind of triangle is ABC? (*b*) Find the lengths of each of its sides.

12. (*a*) Plot the points $A(-5,0)$, $B(3,0)$, $C(-1,4)$. Is the triangle a right triangle? Why? (*b*) Find the lengths of each of its sides.

13. Connect the points $(6,12)$ and $(0,-6)$ by a straight line. (*a*) Does this line pass through point $(1,-3)$? (*b*) Give the coordinates of two points (other than the end points) which lie on this line.

14. Plot the points $A(-6,-3)$, $B(3,-3)$, $C(3,3)$, $D(0,6)$. What is the area of the quadrilateral $ABCD$?

15. Find the lengths of the sides of the triangle ABC formed by points $A(-11,-15)$, $B(12,-8)$, $C(5,15)$. Using the Pythagorean relation, determine whether this triangle is a right triangle.

103. Graphs of Functions of the First Degree. In what is called the *graph* of a function we have a means of picturing the function, and thus of forming a visual idea of its properties.

Suppose, for example, that the dependent variable y is given by the equation

$$y = 2x - 1.$$

In order to graph this equation, we arbitrarily choose any number of different values of x and from the equation compute the corresponding values of y. We arrange them in a table as follows:

When $x =$	0	1	3	$\frac{3}{2}$	-1	-2
$y =$	-1	1	5	2	-3	-5

If we plot these pairs of values on a coordinate system, we have the graph of the equation shown in Fig. 7.

This graph is, as we see, a straight line; furthermore, if we construct the graph of any equation of the form

$$Ax + By + C = 0,$$

which is of the first degree in the variables x and y, we shall find that the graph is a straight line. This is the reason why an equation of the first degree in x and y is often referred to as a *linear* equation.

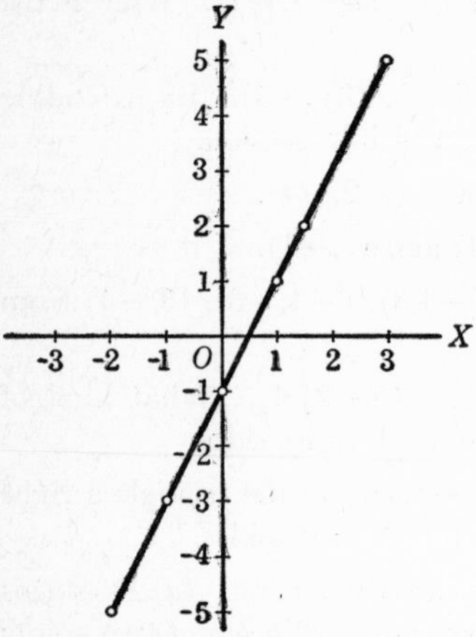

Fig. 7

Since we know that a straight line is definitely determined when two distinct points are known, in graphing a linear equation only two table entries are absolutely necessary. As a general rule, however, it is advisable to obtain three table entries in order to have a check on the correctness of the work.

Example 1. Graph the equation $x + 3y = 6$.

Solution. It is immaterial which variable we consider the independent variable. It is more convenient in this example to let y assume arbitrary values. Thus, we have the table of three entries

When $y =$	0	4	2
$x =$	6	-6	0

Plotting these three points, we have the graph shown in Fig. 8.

NOTE. For the sake of accuracy in actually drawing a line, it is advisable to choose two points that are reasonably far apart.

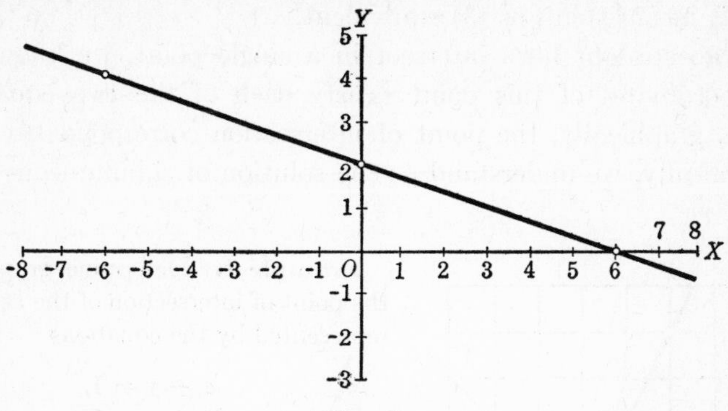

FIG. 8

EXERCISE 80

Draw the graph of each of the following equations. Choose several extra pairs of values which satisfy the equation and see if they lie on the graph:

1. $x + y = 5$ **2.** $2x - y = 4$ **3.** $3x - 2y = 6$
4. $x - 5y = 10$ **5.** $3x + 4y = 12$ **6.** $2x + 3y = 9$

Draw the graph of each of the following equations. Choose several points which lie on the graph and see if the coordinates of these points satisfy the given equation:

7. $x - 2y = 2$ **8.** $3x - y = 1$ **9.** $2x - 2y = 5$
10. $4x - 3y = 12$ **11.** $5x + 2y = 5$ **12.** $x + 6y = 3$

13. On the same coordinate system, draw the graphs of the three equations $x - y = 3$, $x - y = 0$, $x - y = -2$. What have these lines in common?

14. On the same coordinate system, draw the graphs of the three equations $y = x$, $y = 4x$, $y = -2x$. What have these lines in common?

15. Draw the graphs of the three equations $y = 4$, $y = 0$, $y = -2$; let x have any value whatsoever.

16. Draw the graphs of the three equations $x = 5$, $x = 0$, $x = -3$; let y have any value whatsoever.

104. Intersection of Straight Lines. If we plot the graphs of two linear equations on the same system of coordinates, we find that the two straight lines formed either (1) intersect in a single point, (2) are parallel, or (3) are identical. In correspondence with these three situations, we say that their equations are (1) independent, (2) inconsistent, or (3) equivalent.

If two straight lines intersect in a single point, we know that the coordinates of this point satisfy each of the two equations. Hence, graphically, the point of intersection corresponds to what, algebraically, we understand as the solution of simultaneous equations.

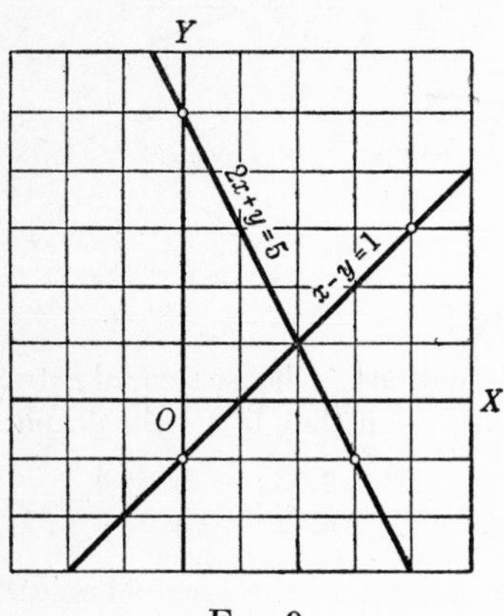

Fig. 9

Example 1. Determine graphically the point of intersection of the two lines represented by the equations

$$x - y = 1,$$
$$2x + y = 5.$$

Solution. Plotting each of the two lines separately on the same system of coordinates, we see that the lines intersect at the point where $x = 2$ and $y = 1$.

These values of x and y evidently satisfy both of the given equations, and hence the point of intersection of the lines is given by (2,1).

Example 2. On the same coordinate system, draw the graphs of the two equations

$$x - y = 2,$$
$$2x - 2y = 9.$$

Solution. Plotting these two equations, we see that their graphs do not intersect. That is to say, the two lines obtained are parallel.

If we multiply the first equation by 2, it becomes $2x - 2y = 4$. It is evidently impossible that $2x - 2y$ can be equal to both 4 and 9 at the same time. It is for this reason that such equations are called inconsistent.

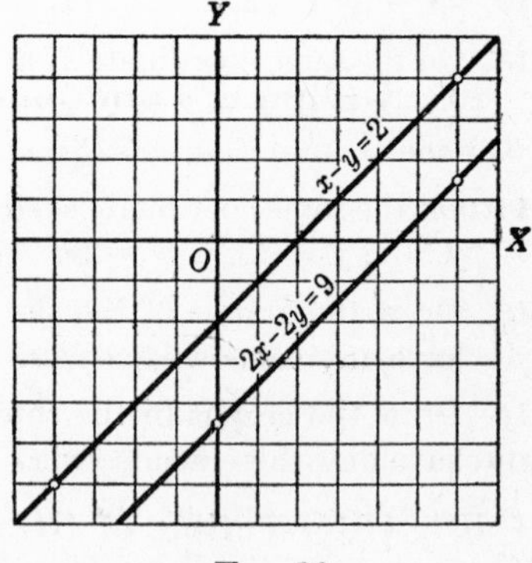

Fig. 10

EXERCISE 81

Plot the following pairs of equations; if they intersect, find their point of intersection:

1. $2x + y = -5$, $x - 2y = 0$
2. $x + y = 4$, $x - y = 2$
3. $3x - y = -5$, $x + 2y = 3$
4. $2x - y = 1$, $6x - 3y = -4$
5. $3x - 2y = 3$, $2x + 3y = 2$
6. $x + 4y = 1$, $x + 6y = 3$
7. $y = 2x + 4$, $x = 1 - y$
8. $x + 3y + 6 = 0$, $y = 4x - 2$
9. $2x + 4y = 3$, $3x + 6y = 10$
10. $3y = 4x + 2$, $2x + 5y = 12$
11. $x + 2y - 2 = 0$, $3x - y - 20 = 0$
12. $2x - y + 6 = 0$, $2x + y - 6 = 0$
13. $3x - y = 3$, $\frac{1}{2}x + \frac{1}{3}y = 2$
14. $4x + y = 3$, $2x + \frac{1}{2}y + 1 = 0$
15. $\frac{1}{4}x + y = -5$, $x - \frac{1}{3}y = 6$
16. $0.2x + 0.7y = 1.5$, $0.4x - 0.3y = 1.3$
17. $0.6x + y = -1$, $x - 0.5y = 7$
18. $0.02y = 0.03x - 0.04$, $1.5x - y = 3$

Plot the following pairs of equations and estimate the coordinates of their point of intersection to the nearest tenth in decimals:

19. $y = x$, $7x + 4y = 24$
20. $x + y = 1$, $7x - 3y + 5 = 0$
21. $x - y - 2 = 0$, $11x + 8y + 6 = 0$
22. $x + 2y = 0$, $7x - 6y = 4$
23. $y = 2x + 1$, $3x + y + 1.2 = 0$
24. $7x + 2y = 4$, $7x - 2y = 2$

105. Graphs of Functions of the Second Degree. A function y of the second degree in the variable x can always be written in the form

$$y = ax^2 + bx + c, \tag{1}$$

where a, b, and c represent known numbers and a is not zero.

To construct the graph of such an equation we proceed just as we did in plotting a linear equation.

Take, for example, the equation

$$y = x^2 - 4x + 3.$$

We first choose an arbitrary set of values for x and compute from the equation the corresponding values for y.

When $x =$	-1	0	1	2	3	4	5
$y =$	8	3	0	-1	0	3	8

When the points corresponding to these pairs of values are plotted on ruled paper and the succession of points are connected by a smooth curve, we obtain the graph shown in Fig. 11.

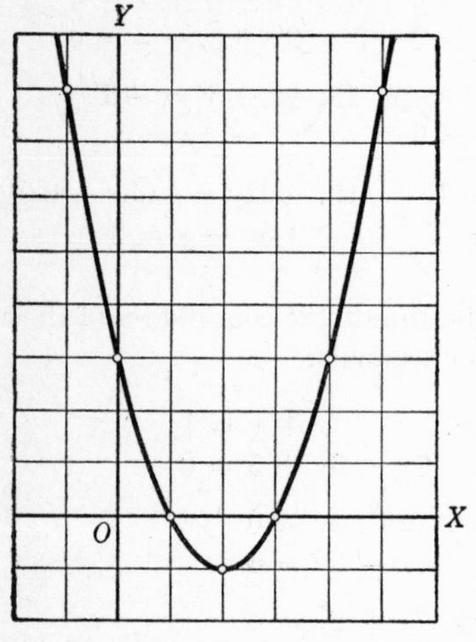

Fig. 11

The curve which is shown here is called a *parabola*. Furthermore, the graph of any equation of the form (1) is always a parabola and its position is such that it is symmetrical about a vertical axis.

If in the general equation the constant a is positive, the parabola is open upward and the lowest point of the curve is then called its *vertex*. If the constant a is negative, the parabola is open downward and its vertex is the uppermost point of the curve.

Thus, for the parabola just drawn, the *axis of symmetry* is the vertical line $x = 2$, and the *vertex* is at the point $(2, -1)$.

Example 1. Graph the equation $y = 60 - 40x - 20x^2$.

Solution. Setting up a table of values which satisfy the given equation, we have

When $x =$	2	1	0	-1	-2	-3	-4
$y =$	-100	0	60	80	60	0	-100

From an inspection of the table we see that it is out of the question to use the same scale of measurement on both axes. Hence, we take the same unit to represent 20 along the y axis as represents 1 along the x axis.

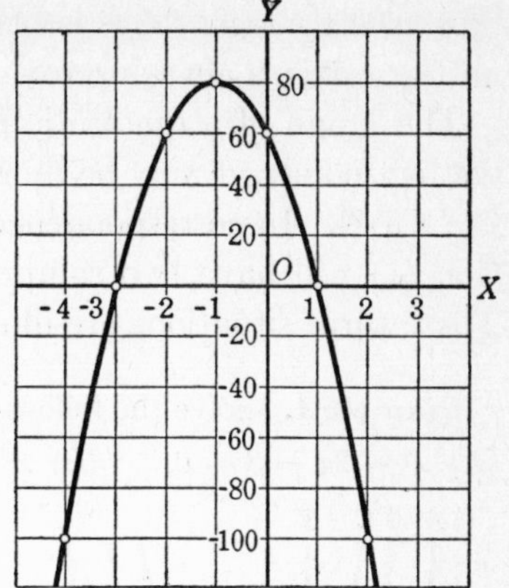

FIG. 12

NOTE. The graphing of an equation differs from the graphing of observational data in that for an equation any number of table entries can be found and the points on the graph can be chosen as closely together as we please. This indicates that the graphs of equations are *smooth* curves and not made up of line segments, as was the case with observational data.

EXERCISE 82

Graph the following equations. Give the equation of the axis of symmetry and the coordinates of the vertex.

1. $y = x^2$ **2.** $y = x^2 - x + 1$ **3.** $y = 4 - x^2$

4. $y = x^2 - 6x + 5$ **5.** $y = 5 + 4x - x^2$ **6.** $y = x^2 + 3x + 3$

7. $y = -2x^2 + 4x - 3$ **8.** $y = 4x^2 - 8x - 5$ **9.** $y = x^2 + 4x + 4$

Graph the following equations:

10. $y = 5x^2 + 5x - 60$ **11.** $y = 30 - 4x - 2x^2$ **12.** $y = \frac{1}{2}x^2 + 7x - 10$

13. Graph the equations $y = x^2 - 2x$ and $y = x^2 + 2x$ on the same coordinate system.

14. Graph the equations $y = x^2 - x - 6$ and $y = -x^2 + x + 6$ on the same coordinate system.

15. Graph the equations $y = x^2 + 5x + 6$ and $y = x^2 + 5x - 6$ on the same coordinate system.

16. Graph the equations $y = x^2 - 2x - 3$ and $y = 2x^2 - 4x - 6$ on the same coordinate system.

106. Solution of Quadratic Equations by Graphing. The solutions of a quadratic equation

$$(1) \qquad ax^2 + bx + c = 0$$

may be found graphically by observing where the graph of the function

$$y = ax^2 + bx + c$$

crosses the x axis. This is true because the value of y at any point on the x axis is always zero.

The graph of a quadratic function may cross the x axis at two distinct points, it may be tangent to the x axis, or it may not touch the x axis. In correspondence to each of these possible cases equation (1) will have two solutions, one solution, or no real solution. These three situations are illustrated in the following example.

Example 1. Solve the following equations graphically:

$(A)\ x^2 - 2x - 3 = 0,\qquad (B)\ x^2 - 2x + 1 = 0,\qquad (C)\ x^2 - 2x + 4 = 0.$

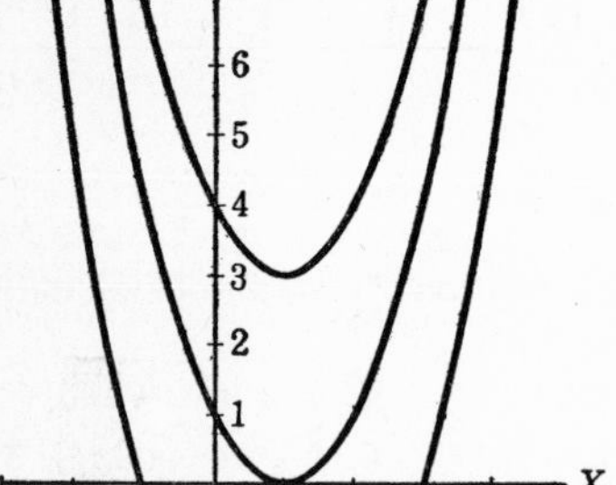

FIG. 13

Solution. Graphing the functions

$$(A)\ y = x^2 - 2x - 3,$$
$$(B)\ y = x^2 - 2x + 1,$$
$$(C)\ y = x^2 - 2x + 4,$$

we have the parabolas indicated by A, B, and C in Fig. 13.

A crosses the x axis at $x = -1$ and $x = 3$. By substitution it is evident that these values do satisfy the given equation.

B touches the x axis at just one point where $x = 1$. Thus the second of the given equations has the single solution $x = 1$.

C does not touch the x axis at all and hence the corresponding equation has no real solution. Later we shall see that an equation such as this does have what are called *imaginary* solutions.

EXERCISE 83

Solve each of the following equations graphically, and check each root:

1. $x^2 + 4x - 5 = 0$
2. $x^2 + 4x + 3 = 0$
3. $x^2 - 8x + 7 = 0$
4. $x^2 - 4x - 12 = 0$
5. $x^2 + 3x - 10 = 0$
6. $x^2 + 6x + 9 = 0$
7. $2x^2 + 5x - 3 = 0$
8. $2x^2 - 13x + 15 = 0$
9. $4x^2 + 12x + 9 = 0$
10. $2x^2 - 7x + 6 = 0$
11. $3x^2 + 8x - 3 = 0$
12. $6x^2 + x - 1 = 0$

Solve each of the following equations graphically, finding each root in decimals to the nearest tenth:

13. $x^2 - 2x - 2 = 0$ **14.** $x^2 - 4x + 1 = 0$

15. $x^2 + 6x + 7 = 0$ **16.** $x^2 + 2x - 4 = 0$

17. $2x^2 - 3x - 1 = 0$ **18.** $3x^2 - x - 1 = 0$

107. Graphs of Miscellaneous Functions. (Optional.) The principles underlying the theory of graphing any functions whatsoever are the same regardless of the nature of the function. In summarized form, they are:

1. Choose arbitrary values of the independent variable.
2. Using the given equation, compute the corresponding values for the dependent variable and arrange in tabular form.
3. Plot the pairs of values thus obtained, and connect the succession of points with a smooth curve.

Example 1. Graph the equation $y = x^3 - x$.

Solution. By substitution, we find the following pairs of corresponding values of x and y:

When $x =$	-2	$-\frac{3}{2}$	-1	$-\frac{1}{2}$	0	$\frac{1}{2}$	1	$\frac{3}{2}$	2
$y =$	-6	$-\frac{15}{8}$	0	$\frac{3}{8}$	0	$-\frac{3}{8}$	0	$\frac{15}{8}$	6

Plotting these values, we have the graph shown in Fig. 14. There is no set rule to indicate what arbitrary values of x to choose in setting up the table of points. Observe in this instance that it was essential to choose fractional values of x. If the four points having fractional abscissas are suppressed, the remaining five points by no means give a clear picture of the nature of the curve. In general, *when in doubt as to the character of the curve in any neighborhood, plot points in that neighborhood to locate the curve definitely.*

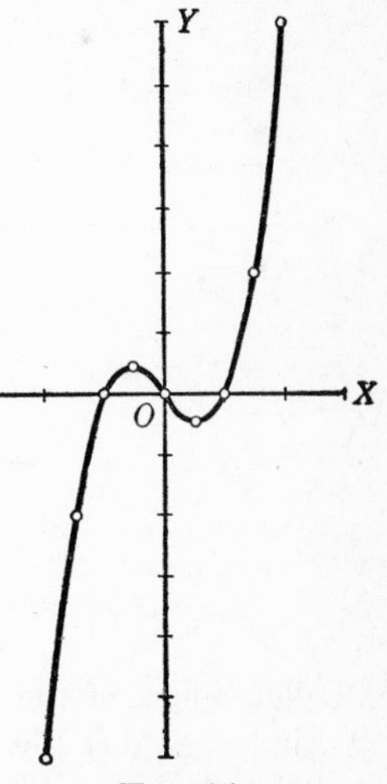

FIG. 14

Example 2. Graph the equation $y = \dfrac{6}{x}$.

Solution. By substitution, we find the following pairs of corresponding values of x and y:

When $x =$	-12	-6	-3	-2	-1	$-\frac{1}{2}$	$\frac{1}{2}$	1	2	3	6	12
$y =$	$-\frac{1}{2}$	-1	-2	-3	-6	-12	12	6	3	2	1	$\frac{1}{2}$

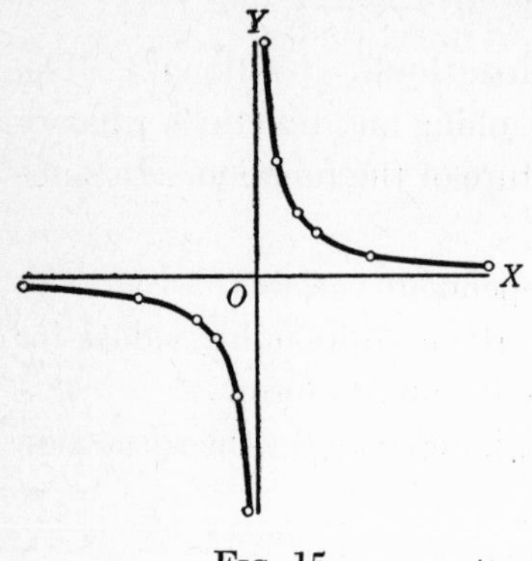

FIG. 15

Plotting these pairs of values, we have the graph shown in Fig. 15. Observe that for $x = 0$, from the equation, we have $y = \frac{6}{0}$, an expression that is meaningless, because division by zero is not allowed in algebra. However, observe on the graph that as x approaches closer and closer to zero, the curve goes farther and farther away from the x axis.

When graphing functions y which are expressed by fractional expressions in the independent variable x, *do not* choose values which make the denominator zero, but *do* choose abscissas which are near these values.

EXERCISE 84

Graph the following functions:

1. $y = x^3$ **2.** $y = x^3 - x^2$ **3.** $y = x^3 - 1$

4. $y = x - x^3$ **5.** $y = x^3 + x^2 + x$ **6.** $y = x^3 - 3x^2 + 2x$

7. $y = \dfrac{4}{x}$ **8.** $y = -\dfrac{6}{x}$ **9.** $y = \dfrac{1}{x - 1}$

10. $y = \dfrac{1}{x^2}$ **11.** $y = 1 + \dfrac{1}{x}$ **12.** $y = x - \dfrac{1}{x}$

13. $y = x^4 - x^2$ **14.** $y = \dfrac{12}{x(x - 5)}$

REVIEW OF CHAPTER IX

A

1. The index of the cost of living is given in the following table. The index number 100 is chosen to represent the average cost of living from 1935 to 1939. Plot the number with respect to time.

Year	1927	'29	'31	'33	'35	'37	'39	'41	'43	'45	'47	'49	'51
Index no.	121	120	109	92	98	103	99	105	124	128	159	170	185

2. The federal income tax receipts for the state of Oregon are given in the following table. Plot these figures with respect to time.

Year	1939	'40	'41	'42	'43	'44	'45	'46	'47	'48	'49	'50	'51
Millions of dollars	6	7	12	34	79	228	225	179	192	210	181	160	157

3. The average cost per passenger-mile for travel on the railways in the United States is given in the following table. Plot the cost with respect to time.

Year	1918	'21	'24	'27	'30	'33	'36	'39	'42	'45	'48	'51
Cost in cents	2.42	3.09	2.99	2.90	2.72	2.02	1.84	1.84	1.92	1.87	2.34	2.45

4. The monthly mean precipitation in inches for Chicago and Los Angeles is given in the following table. Plot each with respect to time on the same chart, and compare.

Month	Jan.	Feb.	Mar.	Apr.	May	June	July	Aug.	Sept.	Oct.	Nov.	Dec.
Chicago	1.9	2.1	2.6	2.8	3.5	3.3	3.3	3.2	3.1	2.5	2.4	2.0
Los Angeles	3.1	3.1	2.8	1.0	0.4	0.1	0.0	0.0	0.2	0.7	1.2	2.6

5. The monthly mean Fahrenheit temperatures for Detroit and St. Louis are given in the following table. Plot each with respect to time on the same chart, and compare.

Month	Jan.	Feb.	Mar.	Apr.	May	June	July	Aug.	Sept.	Oct.	Nov.	Dec.
Detroit	24	25	33	46	58	67	72	70	64	52	39	29
St. Louis	31	35	44	56	67	75	79	78	70	59	45	35

6. The consumers' price indexes (1935–39 = 100) for various items are given in the following table. Plot each with respect to time on the same chart.

Year	1933	'35	'37	'39	'41	'43	'45	'47	'49	'51
Food	84	100	105	95	106	138	139	194	202	227
Apparel	88	97	103	101	106	130	146	186	190	205
Rent	101	94	101	104	106	108	108	111	121	136

B

1. In Fig. 16, give the coordinates of the points A, B, C, D, E.
2. At what value of x does the line segment CD cross the x axis?

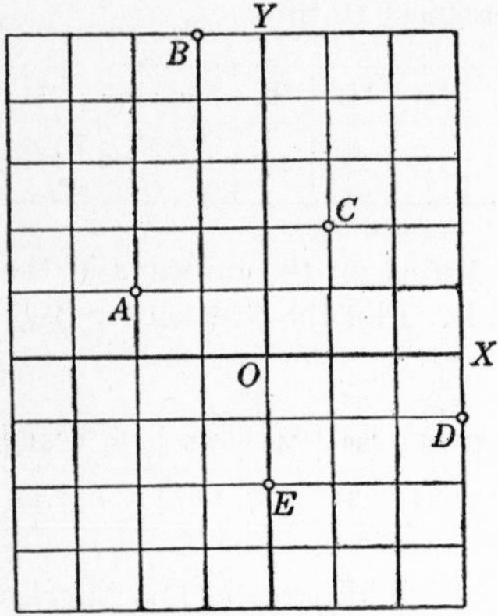

FIG. 16

3. At what value of y does the line segment AC cross the y axis?
4. At what point do the line segments AD and CE intersect?
5. At what point do the extensions of BA and DE intersect?
6. What is the distance between points B and E? Leave your answer in radical form.
7. What is the equation of the horizontal line through point C?
8. What is the equation of the vertical line through point D?
9. (*a*) Plot the points $A(-3,0)$, $B(1,0)$, $C(1,2)$, $D(-2,2)$. (*b*) Connect these points in succession. What kind of figure is $ABCD$? (*c*) What is its altitude? (*d*) Find its area.
10. Plot the points $A(-2,0)$, $B(0,5)$, $C(3,0)$, $D(0,2)$ on three different coordinate systems. Connect the points on the first in the order $ABCD$; on the second, in the order $ADBC$; and on the third, in the order $ABDC$. Find the area of each of the three figures obtained.

Graph the following pairs of linear equations; if they intersect, find their point of intersection:

11. $x + y = 5$
    $2x - y = 4$
12. $x - 3y = 6$
    $3x + y = 3$
13. $2x - 2y = 5$
    $3x - 3y = -2$
14. $5x - 3y = 13$
    $3x + 7y = -1$
15. $2x + 5y = 1$
    $3x - y = -7$
16. $y = 6 - 4x$
    $3x - 2y + 1 = 0$

Graph the following equations and find the coordinates of their point of intersection to the nearest tenth:

17. $4x + y = 1$
    $2y = 2x + 1$
18. $4x - 2y = -5$
    $3x + 3y = -1$
19. Plot the graphs of the three lines $x = 1$, $y = 2$, and $2x + y = 7$. Find the area of the triangle enclosed by these three lines.

20. Plot the graphs of the three lines $y = 0$, $y - x = 2$, and $2x + y = 8$. Find the area of the triangle enclosed by these three lines.

C

Graph the following equations. In each case give the equation of the axis of symmetry and the coordinates of the vertex:

1. $y = x^2 - 2$ **2.** $y = -x^2 + 4x - 3$ **3.** $y = 15 - 2x - x^2$
4. $y = x^2 + 3x - 10$ **5.** $y = x^2 + 2x + 2$ **6.** $y = 3x - x^2$

Solve each of the following equations graphically, and check each root:

7. $x^2 + 4x - 12 = 0$ **8.** $x^2 - 7x + 6 = 0$ **9.** $2x^2 - x - 6 = 0$

Solve each of the following equations graphically, finding each root to the nearest tenth:

10. $2x^2 - 4x + 1 = 0$ **11.** $2x^2 + 2x - 1 = 0$ **12.** $3x^2 - 6x + 1 = 0$

In each of the following, draw the graphs of the two equations on the same coordinate system and give the coordinates of their points of intersection:

13. $y = x^2$
$y = x + 2$

14. $y = 5 + 4x - x^2$
$2x + y = 5$

15. $y = 1 - x^2$
$y = x^2 + 2x + 1$

16. $y = 2x^2 - 3x - 4$
$y = x^2 - x - 1$

Graph the following equations:

17. $y = 4x - x^3$ **18.** $y = x^3 - 6x^2 + 8x$

19. $y = \dfrac{8}{x}$ **20.** $y = \dfrac{1}{2 - x}$

A Leveling Game

Three men play a game with the understanding that the loser is to double the money of each of the other two. After three games each has lost just once and each ends up with \$24. How much did each one start with?

Try ABBA

Show that any symmetric four-digit number is a multiple of 11.

CHAPTER X

QUADRATIC EQUATIONS

108. Quadratic Equations in One Unknown. When an equation is one unknown, after being written in its simplest form, contains the square of the unknown quantity *but no higher power*, it is called a **quadratic equation,** or an **equation of the second degree.**

If the equation contains both the square and the first power of the unknown, it is called a *complete quadratic;* but when it contains only the square of the unknown, it is called an *incomplete quadratic* or a *pure quadratic.*

Thus,

$2x^2 + x = 6$ is a complete quadratic equation,

and $4x^2 - 9 = 0$ is an incomplete quadratic equation.

109. Solution of Incomplete Quadratic Equations. An incomplete quadratic equation may be considered as a linear equation in which the square of the unknown quantity is to be found.

Example 1. Solve $\dfrac{11}{x^2 + 6} = \dfrac{6}{x^2 - 4}$.

Solution.

$$\frac{11}{x^2 + 6} = \frac{6}{x^2 - 4}, \quad \text{given equation;}$$

$$11x^2 - 44 = 6x^2 + 36, \quad \text{clearing fractions;}$$

$$5x^2 = 80, \quad \text{transposing;}$$

$$x^2 = 16, \quad \text{simplifying.}$$

Taking the square root of these equals, we have

$$x = \pm 4.$$

Note. In extracting the square root of the two sides of the equation $x^2 = 16$, it is unnecessary to prefix both sides with the double sign $\pm$. This is evident since $\pm x = \pm 4$ represents the four cases:

$$+x = +4, \ +x = -4, \ -x = -4, \ -x = +4,$$

and the last two of these equations merely duplicate the first two equations.

EXERCISE 85

Solve the following equations:

1. $x^2 = 64$
2. $4x^2 = 49$
3. $3x^2 - 17 = 58$
4. $5x^2 - 13 = x^2 + 23$
5. $9x = \frac{25}{x}$
6. $\frac{x}{2} + \frac{2}{x} = x$
7. $\frac{5}{x^2 + 1} = \frac{3}{x^2 - 3}$
8. $\frac{2}{x^2 + 1} = \frac{3}{2x^2 - 11}$
9. $(x + 2)(x - 2) = 2x^2 - 13$
10. $\frac{x^2 - 1}{4} - \frac{2x^2 - 3}{15} - \frac{x^2 - 4}{5} = 0$

Solve each of the following equations for x:

11. $\frac{5a^2}{2} - x^2 = \frac{3x^2}{2}$
12. $5x^2 - 3a^2 = x^2 + 13a^2$
13. $\frac{x^2 - 3c^2}{2} - \frac{x^2 + 8c^2}{3} = 0$
14. $2(x^2 - 10m^2) = (x + 4m)(x - 4m)$
15. $(x + a)(x + b) + (x - a)(x - b) = 2a(4a + b)$
16. $\frac{x^2 - 9a^2}{16a^2} = \frac{x^2 - 16a^2}{9a^2}$

Solve the following equations and leave answers in terms of simplified radicals:

17. $x^2(x + 1) = (x + 2)(x^2 - 2x + 4)$
18. $(x - 1)(x^2 - 2x - 2) = (x - 2)(x^2 + x + 2)$
19. $\frac{x + 1}{2x + 1} = \frac{2x + 3}{x + 7}$
20. $\frac{1}{x} + \frac{1}{x + 2} = \frac{5}{x + 4}$
21. $\frac{x - 1}{x + 1} + \frac{x + 1}{x - 1} = 6$
22. $\frac{x^2 + 3}{x^2} = \frac{x^2 + 9}{x^2 + 4}$
23. $\frac{1}{4x^2 + 3} + \frac{1}{4x^2 - 3} = \frac{1}{16x^4 - 9}$
24. $\frac{x + 7}{x} - \frac{5}{4} = \frac{x - 1}{4}$

Solve each of the following formulas for the letter indicated:

25. $A = \pi r^2$, for r
26. $S = \frac{1}{2}gt^2$, for t
27. $V = \frac{1}{3}\pi r^2 h$, for r
28. $F = \frac{mM}{d^2}$, for d
29. $\frac{1}{a^2} + \frac{1}{b^2} = \frac{1}{c^2}$, for a
30. $V = \frac{1}{4}\pi d^2 h$, for d

110. Solution of Complete Quadratic Equations by Factoring. The solution of any algebraic equations by factoring depends on the following important principle:

Rule for the Solution of Equations by Factoring

If the product of two or more factors is zero, then at least one of the factors must be equal to zero.

Thus, if $A \cdot B = 0$, then either $A = 0$ or $B = 0$, or both $A = 0$ and $B = 0$.

We may use this principle to solve quadratic equations when such equations can be factored.

Example 1. Solve the equation $2x^2 + 7x - 4 = 0$.

Solution. Factoring the terms on the left side of the equation, we have

$$(2x - 1)(x + 4) = 0.$$

In accordance with the above principle, this equation implies that either

$$2x - 1 = 0, \text{ or } x + 4 = 0.$$

Solving each of these linear equations gives

$$x = \tfrac{1}{2} \text{ or } x = -4.$$

That is, the roots of the given equation are $\frac{1}{2}$ and -4.

Check. To verify our results, we substitute each of these values in turn in the given equation. Thus,

when $x = \frac{1}{2}$;

$$2(\tfrac{1}{2})^2 + 7(\tfrac{1}{2}) - 4 = 0,$$
$$\tfrac{1}{2} + \tfrac{7}{2} - 4 = 0,$$
$$0 = 0.$$

when $x = -4$;

$$2(-4)^2 + 7(-4) - 4 = 0,$$
$$32 - 28 - 4 = 0,$$
$$0 = 0.$$

Example 2. Solve the equation $98x - 6x^2 = 400$ and check results.

Solution. Transposing all terms to the left side of the equation and rearranging terms, we have

$$-6x^2 + 98x - 400 = 0.$$

To facilitate factoring, (1) *always make the coefficient of the squared term positive*, and (2) *always reduce the coefficients to their lowest integral form.*

Thus, in this example, we (1) change signs throughout and (2) divide by 2. This gives

$$3x^2 - 49x + 200 = 0.$$

Factoring, we have

$$(x - 8)(3x - 25) = 0.$$

Hence, from $x - 8 = 0$ and $3x - 25 = 0$, we have

$$x = 8 \text{ or } 8\tfrac{1}{3}.$$

Check.

When $x = 8$,	When $x = \frac{25}{3}$,
$98(8) - 6(8)^2 = 400,$	$98(\frac{25}{3}) - 6(\frac{25}{3})^2 = 400,$
$784 - 384 = 400,$	$\frac{2450}{3} - \frac{1250}{3} = 400,$
$400 = 400.$	$400 = 400.$

EXERCISE 86

Solve the following equations and check the roots:

1. $x^2 - 8x - 9 = 0$ **2.** $x^2 + 4x - 21 = 0$
3. $x^2 - 5x + 4 = 0$ **4.** $x^2 - 8x + 15 = 0$
5. $x^2 + 9x + 20 = 0$ **6.** $x^2 - 11x - 26 = 0$
7. $x^2 = 9x - 18$ **8.** $x^2 = 5x + 6$
9. $x^2 - x = 6$ **10.** $x = x^2 - 12$
11. $2x^2 + 5x - 3 = 0$ **12.** $2x^2 + 5x + 2 = 0$
13. $2x^2 - 11x + 5 = 0$ **14.** $2x^2 - 9x + 9 = 0$
15. $5x^2 + 12x + 4 = 0$ **16.** $8x^2 - 26x + 15 = 0$
17. $2 - x - 3x^2 = 0$ **18.** $-4x^2 + x + 5 = 0$
19. $6x^2 + 17x - 14 = 0$ **20.** $4x^2 + 8x - 45 = 0$
21. $3x^2 - 2x = 5$ **22.** $3x^2 - x = 4$
23. $6x^2 = x + 35$ **24.** $6x^2 = 11x + 10$
25. $12x^2 - 10x - 12 = 0$ **26.** $8x^2 - 22x - 90 = 0$
27. $30x^2 - 85x + 25 = 0$ **28.** $21x^2 - 77x + 42 = 0$
29. $10x^2 + 48x - 10 = 0$ **30.** $12x^2 - 54x - 30 = 0$
31. $18x^2 + 57x + 45 = 0$ **32.** $15x^2 + 20x - 75 = 0$
33. $-x^2 + 10x + 1200 = 0$ **34.** $x^2 - 105x + 2700 = 0$
35. $x^2 - 48x + 540 = 0$ **36.** $x^2 - 43x - 5700 = 0$
37. $6x^2 + 25x - 150 = 0$ **38.** $12x^2 + 43x + 36 = 0$
39. $4x^2 - 10x - 750 = 0$ **40.** $36x^2 + 216x - 576 = 0$

111. Equations Reducing to Quadratic Equations. If any algebraic equation can be reduced to a factorable equation of the form $ax^2 + bx + c = 0$ where the letters a, b, and c represent integral constants, the equation may be solved in accordance with the principles outlined in the preceding article.

Example 1. Solve the equation $(x + 2)(x - 2) = 2(3x - 2)$ and check results.

Solution. Multiply out and transpose all terms to the left side of the equation before factoring:

$$\begin{aligned} x^2 - 4 &= 6x - 4, \\ x^2 - 6x &= 0, \\ x(x - 6) &= 0, \\ x &= 0 \text{ or } 6. \end{aligned}$$

Check.

When $x = 0$,

$$\begin{aligned} (0 + 2)(0 - 2) &= 2(3 \cdot 0 - 2), \\ (+2)(-2) &= 2(-2), \\ -4 &= -4. \end{aligned}$$

When $x = 6$,

$$\begin{aligned} (6 + 2)(6 - 2) &= 2(3 \cdot 6 - 2), \\ (8)(4) &= 2(18 - 2), \\ 32 &= 32. \end{aligned}$$

Example 2. Solve the equation $\dfrac{1}{x - 2} + \dfrac{1}{x + 3} = \dfrac{1}{6}$.

Solution. First clear the fractions:

$$\begin{aligned} 6(x + 3) + 6(x - 2) &= (x - 2)(x + 3), \\ 6x + 18 + 6x - 12 &= x^2 + x - 6, \\ x^2 - 11x - 12 &= 0, \\ (x - 12)(x + 1) &= 0, \\ x &= 12 \text{ or } -1. \end{aligned}$$

Example 3. Solve the equation $\sqrt{2x + 3} - \sqrt{4x - 1} = 1$.

Solution. For convenience we transpose one radical before squaring:

$$\begin{aligned} \sqrt{2x + 3} &= 1 + \sqrt{4x - 1}, \\ 2x + 3 &= 1 + 2\sqrt{4x - 1} + (4x - 1), \\ 2\sqrt{4x - 1} &= 3 - 2x, \\ 4(4x - 1) &= 9 - 12x + 4x^2, \\ 4x^2 - 28x + 13 &= 0, \\ (2x - 1)(2x - 13) &= 0, \\ x &= \tfrac{1}{2} \text{ or } \tfrac{13}{2}. \end{aligned}$$

Checking for extraneous roots, we have

when $x = \frac{1}{2}$,

$$\begin{aligned} \sqrt{2(\tfrac{1}{2}) + 3} - \sqrt{4(\tfrac{1}{2}) - 1} &= 1, \\ \sqrt{4} - \sqrt{1} &= 1, \\ 2 - 1 &= 1, \\ 1 &= 1. \end{aligned}$$

when $x = \frac{13}{2}$,

$$\begin{aligned} \sqrt{2(\tfrac{13}{2}) + 3} - \sqrt{4(\tfrac{13}{2}) - 1} &= 1, \\ \sqrt{16} - \sqrt{25} &= 1, \\ 4 - 5 &= 1, \\ -1 &= 1. \end{aligned}$$

Hence, the only valid solution of the given radical equation is $x = \frac{1}{2}$.

EXERCISE 87

Solve the following equations:

1. $(x + 1)(x + 5) = 96$
2. $(x - 2)(x + 3) = 24$
3. $x(3x + 2) = (x + 2)^2$
4. $(x + 1)(5x + 1) = (2x + 1)(x + 3)$
5. $(x + 1)(x + 2) + (x + 3)(x - 4) = 2$
6. $(4x + 1)(3x + 1) - (2x + 1)(x + 1) = 48$
7. $x^2 + (x + 1)^2 + (x + 2)^2 = 194$
8. $(x - 2)^2 + 7(x - 2) + 10 = 0$
9. $x^3 - (x - 1)^3 = 61$
10. $(x + 2)^3 - x^3 = 98$
11. $\dfrac{2x}{3x + 1} = \dfrac{x - 2}{2x - 1}$
12. $\dfrac{x + 6}{2x + 1} = \dfrac{2x - 3}{x + 2}$
13. $\dfrac{x}{x + 2} = \dfrac{x + 4}{3x}$
14. $\dfrac{1}{x} - \dfrac{1}{x + 1} = \dfrac{1}{6}$
15. $\dfrac{1}{x - 3} + \dfrac{1}{x + 4} = \dfrac{1}{12}$
16. $\dfrac{12}{x + 5} + \dfrac{21}{x + 10} = 1$
17. $\dfrac{9}{x - 1} + \dfrac{4}{x + 3} = 1$
18. $\dfrac{x + 5}{x + 2} + \dfrac{x - 3}{x - 2} = \dfrac{3(x - 5)}{x^2 - 4}$
19. $\dfrac{1}{3x - 4} - \dfrac{3}{3x + 4} - \dfrac{2}{5x} = 0$
20. $\dfrac{1}{2x - 1} + \dfrac{3}{x + 2} + \dfrac{2}{2x + 4} = 1$
21. $\sqrt{3x + 1} = x - 1$
22. $\sqrt{6x - 17} = 2x - 9$
23. $x + \sqrt{x + 6} = 0$
24. $x - \sqrt{4x + 5} = 0$
25. $\sqrt{x + 6} + \sqrt{3x + 7} = 3$
26. $\sqrt{5x - 9} = 1 + \sqrt{x + 4}$
27. $\sqrt{2x - 5} - \sqrt{x - 3} = 1$
28. $\sqrt{2x + 5} + \sqrt{4x + 3} = 3$
29. $\sqrt{2 - x} + \sqrt{5 - x} - \sqrt{x + 8} = 0$
30. $\sqrt{2x - 5} + \sqrt{3 - x} - \sqrt{x - 2} = 0$

112. Literal Equations. The principles of the preceding article apply also to literal equations which are quadratic in the unknown quantity.

Example 1. Solve the equation $6ax^2 - 14a^2x - 40a^3 = 0$ for x.

Solution. We reduce the coefficients to the lowest terms before factoring; thus, dividing throughout the equation by $2a$, we have

$$3x^2 - 7ax - 20a^2 = 0,$$
$$(x - 4a)(3x + 5a) = 0,$$
$$x = 4a \text{ or } -\tfrac{5}{3}a.$$

Example 2. Solve the equation $\frac{x - 2m}{x + 2m} = \frac{x - 3m}{2x - 5m}$ for x.

Solution. We clear of fractions and transpose before factoring:

$$(x - 2m)(2x - 5m) = (x + 2m)(x - 3m),$$
$$2x^2 - 9mx + 10m^2 = x^2 - mx - 6m^2,$$
$$x^2 - 8mx + 16m^2 = 0,$$
$$(x - 4m)(x - 4m) = 0,$$
$$x = 4m \text{ or } 4m.$$

NOTE. When the two roots of a quadratic equation are identical, we say that the equation has a *double root.* Thus, the above quadratic equation has the double root $x = 4m$.

Example 3. Solve the equation $acx^2 - bcx - adx + bd = 0$ for x.

Solution. We factor by grouping terms:

$$(acx^2 - bcx) - (adx - bd) = 0,$$
$$cx(ax - b) - d(ax - b) = 0,$$
$$(ax - b)(cx - d) = 0,$$
$$x = \frac{b}{a} \text{ or } \frac{d}{c}.$$

EXERCISE 88

Solve the following equations for x:

1. $x^2 - 5ax - 14a^2 = 0$
2. $x^2 - 8cx + 12c^2 = 0$
3. $x^2 + 13nx + 42n^2 = 0$
4. $x^2 - 10bx - 56b^2 = 0$
5. $a^2x^2 - 11abx + 18b^2 = 0$
6. $m^2x^2 - 3mnx - 70n^2 = 0$
7. $4p^2x^2 - 10px - 6 = 0$
8. $6ax^2 + 5a^2x + a^3 = 0$
9. $(x - 5a)(x + 2a) = 8a^2$
10. $(x - 3b)(x + b) = 32b^2$
11. $(x + 6m)(x + 2m) = m(3m - 2x)$
12. $(ax - 3b)(ax + b) = (2ax - 7b)(ax - b)$
13. $(ax + 1)(3ax + 7) = (ax + 3)(6ax - 1)$
14. $x(mx + 1) + x(mx + 2) = \frac{6}{m} - x$
15. $x^2 + (x + a)^2 + (x + 2a)^2 = 50a^2$
16. $(a + b)x = x^2 + ab$
17. $\frac{x}{a} + \frac{a}{x} = 2$
18. $\frac{x^2}{ax - 2b} = \frac{3ax - 4b}{a^2}$
19. $\frac{x + c}{2x - 3c} = \frac{3x - 5c}{3x + c}$
20. $\frac{1}{x} + \frac{1}{x + 3a} = \frac{1}{x - a}$

21. $\dfrac{1}{x-y} - \dfrac{1}{x+y} = \dfrac{1}{7y-x}$

22. $\dfrac{1}{x-c} + \dfrac{1}{2x} = \dfrac{1}{c} - \dfrac{1}{x}$

23. $abx^2 + b^2x + acx + bc = 0$

24. $x^2 + ax + bx - cx - ac - bc = 0$

25. $\sqrt{a(x+2a)} = x - 4a$

26. $\sqrt{ax - a^2} = x - 7a$

27. $\sqrt{x^2 + 3b^2} = 3x - b$

28. $\sqrt{m(4x + 7m)} = m + 4x$

113. Stated Problems Leading to Quadratic Equations. The conditions of a stated problem may be such that the equation obtained is of the second degree in one unknown. Such equations, as we have seen, have two solutions. Both solutions may have a reasonable meaning as an answer to the problem, but more often only one of the solutions has any real significance. In giving an answer to such problems, reject any solutions which have absurd meanings.

Example 1. Find a number which when increased by 17 is equal to 60 times the reciprocal of the number.

Solution. Let x represent the number; then

$$\begin{aligned} x + 17 &= \frac{60}{x}, \\ x^2 + 17x &= 60, \\ x^2 + 17x - 60 &= 0, \\ (x+20)(x-3) &= 0, \\ x &= 3, -20. \end{aligned}$$

Answer. The number may be 3 or -20.

Example 2. A alone can do a piece of work in 5 days less than the time required by B alone to do the job. Working together, they complete the work in 6 days. How long does it take each to do the job alone?

Analysis. Let x = the time in days required by A; then
$x + 5$ = the time in days required by B.

	Rate	× Time	= Fraction of Work
A	$\dfrac{1}{x}$	6	$\dfrac{6}{x}$
B	$\dfrac{1}{x+5}$	6	$\dfrac{6}{x+5}$

Solution.
$$\frac{6}{x} + \frac{6}{x+5} = 1,$$
$$6x + 30 + 6x = x^2 + 5x,$$
$$x^2 - 7x - 30 = 0,$$
$$(x - 10)(x + 3) = 0,$$
$$x = 10, -3.$$

Answer. A requires 10 days and B 15 days to do the whole job.

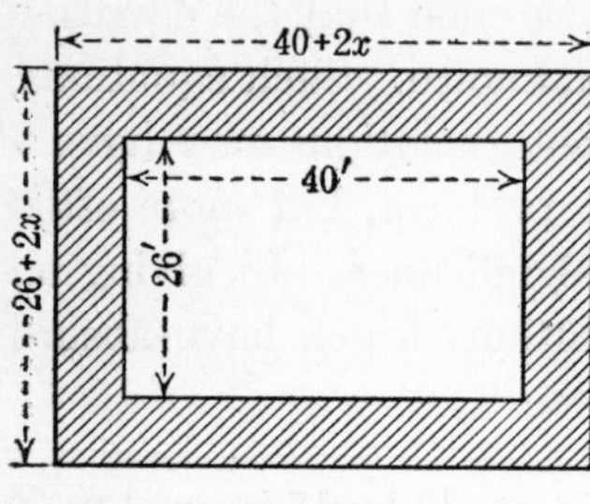

Fig. 17

Example 3. A lawn 40 feet long and 26 feet wide has a path of uniform width around it. If the area of the path is 432 square feet, find its width.

Analysis. Let x = the width of the path.

Solution.
$$(40 + 2x)(26 + 2x) - 40 \cdot 26 = 432,$$
$$4x^2 + 132x - 432 = 0,$$
$$x^2 + 33x - 108 = 0,$$
$$(x - 3)(x + 36) = 0,$$
$$x = 3, -36.$$

Answer. The width of the path is 3 feet.

Example 4. A train travels 150 miles at a uniform rate. If the rate had been 5 miles per hour more, the journey would have taken one hour less. Find the rate of the train.

Analysis. Let x = the rate of the train; then

$$\frac{150}{x} = \text{time of the train for the trip.}$$

When $x + 5$ = the rate of the train,

$$\frac{150}{x+5} = \text{time of the train for the trip.}$$

Solution.
$$\frac{150}{x+5} + 1 = \frac{150}{x},$$
$$150x + x^2 + 5x = 150x + 750,$$
$$x^2 + 5x - 750 = 0,$$
$$(x - 25)(x + 30) = 0,$$
$$x = 25, -30.$$

Answer. The train travels 25 miles per hour.

EXERCISE 89

1. The sum of two whole numbers is 18 and their product is 72. Find the numbers.

2. The sum of the reciprocals of two consecutive even integers is $\frac{7}{24}$. What are the integers?

3. The perimeter of one square exceeds that of another by 12 inches, and its area exceeds twice that of the other by 14 square inches. Find the sides of the squares.

4. A man bought some pigs for \$300. If each pig had cost \$3 more, he would have obtained five less pigs for the same money. How many pigs did he buy?

5. The tens' digit of a two-figured number is 2 more than the units' digit, and the number itself is 1 more than the sum of the squares of its digits. Find the number.

6. An express train makes the run between two cities 250 miles apart in a certain time; and a local train, whose speed averages 10 miles per hour less, makes the same run in 1 hour and 15 minutes longer. Find the average speed of the express train.

7. Harold is four years older than Frank, and the product of their present ages is three times what the product of their ages was four years ago. Find their present ages.

8. A cistern can be filled by two pipes together in 20 minutes; the larger pipe alone will fill the cistern in 9 minutes less than the smaller one. Find the time required by the larger pipe alone to fill the cistern.

9. A man who can row at the rate of 5 miles per hour in still water takes 10 minutes longer to row upstream a distance of 2 miles than he takes to row down the same stream a distance of 2 miles. Find the rate of the current.

10. Find a number whose square diminished by 169 is equal to ten times the excess of the number over 5.

11. The perimeter of a rectangle is 38 inches and its area is 84 square inches. Find its dimensions.

12. A man paid \$990 for some horses. By selling all but 4 of them at a profit of \$10 each, he received the amount he paid for all the horses. How many horses did he buy?

13. The sum of the digits of a two-figured number is 9, and the number itself is equal to twice the product of the digits. Find the number.

14. A cyclist traveling a distance of 8 miles would have arrived 20 minutes earlier if he had ridden two miles per hour faster. What was his rate of speed?

15. Find three consecutive integers the sum of whose squares equals 365.

16. Nancy is one year older than Mildred, and the sum of their present ages is the same as what the sum of the squares of their ages was 9 years ago. Find their present ages.

17. A skilled worker can finish a certain job in 9 days less time than an unskilled worker. Together they can finish the job in 20 days. How long would it take each alone to do the job?

18. A floor can be paved with 720 square tiles of a certain size. If each tile were one inch longer each way, it would take 500 tiles. Find the size of each tile.

19. A motorboat takes one hour longer to make a trip of 48 miles up a stream than it takes on the return trip downstream. If the average rate of the current is 4 miles per hour, what is the rate of the boat in still water?

20. The larger of two numbers exceeds the smaller by 3, and the sum of their squares exceeds their product by 97. Find the two numbers.

21. The longer leg of a right triangle is 2 feet more than twice the shorter leg. If the hypotenuse is 13 feet, find the legs of the triangle.

22. An athletic manager bought some baseballs for $72. Had the price been 10 cents less for each ball, he would have obtained 3 more balls for the same money. How many baseballs did he buy?

23. The units' digit of a number between 10 and 100 is 1 more than its tens' digit. If the number is multiplied by the number with its digits reversed, the result is 736. Find the number.

24. Two trains start at the same time to run 280 miles. One goes 5 miles per hour faster than the other and arrives one hour sooner. Find the rate of each train.

25. A man is six times as old as his son, and the next year his age will be equal to the square of his son's age. Find their present ages.

26. One pipe takes 36 minutes longer to drain a tank than it takes another pipe to fill it. When both pipes are open, it takes 55 minutes to fill the tank. How long would it take to fill the tank if the drain were shut off?

27. A crew that is capable of rowing 12 miles per hour in still water rowed down a stream a distance of 3 miles. If the current had been twice as strong, they would have traveled the course in 2 minutes less time. What was the rate of the current? Are both answers reasonable?

28. The sum of two numbers is 12, and the sum of their cubes is 468. What are the numbers?

29. A lawn 30 feet long and 20 feet wide has a flower border of uniform width around it. If the area of the border is the same as the area of the lawn, find the width of the border.

30. A boy bought some candy bars for \$1.50. He ate 4 bars and sold the rest so as to make 2 cents on each bar. If he made 80 cents on the deal, how many bars did he buy, and what was the cost price per bar?

31. The tens' digit of a two-figured number is 7 more than the units' digit, and the number itself equals the square of the sum of its digits. Find the number.

32. The length of a rectangle is 4 feet more than its width. If both the length and the width were increased by 4 feet, the area of the rectangle would be increased 50 percent. Find the dimensions of the rectangle.

33. George walks at a rate one mile per hour faster than John. They both start walking from the same point at the same time, George going north and John east. At the end of 2 hours they are 10 miles apart. How far apart would they be if they had walked in opposite directions?

34. The sum of the ages of two boys is 11, and last year the sum of the reciprocals of their ages was $\frac{1}{2}$. Find their present ages.

35. It takes a boy 5 days longer to do a certain task than it takes a man. If a man and two boys can do the job in 6 days, how long would it take the man alone to do the job?

36. It takes a boy three hours to row 4 miles down a river with the current and 4 miles back against the current. If the rate of the current is 1 mile per hour, at what rate can the boy row in still water?

37. The sum of a number and its square root is 42. Find the number.

38. The area of a rectangle is 108 square inches and if its dimensions are both decreased by 1 inch the area is decreased by 20 square inches. Find the diagonal of the rectangle.

39. A carpenter worked a certain number of days for a certain wage per day and received \$80 when the work was finished. His helper worked 1 day less, worked at a rate \$6 a day less, and received \$40 for his labor. What was the daily wage of the carpenter? Is the larger answer possible?

40. A number is represented by three digits, the tens' digit being 1 more than the hundreds' digit, and the units' digit being zero. The square of the first two digits considered as a two-figured number is 24 more than the number itself. Find the number.

41. Two cars go from P to Q by different routes, one of which is 150 miles long and the other 100 miles long. The car on the longer route runs five miles per hour faster and takes one hour longer for the trip. How much time was required for the trip by each? Comment on the results you obtain.

42. Robert was born in the year 1937, Sue in 1940, and Charles in 1942. In what calendar year was the sum of the squares of their ages equal to 274?

43. The denominator of a fraction exceeds twice the numerator by 1. If the numerator is increased by 5, and the denominator is decreased by 5, the resulting fraction will be the reciprocal of the given fraction. Find the given fraction.

44. It takes A 3 hours less time than it takes B to do a certain piece of work. After A had worked alone for 5 hours, he was joined by B and they finished the job in 7 hours. How long would it have taken each alone to do the job?

45. A motorboat travels at half speed down a river for a distance of 8 miles; it then finishes the remaining 7 miles of its trip at full speed. If the rate of the current is 3 miles per hour and the total trip lasted just one hour, what is the maximum speed of the boat in still water?

46. A carriage wheel makes a certain number of revolutions in traveling a distance of one mile. If the circumference of the wheel were 2 feet less, the wheel would make 88 more revolutions in traveling the distance. What is the circumference of the wheel?

47. A 2-inch square is cut from each corner of a rectangular piece of cardboard whose length exceeds the width by 4 inches. The sides are then turned up and an open box is formed. If the volume of the box thus formed is 90 cubic inches, find the three dimensions of the box.

48. A man has a certain investment which annually yields \$120 interest. If he had \$1000 less invested at a rate 1% higher, he would receive the same yearly interest. How much has he invested, and at what rate?

49. A tourist receives a certain number of shillings in exchange for \$21. If the exchange value of a shilling were increased 1 cent, the tourist would have received ten less shillings for his money. What is the exchange value of a shilling?

50. The members of a school band were formed into a hollow square, three deep, when it was observed that with the addition of 21 to their number a solid square might be formed in which the number of boys in each side would be 1 greater. How many boys are there in the band?

114. Solution of Quadratic Equations by Completing the Square. We know from the preceding work that the quadratic equation $x^2 = 16$ has as roots either $x = +4$ or $x = -4$. Similarly, by taking the square root of both sides of the equation $(x + 3)^2 = 16$, we find that $(x + 3) = +4$ or $(x + 3) = -4$, from which it follows that $x = 1$ or $x = -7$.

The equation

$$(x + 3)^2 = 16$$

may be written

$$x^2 + 6x + (3)^2 = 16,$$

or

$$x^2 + 6x = 7.$$

It is evident, by retracing our steps, that the equation $x^2 + 6x = 7$ can be solved by first adding $(3)^2$ to both sides in order to make the left member of the equation a *perfect square.*

Whatever the quantity a may be, we know that

$$(x + a)^2 = x^2 + 2ax + a^2$$

and

$$(x - a)^2 = x^2 - 2ax + a^2.$$

That is, if a trinomial is a perfect square, and its *squared term*, x^2, has *unity as a coefficient*, the term not containing x is equal to the *square of half the coefficient of* x.

These considerations may be summarized as a rule for the general solution of quadratic equations. The following procedure is known as the **method of completing the square:**

1. Simplify and write the quadratic equation so that the terms in x^2 and x are on the left side of the equation and the constant term is on the right side of the equation.
2. Make the coefficient of x^2 equal to 1 by dividing through the equation by the coefficient of x^2.
3. Add to each side of the equation the square of half the coefficient of x.
4. Take the square root of each side, prefixing $\pm$ to the right member.
5. Solve the two resulting equations separately.

Example 1. Solve the equation $x^2 + 2x - 15 = 0$.

Solution. Following the steps as outlined, we have

(1) $x^2 + 2x = 15,$

(3) $x^2 + 2x + 1 = 16,$

(4) $x + 1 = \pm 4,$

(5) $x + 1 = +4$ or $x + 1 = -4,$

$x = 3$ or $x = -5.$

Example 2. Solve $2x^2 - 2 = 3x$.

Solution. Following the steps as outlined, we have

$$
\begin{aligned}
&(1) & 2x^2 - 3x &= 2, \\
&(2) & x^2 - \tfrac{3}{2}x &= 1, \\
&(3) & x^2 - \tfrac{3}{2}x + (\tfrac{3}{4})^2 &= 1 + \tfrac{9}{16} = \tfrac{25}{16}, \\
&(4) & x - \tfrac{3}{4} &= \pm\tfrac{5}{4}, \\
&(5) & x - \tfrac{3}{4} = +\tfrac{5}{4} \quad &\text{or } x - \tfrac{3}{4} = -\tfrac{5}{4}, \\
& & x = \tfrac{8}{4} = 2 \text{ or} \quad & x = -\tfrac{2}{4} = -\tfrac{1}{2}.
\end{aligned}
$$

NOTE. We do not multiply out $(\frac{3}{4})^2$ on the left side of the equation.

Example 3. Solve $2x^2 - 8ax + 6a^2 = 0$ for x.

Solution. Following the steps as outlined, we have

$$
\begin{aligned}
&(1) & 2x^2 - 8ax &= -6a^2, \\
&(2) & x^2 - 4ax &= -3a^2, \\
&(3) & x^2 - 4ax + (2a)^2 &= -3a^2 + 4a^2 = a^2, \\
&(4) & x - 2a &= \pm a, \\
&(5) & x - 2a = +a \text{ or } &x - 2a = -a, \\
& & x = 3a \quad \text{or} \quad &x = a.
\end{aligned}
$$

EXERCISE 90

In each of the following, what constant term is needed in order to make the trinomial a perfect square?

1. $x^2 + 4x + (?)$ **2.** $x^2 - 8x + (?)$ **3.** $x^2 - 10x + (?)$
4. $x^2 + 6x + (?)$ **5.** $x^2 + x + (?)$ **6.** $x^2 - 3x + (?)$
7. $x^2 - \frac{1}{3}x + (?)$ **8.** $x^2 + \frac{2}{5}x + (?)$ **9.** $x^2 + 2ax + (?)$
10. $x^2 - 12cx + (?)$ **11.** $x^2 - \frac{4}{7}mx + (?)$ **12.** $x^2 + \frac{3}{4}bx + (?)$

Solve each of the following equations by the method of completing the square:

13. $x^2 + 2x = 8$ **14.** $x^2 - 4x = 21$
15. $x^2 + 8x + 12 = 0$ **16.** $x^2 - 6x + 8 = 0$
17. $x^2 - x - 6 = 0$ **18.** $x^2 - 3x - 4 = 0$
19. $x^2 + 8x + 15 = 0$ **20.** $x^2 - 6x + 5 = 0$
21. $x^2 = 5x - 4$ **22.** $x^2 = x + 30$
23. $2x^2 - 5x - 3 = 0$ **24.** $2x^2 + x - 6 = 0$
25. $3x^2 - 4x - 4 = 0$ **26.** $3x^2 - 8x + 5 = 0$
27. $6x^2 + 7x + 2 = 0$ **28.** $8x^2 - 2x - 3 = 0$
29. $12x^2 + 7x - 12 = 0$ **30.** $24x^2 + 34x - 45 = 0$
31. $8x^2 - 30x - 27 = 0$ **32.** $12x^2 - 7x - 10 = 0$
33. $9x^2 + 12x + 4 = 0$ **34.** $8x^2 - 30x + 27 = 0$

35. $x^2 - 3ax - 10a^2 = 0$
36. $x^2 = ax + 20a^2$
37. $6x^2 + 11cx + 4c^2 = 0$
38. $4x^2 - 7mx - 15m^2 = 0$
39. $4x^2 - 8bx + 3b^2 = 0$
40. $9x^2 - 9px + 2p^2 = 0$

115. Equations with Irrational Roots. The method of completing the square is particularly helpful in determining the roots of nonfactorable quadratic equations.

Example 1. Solve and check: $x^2 - 2x - 1 = 0$.

Solution. Applying the method of completing the square, we have

$$x^2 - 2x = 1,$$
$$x^2 - 2x + 1 = 2,$$
$$x - 1 = \pm\sqrt{2},$$
$$x = 1 + \sqrt{2} \text{ or } 1 - \sqrt{2}.$$

Check. When $x = 1 + \sqrt{2}$,

$$(1 + \sqrt{2})^2 - 2(1 + \sqrt{2}) - 1 = 0,$$
$$1 + 2\sqrt{2} + 2 - 2 - 2\sqrt{2} - 1 = 0,$$
$$0 = 0.$$

When $x = 1 - \sqrt{2}$,

$$(1 - \sqrt{2})^2 - 2(1 - \sqrt{2}) - 1 = 0,$$
$$1 - 2\sqrt{2} + 2 - 2 + 2\sqrt{2} - 1 = 0,$$
$$0 = 0.$$

Example 2. Find the roots of the equation $2x^2 - 6x + 1 = 0$ in decimals, correct to the nearest hundredth.

Solution. Applying the method of completing the square, we have

$$2x^2 - 6x = -1,$$
$$x^2 - 3x = -\tfrac{1}{2},$$
$$x^2 - 3x + (\tfrac{3}{2})^2 = -\tfrac{1}{2} + \tfrac{9}{4} = \tfrac{7}{4},$$
$$x - \tfrac{3}{2} = \pm\frac{\sqrt{7}}{2},$$
$$x = \frac{3 + \sqrt{7}}{2} \text{ or } \frac{3 - \sqrt{7}}{2}.$$

From the table on p. 209 we have $\sqrt{7} = 2.646$; hence,

$$x = \frac{3 + 2.646}{2} = 2.82 \text{ or } x = \frac{3 - 2.646}{2} = 0.18.$$

NOTE. Decimal answers are only approximate and they will not satisfy the given equation exactly. Unless numerical values of the roots are required, it is better to leave the answers in terms of simplified radicals.

EXERCISE 91

Solve and check. Express all radicals in the simplest form.

1. $x^2 - 2x = 5$
2. $x^2 + 4x = 8$
3. $x^2 + 8x = 4$
4. $x^2 - 6x = 6$
5. $x^2 - x - 1 = 0$
6. $x^2 + 3x + 1 = 0$
7. $2x^2 - 4x + 1 = 0$
8. $4x^2 + 4x - 1 = 0$
9. $9x^2 - 12x - 8 = 0$
10. $9x^2 - 18x + 1 = 0$

Solve each of the following equations:

11. $5x^2 - 8x + 1 = 0$
12. $4x^2 - 7x - 2 = 0$
13. $3x^2 + 10x + 6 = 0$
14. $7x^2 + 12x + 4 = 0$
15. $6x^2 - 11x + 4 = 0$
16. $8x^2 - 4x - 3 = 0$
17. $4x^2 + 4x - 5 = 0$
18. $5x^2 - 12x + 5 = 0$
19. $12x^2 - 14x + 3 = 0$
20. $8x^2 - 8x - 3 = 0$

Solve each of the following equations and express the answers numerically correct to the nearest hundredth:

21. $x^2 - 4x - 7 = 0$
22. $x^2 + 10x + 8 = 0$
23. $2x^2 + 3x - 6 = 0$
24. $3x^2 - 2x - 7 = 0$
25. $8x^2 + 13x + 4 = 0$
26. $5x^2 - 7x - 2 = 0$
27. $9x^2 - 12x - 8 = 0$
28. $4x^2 + 6x - 27 = 0$
29. $12x^2 - 9x + 1 = 0$
30. $6x^2 - 18x + 9 = 0$

116. Solution of Quadratic Equations by Formula. We have learned that every quadratic equation can be written in the form

$$ax^2 + bx + c = 0,$$

where a, b, and c may have any numerical values whatever. This equation is called the *general quadratic equation.* If therefore we can solve this quadratic equation, we can solve any quadratic equation.

Applying the method of completing the square, we have, after transposing c,

$$ax^2 + bx = -c.$$

Dividing throughout by a gives

$$x^2 + \frac{b}{a}x = -\frac{c}{a}.$$

Completing the square by adding $\left(\frac{b}{2a}\right)^2$ to each side, we have

$$x^2 + \frac{b}{a}x + \left(\frac{b}{2a}\right)^2 = \frac{b^2}{4a^2} - \frac{c}{a},$$

or
$$\left(x + \frac{b}{2a}\right)^2 = \frac{b^2 - 4ac}{4a^2}.$$

Extracting the square root of both sides we have

$$x + \frac{b}{2a} = \frac{\pm\sqrt{b^2 - 4ac}}{2a};$$

hence,
$$\boldsymbol{x = \frac{-b \pm\sqrt{b^2 - 4ac}}{2a}}.$$

This important result is called the **quadratic formula;** it should be carefully *memorized.*

Instead of going through the process of completing the square in each particular example, we may now make use of this general formula by merely substituting for the values of a, b, and c.

Example 1. Solve $2x^2 + 3x - 9 = 0$.

Solution. Comparing with the general quadratic equation, we see that $a = 2$, $b = 3$, and $c = -9$. Hence, substituting in the quadratic formula,

$$x = \frac{-(3) \pm\sqrt{(3)^2 - 4(2)(-9)}}{2(2)} = \frac{-3 \pm\sqrt{9 + 72}}{4} = \frac{-3 \pm\sqrt{81}}{4},$$

$$= \frac{-3 \pm 9}{4} = \frac{6}{4} \text{ or } \frac{-12}{4} = 1\tfrac{1}{2} \text{ or } -3.$$

Example 2. Solve $9x^2 - 18x + 4 = 0$.

Solution. Here we have $a = 9$, $b = -18$, and $c = 4$; hence,

$$x = \frac{-(-18) \pm\sqrt{(-18)^2 - 4(9)(4)}}{2(9)} = \frac{18 \pm\sqrt{36(9 - 4)}}{18},$$

$$= \frac{18 \pm 6\sqrt{5}}{18} = \frac{3 \pm \sqrt{5}}{3}.$$

Note. To facilitate simplifying the radical expression, we may factor out of $b^2 - 4ac$ such square numbers as are evidently common to both terms.

Example 3. Solve $x^2 - 2x + 2 = 0$.

Solution. Since $a = 1$, $b = -2$, $c = 2$,

$$x = \frac{-(-2) \pm \sqrt{(-2)^2 - 4(1)(2)}}{2(1)} = \frac{2 \pm \sqrt{4 - 8}}{2} = \frac{2 \pm \sqrt{-4}}{2}.$$

The number -4 has no square root, either exact or approximate. Hence no real value of x can be found to satisfy the equation. In such cases the roots are said to be *imaginary*.

EXERCISE 92

Solve the following equations by using the quadratic formula:

1. $x^2 - 2x - 3 = 0$

2. $x^2 - 7x + 12 = 0$

3. $2x^2 + 9x - 5 = 0$

4. $2x^2 - 7x - 30 = 0$

5. $3x^2 + 7x + 2 = 0$

6. $3x^2 + 2x - 8 = 0$

7. $6x^2 = x + 15$

8. $4x^2 - 5 = x$

9. $x = 10x^2 - 21$

10. $8x^2 - 42x + 27 = 0$

11. $8x^2 + 7x - 18 = 0$

12. $12x^2 - 20x = 25$

Solve the following equations by using the quadratic formula and express the roots in the simplest radical form:

13. $5x^2 - 2x - 6 = 0$

14. $3x^2 + 6x + 2 = 0$

15. $4x^2 + 12x - 15 = 0$

16. $8x^2 - 4x - 5 = 0$

17. $12x^2 - 8x - 1 = 0$

18. $20x^2 + 24x + 5 = 0$

19. $2x + 20 - 5x^2 = 0$

20. $16x - 10 = 5x^2$

21. $16x^2 + 12x = 9$

22. $8x^2 + 17 = x$

23. $18x^2 + 5 = 30x$

24. $40x - 16x^2 = 3$

Reduce the following to the standard form and solve, using the formula. Express irrational roots correct to the nearest hundredth:

25. $(x + 1)(2x - 3) = 4x^2 - 9$

26. $(3x - 2)(2x - 3) = x^2 + 2x - 2$

27. $\frac{1}{2}x(x + 1) - \frac{1}{3}x(x - 1) = \frac{1}{2}$

28. $\dfrac{x}{6} + \dfrac{x(x + 3)}{2} = 5\frac{1}{3}$

29. $\dfrac{1}{x} + \dfrac{2}{x + 2} = 1$

30. $\dfrac{1}{x} + \dfrac{2}{x + 1} + \dfrac{3}{x + 2} = 0$

31. $\dfrac{x}{2x + 1} = \dfrac{3x + 2}{4x + 3}$

32. $\dfrac{1}{x^2} + 5 = \dfrac{6x - 7}{x}$

33. $\dfrac{x}{x + 1} + \dfrac{x + 1}{x + 2} = 1$

34. $(x + 2)^3 = x^3 + 218$

35. $\dfrac{x - 4}{2x + 1} - \dfrac{3x - 1}{x - 3} = \dfrac{x^2 - 1}{2x^2 - 5x - 3}$

36. $\dfrac{x + 2}{x - 1} - \dfrac{x + 3}{x + 1} = \dfrac{2x^2 - 13}{1 - x^2}$

117. Quadratic Equations with Imaginary Roots. (Optional.) In Article 93, p. 201, we define an *imaginary number* as the indicated square root of a negative number, and expressed such numbers in terms of an imaginary unit $i = \sqrt{-1}$.

In solving quadratic equations in which an imaginary quantity occurs, it is customary to express the roots in terms of the imaginary unit.

In this sense, we see that *every* quadratic equation has two roots which have the property that they are either (1) real and unequal, (2) real and equal, or (3) imaginary.

Example 1. Show that $1 + i$ is a root of the equation

$$x^2 - 2x + 2 = 0.$$

Solution. Substituting $1 + i$ for x and recalling that $i^2 = -1$, we have

$$\begin{aligned}(1 + i)^2 - 2(1 + i) + 2 &= 1 + 2i + i^2 - 2 - 2i + 2,\\ &= 1 + 2i - 1 - 2 - 2i + 2 = 0.\end{aligned}$$

Since $1 + i$ satisfies the equation, we know that it is a root.

Example 2. Express the roots of the following equation in terms of the imaginary unit:

$$x^2 - 6x + 13 = 0.$$

Solution. Comparing the given equation with the general quadratic equation, we have $a = 1$, $b = -6$, $c = 13$. Hence, substituting in the quadratic formula, we obtain

$$x = \frac{6 \pm \sqrt{36 - 52}}{2} = \frac{6 \pm \sqrt{-16}}{2} = \frac{6 \pm 4i}{2} = 3 \pm 2i.$$

EXERCISE 93

In each of the following, determine by substitution whether the given number is a root of the corresponding quadratic equation:

1. $1 - i$, $x^2 - 2x + 2 = 0$

2. $3 + 2i$, $x^2 - 6x + 13 = 0$

3. $2 + i$, $x^2 - 4x + 4 = 0$

4. $1 + 3i$, $x^2 - 2x + 9 = 0$

5. $2 - 3i$, $x^2 - 4x + 13 = 0$

6. $-1 - 2i$, $x^2 + 2x + 5 = 0$

7. $\frac{1}{2} + \frac{1}{2}i$, $2x^2 - 2x + 1 = 0$

8. $\frac{-1 + i}{3}$, $9x^2 + 6x + 2 = 0$

Express the roots of the following equations in terms of the imaginary unit:

9. $x^2 - 6x + 10 = 0$

10. $x^2 + 2x + 2 = 0$

11. $x^2 - 2x + 17 = 0$

12. $x^2 - 8x + 25 = 0$

13. $x^2 + x + 1 = 0$

14. $x^2 - x + 1 = 0$

15. $9x^2 - 6x + 2 = 0$

16. $2x^2 - 6x + 5 = 0$

17. $4x^2 + 4x + 5 = 0$

18. $9x^2 - 12x + 13 = 0$

19. $2x^2 - 5x + 4 = 0$

20. $3x^2 + 2x + 1 = 0$

21. By the formula, the two roots of a quadratic equation are given by

$$r_1 = \frac{-b + \sqrt{b^2 - 4ac}}{2a}, \quad r_2 = \frac{-b - \sqrt{b^2 - 4ac}}{2a}.$$

(*a*) Show that $r_1 + r_2 = -\frac{b}{a}$. (*b*) Show that $r_1 r_2 = \frac{c}{a}$.

Using the results of Problem 21, find the sum and product of the roots of the following equations without actually solving the equations:

22. $x^2 - 3x + 7 = 0$

23. $3x^2 + 5x - 9 = 0$

24. $6x^2 - 3x - 2 = 0$

25. $4x^2 - 3x + a = 0$

26. $x^2 - sx + p = 0$

27. $2ax^2 - 3x + 4a = 0$

28. $(k + 1)x^2 - kx + 2 = 0$

29. $cx^2 + 2cx + (c + 1) = 0$

30. $5x^2 - (k + 1)x + (k + 2) = 0$

REVIEW OF CHAPTER X

A

Solve by factoring and check results:

1. $x^2 + 5x - 84 = 0$

2. $a^2 + 14a + 45 = 0$

3. $4m^2 - 9 = 0$

4. $2x^2 + 5x - 3 = 0$

5. $2b^2 - 17b + 21 = 0$

6. $8 + 2p - 3p^2 = 0$

7. $3x^2 - 11x - 42 = 0$

8. $25 - 16a^2 = 0$

9. $50 - 5a - a^2 = 0$

10. $4x^2 + 4x - 63 = 0$

11. $-4y^2 + 15y + 4 = 0$

12. $6n^2 + 7n - 10 = 0$

13. $12x^2 + 11x - 36 = 0$

14. $20a^2 - a - 12 = 0$

15. $6c^2 + 20c + 6 = 0$

16. $700 - 15y - y^2 = 0$

17. $48r^2 - 22r - 15 = 0$

18. $16x^2 + 39x - 27 = 0$

19. $160y^2 - 28y - 45 = 0$

20. $36a^2 - 104a + 75 = 0$

B

Solve by the method of completing the square. Express all radicals in the simplest form:

1. $2x^2 + 9x - 5 = 0$

2. $y^2 - 11y + 28 = 0$

3. $a^2 + 4a - 4 = 0$

4. $2c^2 - 6c - 3 = 0$

5. $3m^2 - 8m + 2 = 0$

6. $5x^2 + 10x - 3 = 0$

7. $4y^2 + 3y - 9 = 0$

8. $6a^2 - 10a + 3 = 0$

9. $10 - 6b - 5b^2 = 0$

10. $9d^2 - 13d + 4 = 0$

Solve and express decimal answers correct to the nearest hundredth:

11. $a^2 - 24a + 128 = 0$

12. $x^2 - 9x + 5 = 0$

13. $3y^2 + 7y - 1 = 0$

14. $5c^2 + 17c + 11 = 0$

15. $3 + 16p - 12p^2 = 0$

16. $10z^2 - 5z - 1 = 0$

17. $16b^2 + 8b - 3 = 0$

18. $23 - 17x - 13x^2 = 0$

19. $20y^2 + 7y - 40 = 0$

20. $28a^2 - 32a - 15 = 0$

C

Solve, using the quadratic formula. Express all radicals in the simplest form:

1. $2x^2 - 6x - 9 = 0$

2. $2a^2 + a - 28 = 0$

3. $9m^2 - 3m - 1 = 0$

4. $3d^2 + 2d - 6 = 0$

5. $9y^2 - 6y - 8 = 0$

6. $3 - 2x - 10x^2 = 0$

7. $8c^2 - 16c + 5 = 0$

8. $12q^2 + 6q - 5 = 0$

9. $7x^2 - 3 + 5x = 0$

10. $1 + 2x - 15x^2 = 0$

Solve and express decimal answers correct to the nearest hundredth:

11. $3a^2 - 16a + 4 = 0$

12. $5y^2 - 20y + 4 = 0$

13. $8z^2 - 24z - 3 = 0$

14. $15b^2 - 7b - 36 = 0$

15. $1 - 20p - 10p^2 = 0$

16. $6x^2 - 12x - 5 = 0$

17. $24c^2 - 20c - 15 = 0$

18. $9 - c - 14c^2 = 0$

19. $18y^2 + 14y - 3 = 0$

20. $42x - x^2 - 6 = 0$

D

Solve the following equations:

1. $\frac{1}{5}x^2 - \frac{1}{30}x - \frac{1}{2} = 0$

2. $\frac{1}{3}x^2 + \frac{1}{5}x - \frac{1}{7} = 0$

3. $2.1a^2 - 3.57 = 6.3a$

4. $0.004m^2 = 2.3 - m$

5. $(x - 3)^2 + (x + 3)^2 = 108$

6. $(x + 1)^2 - 7(x + 1) + 12 = 0$

7. $\dfrac{x + 1}{3x - 2} = \dfrac{5x - 4}{3x + 2}$

8. $\dfrac{1}{x - 1} - \dfrac{1}{x + 1} = \dfrac{1}{24}$

9. $\dfrac{3}{x + 1} + \dfrac{2}{2x - 1} = \dfrac{13}{18}$

10. $x^3 - (x - 2)^3 = 98$

11. $(5x + 7)(3x + 10) = 14$

12. $(x - 1)(2x - 1)(3x - 1) = 6x^3 - 82$

13. $\dfrac{x + 3}{x - 1} + \dfrac{x + 4}{x + 1} = \dfrac{8x + 5}{x^2 - 1}$

14. $\dfrac{1}{x} + \dfrac{1}{x - 3} - \dfrac{7}{3x - 5} = 0$

15. $\dfrac{6}{2x - 2} - \dfrac{1}{x + 1} - \dfrac{2}{2x + 2} = 1$

16. $\dfrac{21x^3 - 16}{3x^2 - 4} - 7x = 5$

17. $\sqrt{4x + 1} = x - 5$

18. $\sqrt{10 - x} = 2(x + 4)$
19. $\sqrt{5x - 1} = 4 - \sqrt{x - 1}$
20. $\sqrt{4x + 9} + \sqrt{x} + \sqrt{x + 5} = 0$

E

Solve for x in each of the following:

1. $x^2 - 9abx + 20a^2b^2 = 0$
2. $6m^2x^2 + 5mnx - 25n^2 = 0$
3. $(c + x)(2c - 3x) = 3x^2$
4. $\dfrac{x + a}{x - a} = \dfrac{x - 5a}{2x - 5a}$
5. $\dfrac{1}{x - 2d} + \dfrac{1}{x + d} = \dfrac{1}{x - 3d}$
6. $(x + 3c)^2 - (x + 2c)^2 = (x + c)^2$
7. $a(2a + x) = 2(x^2 + a^2) - bx$
8. $x^2 + a^2 - b^2 = 2ax$
9. $x^2 - a^2 = b(x - a)$
10. $x(x - a) = a(x^2 - a^2)$
11. $x^2 - (a + b)^2x + ab(a + b)^2 = 0$
12. $x^2 - (a + b)cx + abc^2 = 0$
13. $6x^2 - 2x - 4 = ax - a$
14. $x(x - 2a) = \dfrac{2a^2 + 1}{a^2}$
15. $\sqrt{b(ax - 2b)} = ax - 2b$
16. $\sqrt{n(3x + n)} = 3n - \sqrt{nx}$

F

1. Find three consecutive even numbers such that the sum of the squares of the smaller two equals the square of the largest number.

2. The length of a rectangle is 1 foot more than twice the width, and the area of the rectangle is 171 square feet. Find the dimensions of the rectangle.

3. A boy, selling a bicycle for \$24, finds that his loss in percentage is the same as the number of dollars that he paid for the bicycle. What did he pay for the bicycle?

4. The sum of the digits of a two-figured number is 9; and if the number is multiplied by the units' digit, the product obtained is 216. Find the number.

5. A motorist travels 144 miles, and finds that he could have made the trip in 30 minutes less time if he had traveled 4 miles per hour faster. At what rate did he travel?

6. A man can do a certain job in 9 days less time than his son can do it. If they can complete the work in 20 days by working together, how long will it take each alone to do the job?

7. A motorboat capable of traveling 10 miles per hour in still water takes 30 minutes longer to travel upstream a distance of 12 miles than it

takes to travel downstream the same distance. Find the rate of the current.

8. A man is five times as old as his older boy and fifteen times as old as his other son. In two years his age will equal the product of his sons' ages. What are their present ages?

9. The sum of the reciprocals of two consecutive integers is $\frac{17}{72}$. Find the integers.

10. A lawn 60 feet long and 34 feet broad has a path of uniform width around it. If the area of the path is 600 square feet, find its width.

11. A merchant bought a number of yards of cloth for \$250. He kept 5 yards and sold the rest at 70 cents per yard profit. If, in addition to the 5 yards of cloth, he made \$74 on the deal, how many yards of material did he buy?

12. The units' digit of a number between 10 and 100 is 2 more than the tens' digit. If the number is multiplied by the number with its digits reversed, the result is 1855. Find the number.

13. Two motorists start at the same time to travel over the same 120 miles of road. The first travels 80 miles at a certain rate and then increases his speed 10 miles per hour. The second travels at a constant rate of 36 miles per hour, and arrives 20 minutes ahead of the first. Find the original rate of the first motorist.

14. It takes a large pipe 10 minutes less time to fill a tank than a small pipe. If two large pipes and one small pipe can fill the tank in 21 minutes, how long will it take a large pipe alone to fill the tank?

15. Tony is 2 years older than his sister, and next year the product of their ages will be three times what the product of their ages was last year. What are their present ages?

16. The circumference of a rear wheel of a certain wagon is 40 inches longer than the circumference of a front wheel. The rear wheel makes 50 fewer revolutions than the front wheel in traveling a distance of 1000 feet. How large are the wheels?

17. A motorboat whose speed is 12 miles per hour in still water travels upstream a distance of 12 miles and back again in a total time of 2 hours and 8 minutes. What is the rate of the current?

18. A can do a piece of work in 7 days less than B, and they both together can do the work in 9 days less than A alone. Find the time they take when they work together.

19. A group of boys bought a canoe for \$70, dividing the expense equally. If the number of boys had been two less, the expense per person would have been increased \$1.75. How many boys were there in the group?

20. The diagonal and the longer side of a rectangle together are three times the length of the shorter side. If the longer side exceeds the shorter side by 1 foot, what is the area of the rectangle?

Be It Resolved

Determine m, n, and p so that $x^4 + x^3 + px^2 - 4x + 48$ can be resolved into two factors of the form $x^2 + mx + 6$ and $x^2 + nx + 8$.

The Difference Yields

For what value of a do the equations $x^2 + ax - 3 = 0$ and $x^2 - 3x + a = 0$ have a single common root?

A Puzzling Game

Divide a square into 25 smaller squares, and place a penny on each of the nine interior squares. The problem is to remove eight of the pennies and leave the ninth penny in the central square. A penny is removed when another jumps over it; the jumps can be made horizontally, vertically, or diagonally.

One Makes the Difference

Show that $(2 + 1)(2^2 + 1)(2^4 + 1)(2^8 + 1)(2^{16} + 1) = 2^{32} - 1$.

Mind Your P's and Q's

Solve for x when $\dfrac{1}{x + p + q} = \dfrac{1}{x} + \dfrac{1}{p} + \dfrac{1}{q}$.

Simple Division

What divisor of the four numbers 701, 1059, 1417, and 2312 will leave the same remainder in every case?

CHAPTER XI

SYSTEMS INVOLVING QUADRATIC EQUATIONS

118. Quadratic Equations in Two Unknowns. The most general quadratic equation in two unknowns has the form

$$(1) \qquad Ax^2 + Bxy + Cy^2 + Dx + Ey + F = 0,$$

where at least one of the constants A, B, or C is not zero.

It is shown in more advanced courses that the graph of an equation of the form (1) has in general one of the forms shown in Fig. 18, although exceptional cases may arise in which the equation gives rise to no graph at all, or to a graph consisting of just a single point, one line, two intersecting lines, or two parallel lines.

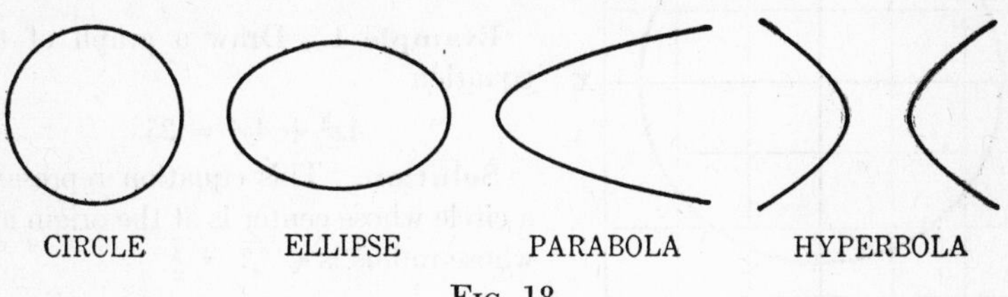

FIG. 18

These curves as a group are often called the *conic sections* inasmuch as they are the curves obtained in taking plane sections of a cone.

A circular cone is generated by revolving one of two intersecting lines about the other as an axis. The fixed line is called the *axis* of the cone. The point of intersection of the lines is called the *vertex* of the cone. The various positions of the rotating line are called the *elements* of the cone.

Let V be the acute angle between the two lines generating a circular cone; then the four curves mentioned above are obtained as the plane sections of the cone where the angle between the plane and the axis of the cone (see Fig. 19) is (A) equal to 90°, (B) less than 90° but greater than V, (C) equal to V, (D) less than V.

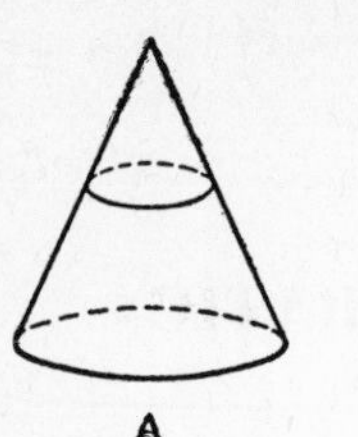

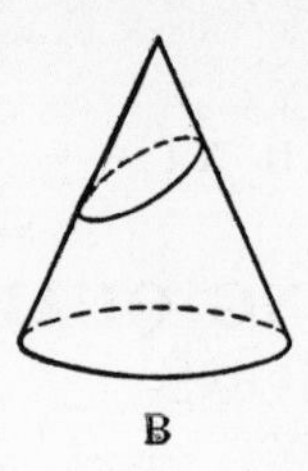

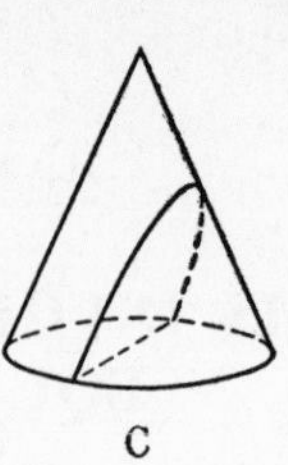

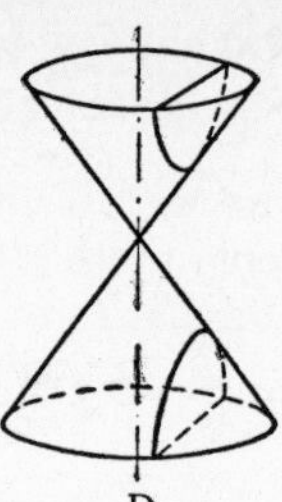

FIG. 19

In the graphical work which follows, we shall restrict ourselves to simpler forms of equation (1).

119. Standard Forms of Quadratic Equations. I. Circle. *An equation of the form* $\mathrm{A}x^2 + \mathrm{A}y^2 = \mathrm{C}$ *in which the constants* A *and* C *are positive represents a circle whose center is at the origin and whose radius is* $\sqrt{\frac{\mathrm{C}}{\mathrm{A}}}$.

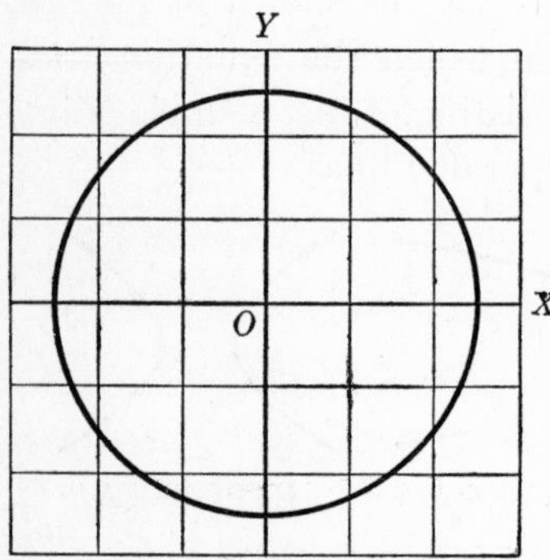

FIG. 20

Example 1. Draw a graph of the equation

$$4x^2 + 4y^2 = 25.$$

Solution. This equation represents a circle whose center is at the origin and whose radius is $\sqrt{\frac{25}{4}} = \frac{5}{2}$.

II. Ellipse. *An equation of the form* $\mathrm{A}x^2 + \mathrm{B}y^2 = \mathrm{C}$ *where the constants are positive and* $\mathrm{A} \neq \mathrm{B}$ *represents an ellipse whose center is at the origin and whose axes lie on the coordinate axes.*

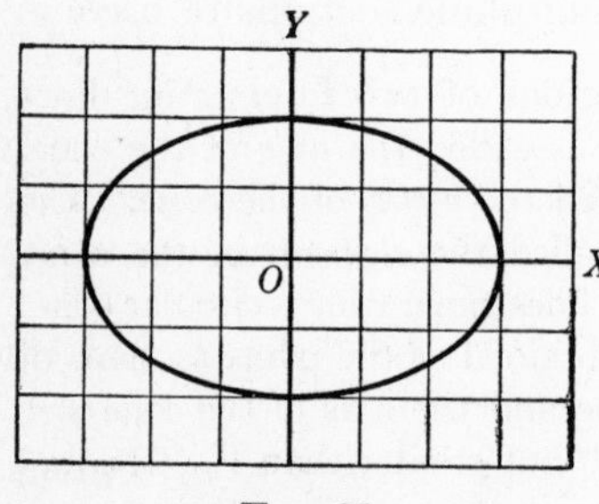

FIG. 21

Example 2. Draw a graph of the equation

$$4x^2 + 9y^2 = 36.$$

Solution. Taking $x = 0$, we obtain the y intercepts, ± 2.

Taking $y = 0$, we obtain the x intercepts, ± 3.

Solving the given equation for y, we use the equivalent equation

$$y = \pm \tfrac{2}{3}\sqrt{9 - x^2}$$

to obtain other points of the curve.

In tabular form, we obtain

$x =$	0	± 3	± 1	± 2
$y =$	± 2	0	± 1.9	± 1.5

NOTE. In plotting irrational values, express them to the nearest tenth in decimals.

III. Hyperbola. *An equation of the form* $Ax^2 - By^2 = C$ *or* $By^2 - Ax^2 = C$ *where the constants are positive and* $C \neq 0$ *represents a hyperbola symmetrical to the coordinate axes.*

Example 3. Draw a graph of the equation

$$4x^2 - 9y^2 = 36.$$

Solution. Taking $y = 0$, we obtain the x intercepts, ± 3.

Solving the given equation for y, we use the equivalent equation

$$y = \pm \tfrac{2}{3}\sqrt{x^2 - 9}$$

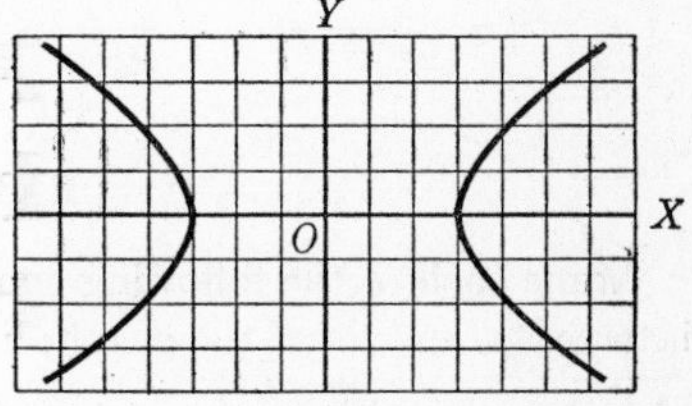

FIG. 22

to obtain other points of the curve. In the above form of the equation it is evident that any values of x numerically less than 3 will give imaginary values for y and hence no points on the curve. Taking $x = \pm 4, \pm 5, \pm 6$, we obtain the table of values

$x =$	± 3	± 4	± 5	± 6
$y =$	0	± 1.8	± 2.7	± 3.5

NOTE. *An equation of the form* $xy = k$, *where the constant* k *is not zero, represents a hyperbola.* This equation is illustrated graphically on p. 232.

IV. Parabola. *An equation of the form* $x = ay^2 + by + c$ *where* $a \neq 0$ *represents a parabola whose axis is horizontal.*

NOTE. *An equation of the form* $y = ax^2 + bx + c$ *represents a parabola whose axis is vertical.* This equation was discussed in Article 105.

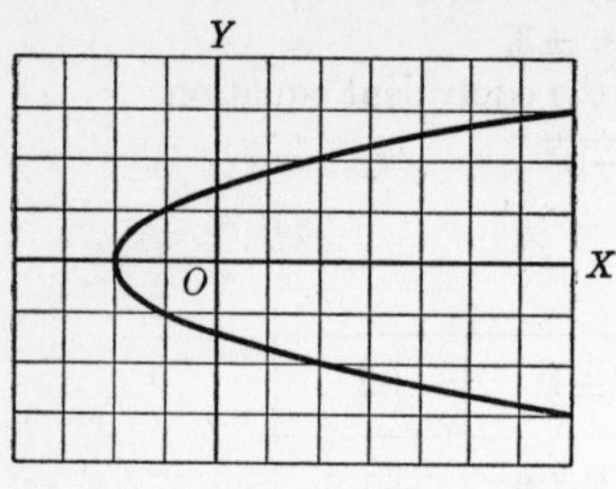

Fig. 23

Example 4. Draw a graph of the equation

$$x = y^2 - 2.$$

Solution. Taking $x = 0$, we obtain the y intercepts, $\pm\sqrt{2}$.

Taking $y = 0$, we obtain the x intercept, -2.

Writing the given equation in the form

$$y = \pm\sqrt{x + 2},$$

we see that the values of y are imaginary for x less than -2. Hence for values of x greater than -2, we obtain the table of values

$x =$	0	-2	2	7
$y =$	± 1.4	0	± 2	± 3

EXERCISE 94

Name each of the following conics, obtain a table of points including the intercepts, and draw the graph:

1. $x^2 + y^2 = 16$ **2.** $9x^2 + 4y^2 = 36$ **3.** $xy = 6$
4. $x^2 - 4y^2 = 16$ **5.** $x = 2y^2 - 2$ **6.** $9x^2 + 9y^2 = 49$
7. $x^2 + 4y^2 = 4$ **8.** $y = x^2 + 2x + 1$ **9.** $y^2 - x^2 = 4$
10. $4x = y^2$ **11.** $xy = -12$ **12.** $9x^2 + 16y^2 = 144$
13. $y^2 = 9 - x^2$ **14.** $9x^2 - 4y^2 = 36$ **15.** $x = y^2 + 2y$
16. $x = y^2 - y - 2$ **17.** $y = x^2 - 3x - 4$ **18.** $4x^2 + 25y^2 = 100$
19. $2xy = 15$ **20.** $4y^2 - 9x^2 = 36$ **21.** $y = 2x^2 - 3x + 1$
22. $2x^2 - 3y^2 = 18$ **23.** $2x^2 + 2y^2 = 13$ **24.** $2x^2 + 3y^2 = 30$
25. $2y^2 - x^2 = 8$ **26.** $x = y^2 - 2y - 3$ **27.** $5x^2 - y^2 = 20$
28. $2y = x^2 - 5x$ **29.** $5x^2 + 4y^2 = 36$ **30.** $2x = y^2 + y + 1$

120. Graphical Solution of a System Involving Quadratic Equations. If we draw the graphs of the two equations

$$x^2 + y^2 = 25$$

and

$$x - y + 1 = 0$$

on the same coordinate system, we see that the graphs intersect at the points (3,4) and (−4,−3). It is evident that each of these pairs

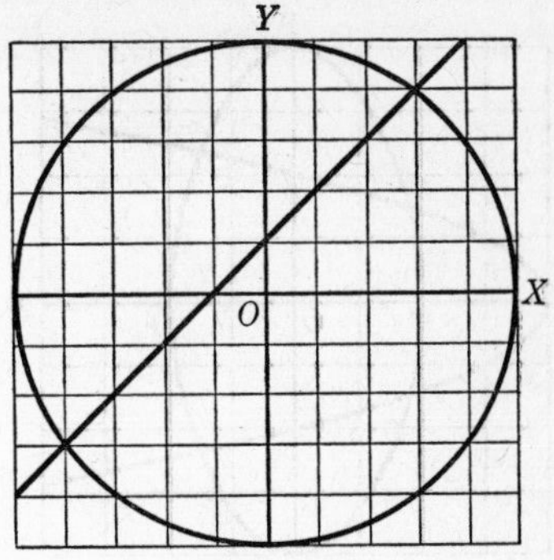

FIG. 24

of values satisfies both of the given equations and hence corresponding to each point of intersection we have a *solution* of those equations.

An inspection of the curves in Fig. 25 indicates that a line can intersect a conic section in two points at most. The curves in Fig. 26 show that there may be only one or possibly no points of intersection. Corresponding to the latter figures the respective systems of equations would have only one or no real solutions. If there are no points of intersection, the system of equations may have imaginary solutions.

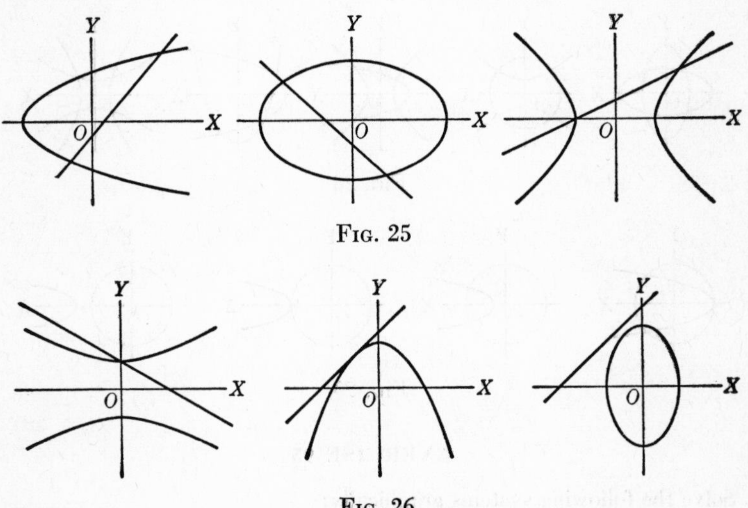

FIG. 25

FIG. 26

Example 1. Solve graphically the equations

$$4x^2 + y^2 = 16,$$
$$y^2 = x + 4.$$

Solution. Plotting the graphs on the same coordinate system, we see in Fig. 27 that they intersect in four points. Each point represents a solu-

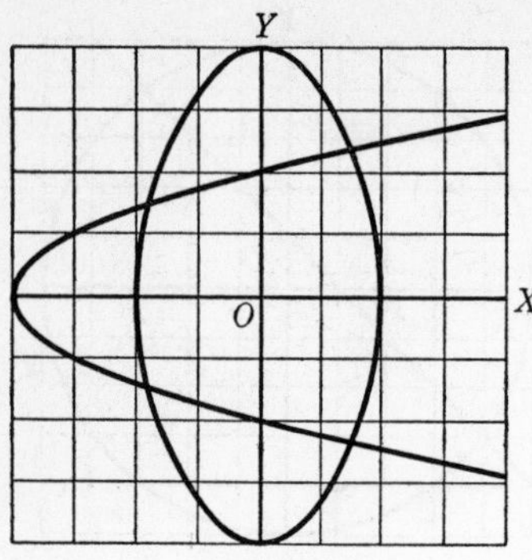

Fig. 27

tion of the given equations and hence the solutions are

$$x = 1.6, \quad 1.6, \quad -1.9, \quad -1.9,$$
$$y = 2.4, \quad -2.4, \quad 1.5, \quad -1.5.$$

Note. The graphs should be drawn carefully on a scale sufficiently large so that the points of intersection may be read correctly to tenths of decimals.

From a study of the various combinations of conic sections it is evident in Fig. 28 that *any two of these curves can intersect in not more than four points.* Hence, the respective systems of equations can have four real solutions at most.

Note. Fig. 29 illustrates that there may be less than four real solutions.

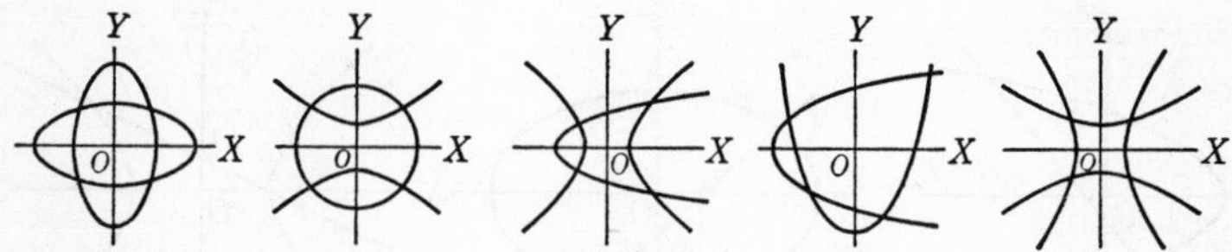

Fig. 28

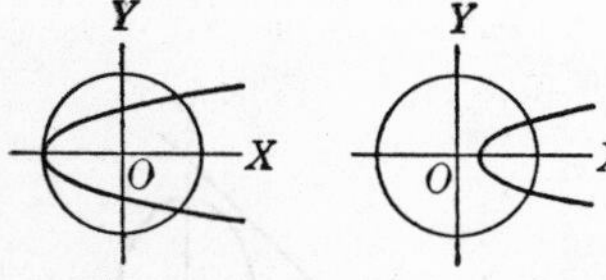

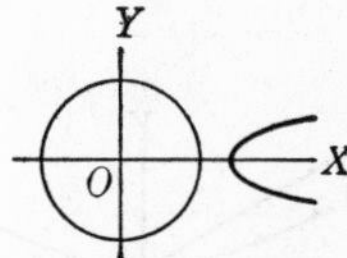

Fig. 29

EXERCISE 95

Solve the following systems graphically:

1. $y^2 = 4x + 4$
   $y = 2x - 2$

2. $x^2 + y^2 = 17$
   $x^2 + 9y^2 = 25$

3. $x^2 + 4y^2 = 25$
   $x - 2y + 1 = 0$

4. $2x + y^2 = 6$
   $x^2 + y^2 = 5$

5. $x^2 - 3y^2 = 1$
   $x^2 + 4y^2 = 8$

6. $y = x^2 - 5x + 4$
   $3x + y = 3$

7. $x^2 + y^2 = 9$
   $x^2 - 3y^2 = 12$

8. $y^2 - x^2 = 4$
   $y = x + 2$

9. $x^2 + 4y^2 = 9$
   $y^2 = 12(x + 3)$

10. $y^2 = 6x + y + 24$, $xy = 6$

11. $3y = x^2$, $9y^2 - 5x^2 = 36$

12. $4x^2 - y^2 = 12$, $4y^2 - x^2 = 12$

13. $x^2 + y^2 = 4$, $y = x + 3$

14. $x^2 - y^2 = 9$, $5x - 4y = 9$

15. $y^2 = 4x - 4$, $x^2 = 4y - 4$

Solve the following systems graphically and express solutions correct to the nearest tenth in decimals:

16. $xy = 5$, $y = x^2$

17. $x^2 + y^2 = 16$, $2x + 3y = 6$

18. $x^2 + y^2 = 25$, $y^2 = 5 - x$

19. $16x^2 + 9y^2 = 144$, $xy = 4$

20. $xy + 6 = 0$, $2x + y = 2$

21. $x^2 + 4y^2 = 4$, $xy = 1$

22. $2x^2 + y^2 = 6$, $5x + y = 9$

23. $y^2 = 3x + 9$, $x^2 + 2x - y = 2$

24. $x^2 = 2x + y + 5$, $xy = 1$

121. Systems Involving One Linear Equation. To solve a system of equations in which one of the equations is of the first degree, the method of substitution is used. The following procedure indicates the method:

1. Solve the linear equation for one of the unknowns in terms of the other.
2. Substitute the value obtained in (1) for that unknown in the second equation.
3. Solve the equation obtained in (2).
4. Substitute the values obtained in (3) in the *linear* equation to find the corresponding values of the second unknown.

Example 1. Solve the following equations and check your results:

$$x + y = 3,$$
$$x^2 + y^2 = 17.$$

Solution. Following the procedure as outlined, we have

$$\begin{aligned}
&(1) & y &= 3 - x,\\
&(2) & x^2 + (3 - x)^2 &= 17,\\
&(3) & x^2 + 9 - 6x + x^2 &= 17,\\
&& 2x^2 - 6x - 8 &= 0,\\
&& 2(x - 4)(x + 1) &= 0,\\
&& x &= 4,\ -1.\\
&(4)\ \text{Since} & y &= 3 - x;\\
&& \text{when } x &= 4,\ y = -1,\\
&& \text{when } x &= -1,\ y = 4.
\end{aligned}$$

Check. Substituting $x = 4$ and $y = -1$ in the given equations, we have

$$(4) + (-1) = 3, \text{ and } (4)^2 + (-1)^2 = 17,$$
$$3 = 3. \qquad 17 = 17.$$

Substituting $x = -1$ and $y = 4$ in the given equations gives

$$(-1) + (4) = 3, \text{ and } (-1)^2 + (4)^2 = 17,$$
$$3 = 3. \qquad 17 = 17.$$

Example 2. Solve the system of equations:

$$\frac{x}{y} + \frac{y}{x} = 2\tfrac{1}{2},$$
$$x + 2y = 4.$$

Solution. Following the procedure outlined, we have

$$(1) \qquad x = 4 - 2y.$$

Clearing fractions in the other equation gives

$$2x^2 + 2y^2 = 5xy.$$

Hence,

$$(2) \qquad 2(4 - 2y)^2 + 2y^2 = 5y(4 - 2y).$$
$$(3) \quad 32 - 32y + 8y^2 + 2y^2 = 20y - 10y^2,$$
$$20y^2 - 52y + 32 = 0,$$
$$5y^2 - 13y + 8 = 0,$$
$$(y - 1)(5y - 8) = 0,$$
$$y = 1, \tfrac{8}{5}.$$

(4) Since $x = 4 - 2y$;

when $y = 1$, $x = 2$,

when $y = \frac{8}{5}$, $x = \frac{4}{5}$.

EXERCISE 96

Solve the following systems of equations for x and y:

1. $x - y = 1$
$x^2 + y^2 = 13$

2. $x + 5y = 2$
$xy + y = -2$

3. $y = 2x - 3$
$2x^2 - xy = 15$

4. $x + y = 2$
$x^2 - 2y^2 = 8$

5. $x - 2y = 1$
$xy = 10$

6. $x - y = 2$
$x^2 + xy + y^2 = 13$

7. $x + 4y = 9$
$x^2 + 4y = 9$

8. $5x - y = 4$
$y = 4x^2 + x - 12$

9. $y = \frac{1}{2}x$
$x^2 + y^2 = 16 - x$

10. $2x + y = 5$
$x^2 - y^2 = 7$

11. $y = 3(x - 1)$
$3x^2 + xy = 3(x + 2y)$

12. $2y = 5x$
$x^2 + 2y^2 = 54$

13. $2x = y + 3$
$x^2 + y^2 = 2xy + 4$

14. $y = \frac{3}{4}x$
$2x^2 - 3xy + 4 = 0$

15. $5x - 3y = 6$
$2x^2 - 3x = y^2$

16. $4x - 3y = 25$
$x^2 + y^2 = 25$

17. $3x + 4y = 7$
$x^2 + x = 2y^2$

18. $2x - 3y = 5$
$x^2 + xy + y^2 = 21$

19. $x + y - 1 = 0$
$\frac{1}{x} + \frac{1}{y} = 4$

20. $x + y = 7$
$\frac{x}{y} + \frac{y}{x} = \frac{25}{12}$

21. $\frac{x - 3}{y - 2} = \frac{x + 1}{y + 2}$
$x^2 + y^2 = 41$

22. $x + \frac{2}{x} = y$
$5x - 2y = 4$

23. $\frac{1}{x - 1} + \frac{1}{y - 1} = 2$
$3x - y = 4$

24. $\frac{x + 3}{y + 2} = 2$
$x^2 - 5y^2 = 4$

25. $x - y = a$
$x^2 + y^2 = 5a^2$

26. $x - y = b$
$xy = a^2 + ab$

27. $x + y = a$
$4xy = a^2 - b^2$

28. $x + y = 2m$
$x^2 - y^2 = 4mn$

29. $x = ay$
$3x - 2y^2 = a^2$

30. $x - y = 2$
$xy + 1 = c^2$

122. Systems Involving Two Quadratic Equations. The algebraic solution of a system of two quadratic equations can often be found by applying methods of elimination similar to those previously studied. In all cases it is important to indicate clearly the pairs of values which together serve as solutions of the problem.

Example 1. Solve the system of equations

$$(1) \qquad x^2 + y^2 = 13,$$
$$(2) \qquad xy = 6.$$

Solution. Solving equation (2) for y, we have

$$y = \frac{6}{x}.$$

Substituting this value for y in equation (1) gives

$$x^2 + \frac{36}{x^2} = 13,$$

$$x^4 + 36 = 13x^2,$$

$$x^4 - 13x^2 + 36 = 0.$$

Factoring, $$(x^2 - 4)(x^2 - 9) = 0.$$

Hence, $$x^2 = 4, 9,$$

$$x = \pm 2, \pm 3.$$

Substituting each of these four values of x in (2), we obtain the corresponding values ± 3, ± 2 for y. Thus the four distinct solutions of the given equations are:

$$x = +2, -2, +3, -3,$$
$$y = +3, -3, +2, -2.$$

Example 2. Solve for x and y:

$$(1) \quad 3x^2 - 2y^2 = 10a^2,$$
$$(2) \quad 2x^2 + 3y^2 = 11a^2.$$

Solution. Multiplying both members of equation (1) by 3, and both members of equation (2) by 2, we have

$$9x^2 - 6y^2 = 30a^2,$$
$$4x^2 + 6y^2 = 22a^2.$$

Adding these equations and solving the resulting equation, we find that $x = \pm 2a$.

On substituting either $+2a$ or $-2a$ for x in equation (2), we find that $y = \pm a$. Hence, we have the four distinct solutions:

$$x = 2a, \quad 2a, \quad -2a, \quad -2a,$$
$$y = a, \quad -a, \quad a, \quad -a.$$

EXERCISE 97

Solve algebraically:

1. $x^2 + y^2 = 13$, $4x^2 - y^2 = 7$
2. $x^2 + 4y^2 = 40$, $4x^2 + y^2 = 25$
3. $x^2 + y^2 = 25$, $xy = 12$
4. $x^2 - xy + y^2 = 4$, $xy = 4$
5. $x^2 - 6y^2 = 3$, $3x^2 + 2y^2 = 29$
6. $y^2 - 3x^2 = 4$, $y + 2x^2 = 4$
7. $x^2 + 3 = xy$, $2x^2 - 6 = xy$
8. $4x^2 + y^2 = 16$, $xy + 4 = 0$
9. $4x^2 + 3y^2 = 19$, $7x^2 - 2y^2 = 26$

10. $3x^2 + 2 = 2y^2$
$y^2 - x^2 = 9$

11. $y^2 = 4 - 4x + y$
$xy = 1$

12. $4x^2 - y = 15$
$2x^2 + 3y = 11$

13. $5x^2 + 19 = 4y^2$
$8x^2 - 20 = 5y^2$

14. $x^2 + y^2 = xy + 19$
$x^2 - y^2 = xy + 1$

15. $3x^2 - 2y^2 = 0$
$5x^2 - 3y^2 = 1$

16. $x^2 + 2y^2 = 4$
$x^2 + y^2 = 2$

17. $(x + y)^2 + (x - y)^2 = 10$
$xy = 2$

18. $(x - 2y)^2 = 4y^2 + 8$
$2x^2 + xy = 34$

19. $\frac{4}{x^2} + \frac{9}{y^2} = 2$
$\frac{12}{x^2} - \frac{18}{y^2} = 1$

20. $\frac{1}{x + 1} + \frac{1}{y + 1} = 1$
$\frac{x}{y} + \frac{y}{x} = \frac{17}{4}$

Solve for x and y:

21. $2x^2 + 3y^2 = 7a^2$
$x^2 - 5y^2 + 3a^2 = 0$

22. $x^2 + 4y^2 = 20a^2$
$xy = 4a^2$

23. $x^2 + y^2 = 2m^2 + 2n^2$
$x^2 - y^2 = 4mn$

24. $4x^2 - y^2 = 4mn - n^2$
$x^2 + 2y^2 = 9m^2 - 8mn + 2n^2$

123. Reduction to Simpler Systems. (Optional.) In order for a product of two or more quantities to be equal to zero, we know that one or more of the factors of that product must be equal to zero. That is, for example, if

$$(x - y)(x - 2y) = 0,$$

then either

$$(x - y) = 0 \text{ or } (x - 2y) = 0.$$

We may use this fact to reduce a particular system of equations to two or more simpler systems of equations.

Example 1. Solve

$$(1) \quad x^2 + y^2 = 8,$$
$$(2) \quad x^2 - 3xy + 2y^2 = 0.$$

Solution. Factoring (2), we have $(x - y)(x - 2y) = 0$. Hence (2) is satisfied if either $x - y = 0$ or $x - 2y = 0$.

Thus any solution of (1) and (2) is a solution of one of the following systems, and conversely any solution of either of the following systems is a solution of the given system:

$$\text{I.} \begin{cases} x^2 + y^2 = 8, \\ x - y = 0. \end{cases} \qquad \text{II.} \begin{cases} x^2 + y^2 = 8, \\ x - 2y = 0. \end{cases}$$

Solving each of the above systems separately, we have the solutions

$$\text{I}'.\ \begin{cases} x = 2, -2, \\ y = 2, -2. \end{cases} \qquad \text{II}'.\ \begin{cases} x = \frac{4}{5}\sqrt{10}, -\frac{4}{5}\sqrt{10}, \\ y = \frac{2}{5}\sqrt{10}, -\frac{2}{5}\sqrt{10}. \end{cases}$$

Thus the complete solution of the given equations consists of the four solutions I′ and II′.

NOTE. The above procedure is known as the method of **reduction by factoring.**

A system of equations in which all the terms containing variables are of the second degree can be reduced to an equivalent system which in turn can be reduced by factoring.

Example 2. Solve:

$$(1) \qquad x^2 + xy = 12,$$
$$(2) \qquad y^2 + 4xy = 48.$$

Solution. Multiply equation (1) by 4 and subtract equation (2) from this product. This gives

$$4x^2 - y^2 = 0 \quad \text{or} \quad (2x - y)(2x + y) = 0.$$

Hence, we solve the systems

$$\text{I}.\ \begin{cases} 2x - y = 0, \\ x^2 + xy = 12. \end{cases} \quad \text{and} \quad \text{II}.\ \begin{cases} 2x + y = 0, \\ x^2 + xy = 12. \end{cases}$$

The solutions of these systems are

$$\text{I}'.\ \begin{cases} x = 2, -2, \\ y = 4, -4. \end{cases} \quad \text{and} \quad \text{II}'.\ \begin{cases} x = 2i\sqrt{3}, -2i\sqrt{3}, \\ y = -4i\sqrt{3}, 4i\sqrt{3}. \end{cases}$$

Thus we see that the given equations in this example have two real and two imaginary solutions.

NOTE. The above procedure is known as the method of **eliminating the constant.**

EXERCISE 98

Reduce to simpler systems and solve:

1. $x^2 + y^2 = 25$
$(3x - 4y)(x + y + 5) = 0$

2. $x^2 - y^2 = 16$
$(3x + 5y)(x + y - 4) = 0$

3. $8x^2 - 3xy = 14$
$(2x + y)(3x - 2y) = 0$

4. $2xy - y^2 = 3$
$(x + y)(2x - y - 3) = 0$

Reduce to simpler systems by factoring and solve:

5. $2xy - y^2 = 9$
$x^2 - y^2 = 0$

6. $x^2 + xy + y^2 = 49$
$x^2 - xy - 2y^2 = 0$

7. $x^2 - 2xy + 3y^2 = 9$
$2x^2 - 13xy + 15y^2 = 0$

8. $x^2 + 4y^2 = 9$
$x^2 + 2xy = 0$

9. $3xy + y^2 = 4$
$(x + y)^2 - 4 = 0$

10. $x^2 - 3y^2 = 9$
$x^2 - 4xy + 4y^2 = 0$

Solve by eliminating constants:

11. $x^2 + 2xy = 8$
$3x^2 - 4y^2 = 8$

12. $x^2 + xy = 6$
$2x^2 + y^2 = 27$

13. $x^2 - xy + y^2 = 21$
$2xy - y^2 = 15$

14. $x^2 - xy + y^2 = 3$
$2x^2 + y^2 = 6$

15. $x^2 + xy = 6$
$xy + y^2 = 3$

16. $x^2 - 2xy + 2y^2 = 5$
$7x^2 + 2y^2 = 15$

17. $x^2 - 2xy + y^2 = 9$
$2x^2 - 3xy + y^2 = 15$

18. $2x^2 - xy + 3y^2 = 18$
$x^2 + 2xy - 3y^2 = 12$

Reduce to four systems, each consisting of two linear equations:

19. $(x + y)(x - y - 2) = 0$
$(x - 3y)(2x - y - 3) = 0$

20. $(x + y + 5)(x + y - 5) = 0$
$12x^2 - 16xy - 3y^2 = 0$

Factor and divide the first equation by the second equation:

21. $x^3 + y^3 = 9$
$x^2 - xy + y^2 = 3$

22. $x^2 - xy = 6$
$xy - y^2 = 2$

Solve by factoring the second equation:

23. $x^2 + xy = 6$
$x^2 + 2x + 1 - y^2 = 0$

24. $x^2 - 5xy = 4$
$x^2 - y^2 = x - y$

REVIEW OF CHAPTER XI

A

1. Find y for $x = 0, \pm 0.2, \pm 0.4, \pm 0.6, \pm 0.8, \pm 1$, and graph the equation $2x^2 + y^2 = 2$.

2. Find y for $x = 0, 1, 2, 3, 4$, and graph the equation $x^2 + y^2 = 4x$. What conic section is this?

3. Find y for $x = \pm\frac{1}{10}, \pm\frac{1}{2}, \pm 1, \pm 2, \pm 3$, and graph the equation $x^2 + xy = 1$. What conic section is this?

4. Draw a graph of the equation $xy + x = 6$.

Solve the following systems graphically:

5. $x^2 + 9y^2 = 25$
$4x + 9y = 25$

6. $x - 2y + 6 = 0$
$xy = 8$

7. $4x^2 - y^2 = 16$
$4x^2 + y^2 = 16$

8. $9x^2 + y^2 = 25$
$xy = 4$

9. $x^2 + 4y^2 = 20$
$4x^2 + y^2 = 20$

10. $y^2 - x^2 = 9$
$5x - 2y = 10$

11. $2x^2 - 3y^2 = 8$
$y^2 = x + 2$

12. $6y = 3x^2 + 2x$
$2x = y^2 - y$

B

Solve algebraically:

1. $x + 2y = 3$
$xy + y^2 + 3x + 1 = 0$

2. $x^2 + 2y^2 = 17$
$3x^2 - 4y^2 = 11$

3. $x^2 - 4y^2 = 15$
$xy = 2$

4. $2x + 3y = 9$
$4x^2 - 5y^2 = 31$

5. $2x^2 - 3y = 2$
$3x^2 + 2y = 16$

6. $4x^2 + 3y^2 = 19$
$3x^2 + 5y^2 = 17$

7. $3x - 2y = 10$
$xy + y^2 = 5$

8. $2x^2 + xy - y^2 = 14$
$xy = 12$

Solve and express results correct to two decimal places:

9. $x = 2y^2 - 3y - 5$
$2x - 5y + 1 = 0$

10. $x^2 + 2y^2 = 21$
$2x^2 - 3y^2 = 6$

11. $3x^2 + 2y^2 = 10$
$xy + 2 = 0$

12. $3x - 7y + 4 = 0$
$3xy - 8 = 0$

13. $5x^2 + 2y^2 = 14$
$y^2 = 2x + 1$

14. $5x + 2y - 7 = 0$
$3x^2 - 2xy + 4y = 6$

Solve for x and y:

15. $x - 2y = a$
$2x^2 - y^2 = 17a^2$

16. $x + a^2y = 2a$
$xy = 1$

C (Optional)

Reduce to simpler systems and solve:

1. $10x^2 + 6xy = 25$
$x^2 - y^2 = 0$

2. $3x^2 + 5xy - 2y^2 = 5$
$2x^2 - 7xy + 6y^2 = 0$

3. $(x + 2y - 10)(2x + y - 5) = 0$
$2x^2 - 5xy + 2y^2 = 0$

4. $3x^2 + xy = 9$
$y^2 + 2xy = 0$

Eliminate the constant and solve:

5. $4x^2 - xy + 2y^2 = 10$
$3x^2 + xy = 5$

6. $2x^2 - xy = 24$
$y^2 + xy = 12$

7. $x^2 + 3xy + 8 = 0$
$y^2 - xy = 12$

8. $x^2 + 16 = 2xy$
$xy + 2y^2 = 60$

9. $(x + y)^2 + (x + 2y)^2 = 1$
$(x + 2y)^2 + (x + 3y)^2 = 5$

10. $(x + y)^2 + x^2 = 5$
$(x + y)^2 + y^2 = 10$

Solve for x and y:

11. $xy + 2y^2 = 24a^2$
$2xy - x^2 = 8a^2$

12. $x^2 - a^2y^2 = 0$
$y^2 = y - x$

Solve the following systems by any method:

13. $x^2y^2 - 4 = 0$
$x^2 + y^2 = 5$

14. $(x + y)(2x + 3y) = 1$
$(x + y)(3x + 2y) = 4$

15. $x^4 - y^4 = 65$
$x^2 - y^2 = 5$

16. $xy^2 + x^2y = 6$
$x + y = 3$

D

Solve each problem, using two unknowns:

1. The area of a rectangle is 75 square feet and its perimeter is 35 feet. Find the sides of the rectangle.
2. The sum of the reciprocals of two numbers is 7 and the product of the numbers is $\frac{4}{45}$. Find the numbers.
3. The hypotenuse of a right triangle is 17 feet long and the area of the triangle is 60 square feet. Find the other sides.
4. In a two-figured number the sum of the squares of the digits is 25, and the product of the number by the number with digits reversed is 1462. Find the number.
5. The product of two integers exceeds their sum by 55, and the quotient of the two integers is 7 less than their difference. Find the integers.
6. A certain fraction is 5% of the sum of its numerator and denominator. If the reciprocal of the fraction is 80% of the sum of the numerator and denominator, find the fraction.
7. A milkman covers his route, which is 21 miles long, by 11 A.M. each day. If his average rate of travel were half a mile faster each hour, he would cover the route by 10 A.M. What time does he start in the morning?
8. The hypotenuse of a right triangle is 17. If one leg is increased by 1 and the other leg by 4, the hypotenuse becomes 20. Find the sides of the triangle.

9. A and B plan to make 80 boxes in two days, making 40 on each day. On the first day A works alone for 3 hours and then B working alone finishes the 40 boxes in 2 hours and 45 minutes. On the second day, A makes 12 boxes before B arrives and then the two working together finish the other 28 boxes 4 hours after A began. Find the number of boxes each makes per hour.

10. On the same day A, B, and C start to solve a certain number of problems. A solved 6 a day and finished them 4 days after B. C solved 3 more a day than B and finished 2 days before he did. Find the number of problems and the number of days each worked.

Play Ball

It takes a seconds for a ball to go from the pitcher to the catcher, and b seconds for the catcher to handle it and get off a throw to second base. It is 90 feet from first base to second, and 130 feet from the catcher's position to second. A runner stealing second has a start of 13 feet from first base when the ball leaves the pitcher's hand; he beats the throw to second base by $\frac{1}{8}$ of a second. The next time he tries it, he gets a start of only $3\frac{1}{2}$ feet, and the ball reaches second base when he is still 6 feet away. Find the runner's rate of travel.

Really?

Prove that the roots of $(x - a)(x - b) = k^2$ are always real.

In Memoriam, John Doe, 18??–1952

John Doe lived a sixth of his life as a child, a twelfth as an adolescent, and a seventh as a bachelor. Five years later, John became the father of a son who lived half as long as John and died four years before him. When was John born?

More Arithmetic

Show that $2^{100} - 1$ is a multiple of 15.

A Radical Problem

If $(5 + 2\sqrt{6})^n = a + b\sqrt{6}$, where n, a, and b are positive integers, what is the value of $a^2 - 6b^2$?

CHAPTER XII

RATIO, VARIATION, AND BINOMIAL THEOREM

124. Comparing Numbers; Ratio. Similar quantities may be compared (1) by giving the difference in magnitude of the quantities, or (2) by stating what multiple part one quantity is of the other.

Thus, in a group composed of 20 men and 10 women, we may say (1) that there are 10 more men than there are women, or (2) that there are twice as many men as there are women.

The second type of comparison illustrated above is obtained by *dividing* the quantities being compared.

Definition. *When two like quantities are compared by division, the numerical result of the division is called the* ***ratio*** *of the two quantities.*

Illustrations. The ratio of \$6 to \$8 is $\frac{3}{4}$.
The ratio of 3 inches to 2 feet is $\frac{3}{24}$ or $\frac{1}{8}$.

The ratio of a with respect to b is sometimes expressed by the notation $a : b$ (read, "a *is to* b"), or otherwise by the division notation $\frac{a}{b}\cdot$

An equality of ratios, such as

$$a : b = 2 : 3$$

is read "a *is to* b *as two is to three.*" When we write the ratios in their equivalent form

$$\frac{a}{b} = \frac{2}{3},$$

we see that this equation implies that a is two-thirds as large as b.

When we wish to compare three or more quantities in the above sense, we may write a relation such as the following:

$$a : b : c = 2 : 3 : 5.$$

Each side of this equality is called a **continued ratio** and the above is read "a *is to* b *is to* c *as two is to three is to five.*" This single relation is an abbreviated notation for expressing the following ordinary ratios:

$$a : b = 2 : 3,$$
$$a : c = 2 : 5.$$

Example 1. The $12,000 profits of a firm are to be divided among three partners in the ratio 2 : 3 : 5. How much does each receive?

Solution. Let $2x$ represent the smallest of the three shares; then $3x$ must represent the next larger share. This is true, since evidently

$$\frac{2x}{3x} = \frac{2}{3} \text{ or } 2 : 3.$$

Similarly, the largest share will be represented by $5x$. Adding, we have

$$2x + 3x + 5x = 12{,}000,$$
$$x = 1200.$$

Answer. The profits are divided in the three shares: $2400, $3600, and $6000.

EXERCISE 99

Express each of the following ratios in its simplest form:

1. $3 : 6$ **2.** $34 : 17$ **3.** $35 : 56$ **4.** $162 : 243$

5. $\frac{3}{4} : \frac{1}{2}$ **6.** $\frac{2}{3} : \frac{1}{5}$ **7.** $0.06 : 0.3$ **8.** $0.6 : 0.15$

9. $5a : 7a$ **10.** $16x^2 : 24x^2$

11. $(x + 1) : (2x + 2)$ **12.** $(x - 1) : (x^2 - 1)$

13. Find the ratio of 2 feet to $5\frac{1}{2}$ yards.

14. Find the ratio of 3 pints to 2 quarts.

15. Find the ratio of 12 ounces to 3 pounds.

16. Find the ratio of 440 yards to 1 mile.

17. What is the ratio of the area of a circle whose radius is 2 inches to that of one whose radius is 3 inches?

18. What is the ratio of the area of a 6-inch square to that of a 3-inch square?

19. Divide 143 into two parts whose ratio is 4 : 7.

20. Find two numbers whose sum is 91 and whose ratio is 3 : 4.

21. Two numbers are in the ratio 3 : 5, and their difference is 34. What are the numbers?

22. Find three numbers whose sum is 153 and whose ratio is 2 : 3 : 4.

23. A line 45 inches long is to be divided into three parts having the ratio 3 : 5 : 7. Find the length of each part.

24. Find four numbers whose sum is 370 and whose ratio is 1 : 2 : 3 : 4.

25. Two numbers are in the ratio of 3 : 4; and if 5 is subtracted from each, the remainders are in the ratio of 2 : 3. Find the numbers.

26. What number must be taken from each term of the ratio 40 : 57 in order to have the ratio 2 : 3?

27. Two numbers are in the ratio of 3 : 5. If 10 is added to the smaller and 12 is taken from the larger, the resulting numbers are equal. Find the numbers.

28. If $x : y = 3 : 4$, find the numerical value of the ratio $(x + y) : (x + 2y)$. (Note that $x = \frac{3}{4}y$.)

29. The ratio of the dimensions of a rectangle is 3 : 5. If 4 feet is added to each dimension, the area of the rectangle is increased 208 square feet. Find the dimensions.

30. Two positive integers are in the ratio of 4 : 7. If 18 is added to the smaller and 6 is taken from the larger, the resulting numbers have the ratio 2 : 3. Find the integers. Are two answers possible? Why?

125. Variation. One variable y is said to *vary directly* as another variable x when they change in such a way that their ratio is constant.

NOTE. The word *directly* is often omitted, and y is said to vary as x.

For example, an automobile traveling at a constant rate of 20 miles an hour will travel 40 miles in 2 hours, 60 miles in 3 hours, 80 miles in 4 hours, and so on. The relation between the distance and the time is simply stated by saying "When the velocity is uniform, the distance varies as the time."

If y varies as x, by definition, we have the algebraic relation

$$\frac{y}{x} = k \text{ or } y = kx,$$

where k is a constant, usually called the *constant of variation.* The numerical value of this constant depends on the corresponding values which the variables assume in any particular problem. Thus, if we know that $y = 6$ when $x = 2$, it follows that $k = 3$.

As another example, we know that the circumference of a circle varies as the diameter, and from the formula $c = \pi d$ we see that the constant of variation in this case is the absolute constant π.

Example 1. If y varies directly as x, and $y = 8$ when $x = 2$, find y when $x = 4$.

Solution. Since y varies directly as x, we have $y = kx$. When $y = 8$ and $x = 2$, by substitution we have $8 = k \cdot 2$. Hence, $k = 4$ and the exact relation between the variables x and y is

$$y = 4x.$$

Now substituting 4 for x, we find that $y = 16$.

Example 2. The weight of a solid spherical ball varies as the cube of its diameter. If a ball 2 inches in diameter weighs 1.6 pounds, how much will a ball 3 inches in diameter weigh?

Solution. Since the weight (W) varies as the *cube* of the diameter (D), we have

$$W = kD^3.$$

When $W = 1.6$, $D = 2$,

$$1.6 = k \cdot 2^3 \text{ or } k = 0.2.$$

Thus, $W = 0.2D^3$; and when $D = 3$,

$$W = 0.2(3)^3 = 5.4 \text{ pounds.}$$

EXERCISE 100

1. If y varies directly as x, and $y = 35$ when $x = 14$, find y when $x = 16$.
2. If m varies directly as n, and $m = 15$ when $n = 20$, find m when $n = 32$.
3. If a varies directly as b, and $a = 35$ when $b = 42$, find a when $b = 78$.
4. If p varies directly as q, and $p = 21$ when $q = 24$, find p when $q = 56$.
5. If y varies directly as x^2, and $y = 8$ when $x = 2$, find y when $x = 7$.
6. If b varies directly as c^2, and $b = 3$ when $c = 4$, find b when $c = 16$.
7. If y varies directly as x^3, and $y = 24$ when $x = 2$, find y when $x = 3$.
8. If p varies directly as q^3, and $p = 54$ when $q = 3$, find p when $q = 4$.
9. If y varies directly as $2x + 3$, and $y = 14$ and $x = 2$, find y when $x = 5$.

10. If b varies directly as $a^2 - 1$, and $b = 15$ when $a = 2$, find b when $a = 3$.

11. If x varies directly as $m + n$, and $x = 12$ when $m = 1$ and $n = 2$, find x when $m = 3$ and $n = 4$.

12. If z varies directly as $2x - 3y$, and $z = 8$ when $x = 5$ and $y = 2$, find z when $x = 7$ and $y = 3$.

13. The weight of a certain copper wire varies directly as its length. If 20 feet of the wire weighs $1\frac{1}{4}$ pounds, how much will 92 feet of it weigh?

14. A man's wages varies as the time he works. If his salary for 20 hours of work is \$35, how much will he make in a week of 44 hours?

15. If the weight of a statue varies directly as the cube of its height, what will a statue 5 feet high weigh, if a corresponding one $1\frac{2}{3}$ feet high weighs 15 pounds?

16. The distance through which a body falls in a given time varies directly as the square of the time. If it is observed that a body falls 64 feet in 2 seconds, how far will it fall in 5 seconds?

17. The amount of paint needed to paint a spherical ball varies as the square of the diameter of the ball. If one pint of paint is needed to paint 100 balls of diameter 3 inches, how much paint is needed to paint 150 balls of diameter 5 inches?

18. The volume of a circular cone varies as the product of its base and its height. If the volume is 45 cubic inches when the base is 27 square inches and the height is 5 inches, find the volume when the base is 37 square inches and the height is 6 inches.

126. Inverse Variation. One variable y is said to *vary inversely* as another variable x when y varies *directly* as the *reciprocal* of x.

Algebraically, this definition states that

$$y = k \cdot \frac{1}{x} \text{ or } xy = k.$$

Hence, if the product of two variables remains constant, the variables vary inversely with each other.

For example, an automobile traveling a distance of 120 miles will take 6 hours when traveling 20 miles per hour, 5 hours when traveling 24 miles per hour, 4 hours when traveling 30 miles per hour, and so on. The relation between the velocity and time is simply stated by saying "When the distance is constant, the velocity varies inversely as the time."

Example 1. If a varies inversely as b, and $a = 6$ when $b = 4$, find a when $b = 3$.

Solution. Since a varies inversely as b, we have $ab = k$. Substituting $a = 6$ and $b = 4$, we find $k = 24$. Hence the exact relation satisfied by a and b is

$$ab = 24.$$

Now substituting 3 for b, we find that $a = 8$.

Example 2. If z varies directly as x and inversely as y, and $z = 4$ when $x = 2$ and $y = 3$, find z when $x = 3$ and $y = 6$.

Solution. Since z varies directly as x and inversely as y, we have

$$z = k\frac{x}{y} \text{ or } \frac{zy}{x} = k.$$

Substituting $x = 2$, $y = 3$, and $z = 4$, we have

$$k = \frac{zy}{x} = \frac{4 \cdot 3}{2} = 6.$$

Hence, substituting $x = 3$ and $y = 6$, we find

$$z = 6 \cdot \frac{x}{y} = 6 \cdot \frac{3}{6} = 3.$$

EXERCISE 101

1. If y varies inversely as x, and $y = 8$ when $x = 9$, find y when $x = 6$.
2. If p varies inversely as q, and $p = 9$ when $q = 12$, find p when $q = 8$.
3. If m varies inversely as n, and $m = 51$ when $n = 42$, find m when $n = 34$.
4. If a varies inversely as b, and $a = 3\frac{1}{4}$ when $b = 4\frac{4}{5}$, find a when $b = 2\frac{3}{5}$.
5. If y varies inversely as x^2, and $y = 9$ when $x = 2$, find y when $x = 3$.
6. If s varies inversely as t^2, and $s = 32$ when $t = 6$, find s when $t = 8$.
7. If p varies inversely as q^3, and $p = 54$ when $q = 4$, find p when $q = 3$.
8. If y varies inversely as x^3, and $y = 15$ when $x = 3$, find y when $x = 5$.
9. If y varies inversely as $3x + 5$, and $y = 8$ when $x = 2\frac{1}{2}$, find y when $x = 1\frac{2}{3}$.
10. If u varies inversely as $v^2 + 1$, and $u = 2$ when $v = 7$, find u when $v = 3$.
11. On a boat having a given amount of rations, the number of days the rations will last varies inversely with the number of people on the boat.

If there are rations enough to last 120 people 8 days, how long will the rations last if 40 shipwrecked individuals are taken aboard?

12. The temperature being constant, the pressure which a gas exerts varies inversely as its volume. If a gas has a volume of 76 cubic inches when the pressure is 16 pounds per square inch, find the volume when the pressure is 64 pounds per square inch.

13. The weight of a body above the surface of the earth varies inversely as the square of its distance from the center of the earth. How much would a body weigh at a distance of 1000 miles above the earth's surface, if the body weighed 100 pounds on the earth's surface? Assume the radius of the earth to be 4000 miles.

14. For a circular cone of given volume, the height of the cone varies inversely as the square of the diameter of its base. If the height is 2 inches when the diameter is 3 inches, what is the height when the diameter is 2 inches?

15. If z varies directly as x and inversely as y, and $z = 20$ when $x = 8$ and $y = 12$, find z when $x = 6$ and $y = 18$.

16. If p varies directly as m and inversely as n, and $p = 10$ when $m = 10$ and $n = 12$, find p when $m = 9$ and $n = 6$.

17. The amperage (I) of electricity passing through a wire varies directly as the potential (E) and inversely as the resistance (R) of the wire. When $E = 110$ volts and $R = 10$ ohms, $I = 11$ amperes. Find I when $E = 550$ volts and $R = 22$ ohms.

18. The length of a piece of wire varies directly as the weight and inversely as the square of its cross section. If a wire 100 feet long with a diameter of $\frac{1}{8}$ inch weighs 6 pounds, how long a wire of the same material with a diameter of $\frac{1}{16}$ inch must be taken to weigh 9 pounds?

127. Proportion. Four quantities are said to be in **proportion** when the ratio of the first to the second is the same as the ratio of the third to the fourth. Thus, if $\dfrac{a}{b} = \dfrac{c}{d}$, then a, b, c, d are *terms of a proportion* or *proportionals*. Written in the ratio notation the equality has the form

$$a : b = c : d.$$

The terms a and d are called the *extremes*; b and c are the *means*. Since $\dfrac{a}{b} = \dfrac{c}{d}$, by clearing of fractions we see that $ad = bc$, that is

Rule of Proportion

In a proportion, the product of the extremes equals the product of the means.

Example 1. Find the fourth proportional to 2, 30, 3.

Solution. Let x be the fourth proportional; then

$$2 : 30 = 3 : x,$$

or

$$\frac{2}{30} = \frac{3}{x}.$$

Hence, $x = 45$ is the fourth proportional.

Example 2. If $\frac{a}{b} = \frac{c}{d}$, prove that $\frac{a+b}{b} = \frac{c+d}{d}$.

Solution. Adding 1 to both sides of the given proportion, we have

$$\frac{a}{b} + 1 = \frac{c}{d} + 1.$$

Hence,

$$\frac{a+b}{b} = \frac{c+d}{d}.$$

EXERCISE 102

Find the fourth proportional to the following:

1. 15, 5, 30 **2.** 18, 12, 21 **3.** 10, 14, 15
4. 2, 3, 4 **5.** 3, 6, 9 **6.** $\frac{1}{2}, \frac{1}{3}, \frac{1}{4}$
7. a, ab, c **8.** a, b, c **9.** a^2, ab, b^2
10. a, a^2, a^3 **11.** $a^2 - b^2, a - b, a + b$ **12.** $a^2, a, 1$

If $a : b = c : d$, prove that:

13. $(a - b) : b = (c - d) : d$ **14.** $(a + 2b) : b = (c + 2d) : d$
15. $a : (a + b) = c : (c + d)$ **16.** $a : c = b : d$
17. $a^2d^2 : b^2c^2 = ad : bc$ **18.** $(ad + bc) : ac = 2b : a$
19. $(a + b) : (c + d) = b^2(c - d) : d^2(a - b)$
20. $(a - b) : (c - d) = b^2(c + d) : d^2(a + b)$

Solve the equations:

21. $x : (x + 2) = (2x + 1) : 5x$ **22.** $(x + 2) : x = (x + 5) : (x + 1)$
23. $(x - 1) : (x + 5y) = (x + 1) : (2x + 5y) = 2 : 5$
24. $(x + 2) : (y + 2) = (5x - 2) : (3y - 1) = 4 : 7$

128. The Binomial Theorem. In many branches of mathematics it is important to be able to express any power of a binomial in terms of its component parts. Thus, in multiplying the binomial $(A + B)$ by itself successively we find that

$$(A + B)^1 = A + B,$$
$$(A + B)^2 = A^2 + 2AB + B^2,$$
$$(A + B)^3 = A^3 + 3A^2B + 3AB^2 + B^3,$$
$$(A + B)^4 = A^4 + 4A^3B + 6A^2B^2 + 4AB^3 + B^4.$$

We observe in each of the above examples that the expansion of the expression $(A + B)^n$ for $n = 1, 2, 3, 4$ can be obtained by the following rule:

1. The first term is A^n.
2. The exponent of A decreases by 1 in each successive term and the exponent of B increases by 1.
3. If the coefficient of any term is multiplied by the exponent of A in that term and divided by 1 more than the exponent of B, the result is the coefficient of the next term.

The above rule is known as the **Binomial Theorem.** We have verified that this theorem is true for $n = 1, 2, 3, 4$ and we shall assume that it is true when n is any positive integer.

Writing the Binomial Theorem symbolically, we have the following formula:

Binomial Formula

$$(A + B)^n = A^n + nA^{n-1}B + \frac{n(n-1)}{1 \cdot 2}A^{n-2}B^2 + \frac{n(n-1)(n-2)}{1 \cdot 2 \cdot 3}A^{n-3}B^3 + \cdots + B^n$$

Example 1. Expand $(x - 2)^5$ by the Binomial Theorem.

Solution. Since $x - 2 = x + (-2)$, $A = x$ and $B = -2$. Substituting these values in the binomial formula gives the equation

$$(x - 2)^5 = x^5 + 5x^4(-2) + \frac{5 \cdot 4}{1 \cdot 2}x^3(-2)^2 + \frac{5 \cdot 4 \cdot 3}{1 \cdot 2 \cdot 3}x^2(-2)^3 + \frac{5 \cdot 4 \cdot 3 \cdot 2}{1 \cdot 2 \cdot 3 \cdot 4}x(-2)^4 + \frac{5 \cdot 4 \cdot 3 \cdot 2 \cdot 1}{1 \cdot 2 \cdot 3 \cdot 4 \cdot 5}(-2)^5$$
$$= x^5 - 10x^4 + 40x^3 - 80x^2 + 80x - 32.$$

Example 2. Write the first four terms in the expansion of

$$\left(2x + \frac{x^2}{2}\right)^{10}.$$

Solution. Here $A = 2x$ and $B = \frac{x^2}{2}$; hence,

$$\left(2x + \frac{x^2}{2}\right)^{10} = (2x)^{10} + 10(2x)^9\left(\frac{x^2}{2}\right) + \frac{10 \cdot 9}{1 \cdot 2}(2x)^8\left(\frac{x^2}{2}\right)^2$$
$$+ \frac{10 \cdot 9 \cdot 8}{1 \cdot 2 \cdot 3}(2x)^7\left(\frac{x^2}{2}\right)^3 + \cdots$$
$$= 1024x^{10} + 2560x^{11} + 2880x^{12} + 1920x^{13} + \cdots.$$

Example 3. Find the fifth term in the expansion of $(2x - y)^{12}$.
Solution. The fifth term of the binomial formula is

$$\frac{n(n-1)(n-2)(n-3)}{1 \cdot 2 \cdot 3 \cdot 4} A^{n-4}B^4.$$

Hence,

$$\text{fifth term} = \frac{12 \cdot 11 \cdot 10 \cdot 9}{1 \cdot 2 \cdot 3 \cdot 4}(2x)^8(-y)^4 = 126{,}720x^8y^4.$$

EXERCISE 103

Expand each of the following by the Binomial Theorem:

1. $(a + x)^6$ **2.** $(2x - a)^7$ **3.** $(x - 2y)^4$
4. $(3a + 2b)^5$ **5.** $(ax - b)^8$ **6.** $(x - \frac{1}{2})^5$
7. $(x^2 + x)^7$ **8.** $(\frac{2}{3}x - \frac{3}{2})^4$ **9.** $\left(\frac{x}{2} + 2\right)^5$
10. $(x^2 - 3)^6$

Write and simplify the first four terms in the expansion of the following:

11. $(x - 3y)^{10}$ **12.** $(a + \frac{1}{2})^{16}$ **13.** $(2x^2 - \frac{1}{2})^9$
14. $\left(1 - \frac{x}{3}\right)^{27}$ **15.** $(m - 3)^{12}$ **16.** $(x - \frac{2}{3})^{18}$
17. $\left(2x + \frac{1}{x}\right)^8$ **18.** $\left(1 + \frac{x}{y}\right)^{20}$ **19.** $\left(\frac{x}{2} - \frac{y}{3}\right)^{12}$
20. $\left(1 + \frac{5x}{2y}\right)^{24}$

In each of the following, give only the indicated term:

21. $(x - 3)^{15}$; 5th term. **22.** $\left(\frac{x}{2} + 2\right)^{12}$; 6th term.

23. $\left(a + \frac{b}{2}\right)^{16}$; 4th term.

24. $(x^3 - 4)^{10}$; 4th term.

25. $(2m - 3)^9$; 6th term.

26. $\left(1 - \frac{x}{2}\right)^{32}$; 5th term.

Evaluate each of the following by the Binomial Theorem:

27. $(10 - 1)^5$
28. $(100 + 2)^4$
29. $(1 + 0.01)^3$
30. $(1 - 0.02)^4$
31. $(1 + \sqrt{2})^3$
32. $(\sqrt{3} - 1)^4$
33. $(1.06)^4$
34. $(98)^3$
35. $(\sqrt{3} + \sqrt{2})^4$

REVIEW OF CHAPTER XII

A

Reduce the following ratios to simplest terms:

1. 123 : 205
2. 3 inches : 2 yards
3. 4 pints : 1 gallon
4. $(x^2 - 1) : (x^2 - x)$
5. $78ab : 117ab$
6. 8 square inches : 3 square feet

7. What is the ratio of the volume of a 4-inch cube to that of a 2-inch cube?

8. What is the ratio of the area of a circle whose diameter is 2 feet to the area of a square whose sides are 2 feet? Use $\pi = \frac{22}{7}$.

9. Divide 333 into two parts whose ratio is 4 : 5.

10. Two numbers are in the ratio 3 : 7 and their difference is 76. Find the numbers.

11. Find three numbers in the ratio 3 : 4 : 8, whose sum is 285.

12. A sum of money amounting to $727.80 is to be divided into three parts which are in the ratio 17 : 20 : 23. How should it be divided?

13. A wire 6 feet long is to be divided into four parts having the ratio 3 : 5 : 7 : 9. Find the length of each part.

14. At a boxing match the contestants agree to split the proceeds in the ratio 5 : 3, the winner getting the larger share. If the net gate receipts are $109,147.20, how much does each receive?

15. According to the terms of a will, a man leaves $5000 to charity; $\frac{1}{2}$ and $\frac{1}{3}$ of the remainder are willed to his wife and son respectively. What is left is to be given to his niece. If the man's property amounted to $95,000, in what ratio was the money divided?

16. If $x : y = 5 : 3$, find the numerical value of the ratio $(x - y) : (x + y)$.

17. What number must be subtracted from each member of the ratio 71 : 82 in order to have the ratio 3 : 4?

18. Two numbers are in the ratio 1 : 2. If the smaller number is decreased by 1 and the larger is increased by 1, the ratio of the resulting numbers is 3 : 7. What are the numbers?

19. The ratio of the dimensions of a rectangle are 5 : 7. If each dimension is increased by 5 feet, the area of the rectangle is increased 445 square feet. What are the dimensions of the rectangle?

20. A merchant maintains a "selling price to cost" ratio of 7 : 5 on all his merchandise. If, at a sale, he sells his goods at "$\frac{1}{4}$ off," what is the ratio of the new selling price to the cost?

B

1. If y varies directly as x, and $y = 18$ when $x = 27$, find y when $x = 57$.

2. If p varies directly as q, and $p = 25$ when $q = 15$, find p when $q = 21$.

3. If m varies directly as n, and $m = 0.02$ when $n = 0.7$, find m when $n = 0.035$.

4. If b varies directly as $a + 5$, and $b = 12$ when $a = 2\frac{2}{5}$, find b when $a = 7\frac{1}{3}$.

5. If x varies directly as $a^2 + b^2$, and $x = 25$ when $a = 3$ and $b = 4$, find x when $a = 4$ and $b = 5$.

6. If s varies inversely as t, and $s = 5$ when $t = 24$, find s when $t = 16$.

7. If x varies inversely as y, and $x = 2\frac{1}{2}$ when $y = 2\frac{1}{3}$, find x when $y = 1\frac{2}{3}$.

8. If c varies inversely as d, and $c = 0.2$ when $d = 0.027$, find c when $d = 0.06$.

9. If p varies inversely as q^2, and $p = 50$ when $q = 3$, find p when $q = 2\frac{1}{2}$.

10. If y varies inversely as $m + n$, and $y = 20$ when $m = 2$ and $n = 3$, find y when $m = 4$ and $n = 4$.

11. If z varies directly as x and inversely as y, and $z = 4$ when $x = 2$, and $y = 6$, find z when $x = 4$ and $y = 8$.

12. If p varies directly as a and inversely as b, and $p = 9$ when $a = 6$ and $b = 12$, find p when $a = 8$ and $b = 8$.

13. The cost of a certain copper wire varies directly as its length. If 348 feet of it costs \$4.06, how much will 1200 feet of it cost?

14. If the weight of a hollow statue varies directly as the square of its height, what will a statue 6 feet high weigh, if a similar one 10 inches high weighs $12\frac{1}{2}$ pounds?

15. The brightness of illumination from a lamp varies inversely as the square of the distance from the lamp. At what distance from a lamp must a page be held so that it may be illuminated twice as brightly as a page which is held 4 feet away?

16. In a motorboat, the power of the motor varies directly as the cube of the velocity of the boat. If a 32-horsepower motor can run a boat at a speed of 10 miles per hour, how many horsepower are needed to run the boat at a speed of 15 miles per hour?

17. The cost of a certain article varies inversely as the square root of the demand for the article. If the article costs 36 cents when the company is producing 200,000, what should the cost be if the production jumps to 450,000?

18. The strength of a beam varies inversely as the cube of its length. If a beam 10 feet long can support 2025 pounds, how much can a similar 15-foot beam support?

19. The electrical resistance of a wire varies directly as the length and inversely as the square of the diameter. If a wire 2000 feet long and 0.09 inch in diameter has a resistance of 2 ohms, what is the resistance of a wire of the same material which is 3000 feet long and 0.16 inch in diameter?

20. The volume of a gas varies directly as its absolute temperature and inversely as its pressure. If a gas whose temperature is 290° and whose pressure is 15 pounds per square inch occupies a space of 1000 cubic inches, how much space would the gas occupy if the temperature were raised to 300° and the pressure increased to 20 pounds per square inch?

C

Expand each of the following, using the binomial formula:

1. $(3x + 2)^4$ **2.** $(x^2 + 4)^5$ **3.** $(2m - 1)^7$

4. $(1 - 4x)^6$ **5.** $\left(\frac{x}{3} - 3\right)^4$ **6.** $(5 + 2p)^5$

Write and simplify the first four terms in the expansion of the following:

7. $(x + 5y)^{13}$ **8.** $(1 - 2x)^{20}$ **9.** $\left(\frac{x}{2} - \frac{2}{x}\right)^{10}$

10. $(x^2 + 4)^{17}$ **11.** $\left(x^2 - \frac{3}{x}\right)^9$ **12.** $\left(3x^2 + \frac{2}{x}\right)^8$

Give the term indicated in the expansion of each of the following:

13. $(a^2 + 4)^{14}$; 6th term. **14.** $\left(3x^2 - \dfrac{1}{3x}\right)^{12}$; 6th term.

15. $\left(1 - \dfrac{x}{3}\right)^{21}$; 5th term. **16.** $\left(\dfrac{x^2}{2} + \dfrac{4}{x}\right)^{10}$; 4th term.

Evaluate each of the following by the Binomial Theorem:

17. $(1.04)^5$ **18.** $(101)^4$ **19.** $(\sqrt{2} - 1)^4$ **20.** $(\sqrt{6} - \sqrt{2})^4$

Whatsis Is So?

If the whatsis varies directly as the whosis and inversely as the so-and-so, and if the whatsis is so when the whosis is is and the so-and-so is is so, what is the whatsis when the whosis is so and the so-and-so is so so?

The Binomial Coefficients

If $(x + y)^n = c_0x^n + c_1x^{n-1}y + \cdots + c_ny^n$, find the sums

$$(a)\ c_0 + c_1 + \cdots + c_n,$$
$$(b)\ c_0 - c_1 + \cdots + (-1)^nc_n.$$

Dear Father

What digits satisfy the following sum?

SEND + MORE = MONEY

The Golden Mean

A rectangle whose area is equal to the square of its length less the square of its width is said to have the most pleasing proportions; and the ratio of its sides is called the "golden mean." Find the golden mean.

Let George Do It

If the quantity of work done by George in an hour varies directly as his pay per hour and inversely as the square root of the number of hours he works per day, and if he can finish a job in 6 days when working 9 hours a day at $1.00 per hour, how long will it take him to finish the same job when working 16 hours a day at $1.50 per hour?

CHAPTER XIII

LOGARITHMS

129. Introduction. Many of the problems arising in mathematics and in its applications to other fields of knowledge involve arithmetic computations of considerable length. We shall see that the computation required in extensive and complicated problems may be greatly lessened by the use of *logarithms.*

130. Definition of a Logarithm. *If* $N = b^x$ $(b > 0, b \neq 1)$, *the exponent* x *is called the logarithm of* N *to the base* b, *and is written* $x = \log_b N$.

Thus, by definition, the expressions

$$\text{(E)} \quad N = b^x \quad \text{and} \quad \text{(L)} \quad x = \log_b N$$

represent exactly the same relationship; (E) is called the exponential form of the relation and (L) the logarithmic form.

Illustrations. The following table gives several equivalent relations:

Exponential form	$3^4 = 81$	$4^{\frac{1}{2}} = 2$	$2^{-3} = \frac{1}{8}$	$5^0 = 1$
Logarithmic form	$\log_3 81 = 4$	$\log_4 2 = \frac{1}{2}$	$\log_2 \frac{1}{8} = -3$	$\log_5 1 = 0$

To appreciate fully the significance of any logarithm, such as $\log_2 8$, it is advisable at first to write the expression $x = \log_2 8$ in its equivalent exponential form. Thus $2^x = 8$. In this form it is evident that $x = 3$, inasmuch as $2^3 = 8$.

Illustration. To find the value of $\log_9 3$, we write $9^x = 3$. Hence $x = \frac{1}{2}$, since $9^{\frac{1}{2}} = \sqrt{9} = 3$.

NOTE. For any base b, we have $b^0 = 1$ and $b^1 = b$. Hence

$$\mathbf{\log_b 1 = 0 \text{ and } \log_b b = 1.}$$

Example 1. Find N, if $\log_2 N = 5$.

Solution. Write the given equation in its equivalent exponential form; that is, $N = 2^5$. Evaluating, we have $N = 32$.

Example 2. Find x, if $\log_3 \frac{1}{9} = x$.

Solution. Writing in the exponential form, we have $3^x = \frac{1}{9}$. Since by the laws of exponents,

$$\frac{1}{9} = \frac{1}{3^2} = 3^{-2},$$

it follows from $3^x = 3^{-2}$, that $x = -2$.

Example 3. Find b, if $\log_b 4 = \frac{2}{3}$.

Solution. Writing in exponential form, we have $b^{\frac{2}{3}} = 4$. Raising both sides to the power $\frac{3}{2}$ and recalling that b must be positive, we have

$$(b^{\frac{2}{3}})^{\frac{3}{2}} = 4^{\frac{3}{2}},$$
$$b^1 = b = (\sqrt{4})^3 = 8.$$

EXERCISE 104

Express the following equations in logarithmic form:

1. $4^2 = 16$ **2.** $10^3 = 1000$ **3.** $7^2 = 49$ **4.** $3^4 = 81$
5. $6^{-2} = \frac{1}{36}$ **6.** $9^{\frac{1}{2}} = 3$ **7.** $8^{-\frac{2}{3}} = \frac{1}{4}$ **8.** $16^{\frac{3}{4}} = 8$

Express the following equations in exponential form:

9. $\log_2 16 = 4$ **10.** $\log_{10} 100 = 2$ **11.** $\log_5 125 = 3$
12. $\log_2 64 = 6$ **13.** $\log_8 4 = \frac{2}{3}$ **14.** $\log_5 1 = 0$
15. $\log_{10} 0.001 = -3$ **16.** $\log_9 \frac{1}{3} = -\frac{1}{2}$

17. When the base is 2, what are the logarithms of the numbers 1, 2, $\frac{1}{2}$, 4, $\frac{1}{4}$, 8?

18. When the base is 10, what are the logarithms of the numbers 1, 10, $\frac{1}{10}$, 100, $\frac{1}{100}$, 1000?

19. When the base is 3, what are the numbers whose logarithms are 0, 1, 2, 3, -1, -2, -3?

20. When the base is 10, what are the numbers whose logarithms are 0, 1, 2, 3, -1, -2, -3?

Find the value of each of the following logarithms:

21. $\log_2 4$ **22.** $\log_4 2$ **23.** $\log_{10} 0.01$ **24.** $\log_3 81$
25. $\log_5 5$ **26.** $\log_2 32$ **27.** $\log_{27} 9$ **28.** $\log_4 \frac{1}{4}$

Find the unknown b, x, or N in each of the following:

29. $\log_2 \frac{1}{8} = x$ **30.** $\log_2 N = \frac{1}{2}$ **31.** $\log_b 27 = 3$
32. $\log_5 N = -3$ **33.** $\log_b 9 = \frac{2}{3}$ **34.** $\log_b 0.1 = -1$
35. $\log_{\frac{1}{4}} N = 2\frac{1}{2}$ **36.** $\log_b 16 = \frac{1}{2}$

Evaluate each of the following, finding each logarithm separately and then combining arithmetically:

37. $\log_2 4 + \log_4 2 - \log_8 1$ **38.** $\log_3 9 - \log_9 3 + \log_3 3$
39. $\log_2\sqrt{2} + \log_2\sqrt[3]{2} + \log_2\sqrt[6]{2}$ **40.** $\log_{10} 0.1 - \log_{10} 0.001 + \log_{10} 10$

131. Common Logarithms. Although any positive number except unity may be taken as the base for a system of logarithms, for purposes of computation in our decimal system of numbers the logarithms that have 10 as a base are most convenient to use.

The system of logarithms with 10 as a base is called the system of *common logarithms.* In the work that follows, when the base is not explicitly written it is understood to be 10. That is, log 2 means $\log_{10} 2$.

In the following table we have listed some of the numbers that have integers for logarithms in the common system.

Exponential Form	Logarithmic Form
$10^4 = 10{,}000$	$\log 10{,}000 = 4$
$10^3 = 1000$	$\log 1000 = 3$
$10^2 = 100$	$\log 100 = 2$
$10^1 = 10$	$\log 10 = 1$
$10^0 = 1$	$\log 1 = 0$
$10^{-1} = 0.1$	$\log 0.1 = -1$
$10^{-2} = 0.01$	$\log 0.01 = -2$
$10^{-3} = 0.001$	$\log 0.001 = -3$
$10^{-4} = 0.0001$	$\log 0.0001 = -4$

Since, if its base is greater than 1, a logarithm increases as its number increases, any number between 100 and 1000 has a logarithm between 2 and 3. Similarly a number between 0.001 and 0.01 has a logarithm between -3 and -2.

In general, for any number that is not an exact power of 10, the logarithm consists of a whole-number part (positive or negative) increased by a decimal part (less than 1). We call the integral part the *characteristic* of the logarithm and the decimal part the *mantissa* of the logarithm.

Thus, since 3694 is a number lying between 10^3 and 10^4, we have

$$\log 3694 = 3 + \text{a decimal.}$$

Similarly, since 0.0052 is a number lying between 10^{-3} and 10^{-2}, we have

$$\log 0.0052 = -3 + \text{a decimal.}$$

Note 1. The mantissa of a logarithm is always positive. Thus if $\log N = -1.6108$, the mantissa is not 0.6108. Writing -1.6108 with a positive decimal, we have $-2 + 0.3892$. In this form we see that the mantissa is 0.3892.

Note 2. Logarithms with negative characteristics like $-2 + 0.3892$ are sometimes written $\bar{2}.3892$, the negative sign being placed above the characteristic. For purposes of computation it is more convenient to write characteristics in the binomial form. That is,

$$-1 = 9 - 10 \qquad -4 = 6 - 10$$
$$-2 = 8 - 10 \qquad -5 = 5 - 10$$
$$-3 = 7 - 10 \qquad \text{etc.}$$

Illustration 1. To write $\log N = -1.6108$ in the binomial form, we subtract from zero in the following manner:

$$\begin{array}{rr} & 10.0000 - 10 \\ & 1.6108 \\ \hline \log N = & 8.3892 - 10 \end{array}$$

Illustration 2. The following table lists several logarithms and gives the characteristic and mantissa of each:

Logarithm	Characteristic	Mantissa
0.7782	0	0.7782
2.0253	2	0.0253
9.0000 − 10	−1	0.0000
7.9518 − 10	−3	0.9518
−0.7118	−1	0.2882

132. Rule for Determining Characteristics. By direct computation we know that

$$10^{0.25} = 10^{\frac{1}{4}} = \sqrt[4]{10} = \sqrt{\sqrt{10}} = 1.778.$$

If we multiply and divide this relation successively by 10, we obtain

$10^{7.2500-10} = 0.001778$	hence $\log 0.001778 = 7.2500 - 10$
$10^{8.2500-10} = 0.01778$	$\log 0.01778 = 8.2500 - 10$
$10^{9.2500-10} = 0.1778$	$\log 0.1778 = 9.2500 - 10$
$10^{0.2500} = 1.778$	$\log 1.778 = 0.2500$
$10^{1.2500} = 17.78$	$\log 17.78 = 1.2500$
$10^{2.2500} = 177.8$	$\log 177.8 = 2.2500$
$10^{3.2500} = 1778$	$\log 1778 = 3.2500$

From the above table it is evident that:

1. The mantissas are the same when the numbers differ only in the position of the decimal point.
2. The characteristic is determined by the position of the decimal point in accordance with the following rule:

Rule for Characteristics

1. ***If a number is greater than or equal to 1, the characteristic of its logarithm is positive and is 1 LESS than the number of figures to the left of the decimal point.***
2. ***If a number is less than 1, the characteristic of its logarithm is negative and is 1 MORE than the number of zeros immediately following the decimal point.***

Illustration. The characteristics of the logarithms of the numbers 472.5, 3.023, 0.0007, 15000, and 0.1006 are respectively 2, 0, −4, 4, and −1.

Example 1. If log 7.42 = 0.8704, find the logarithms of the numbers 0.742, 742, 0.00742, 742,000.

Solution. If numbers differ only in the position of their decimal point, their mantissas are the same. Hence the mantissa for each of the given numbers is 0.8704. By the above rule their respective characteristics are −1, 2, −3, 5. Thus,

$$\log 0.742 = 9.8704 - 10, \qquad \log 742 = 2.8704,$$
$$\log 0.00742 = 7.8704 - 10, \qquad \log 742{,}000 = 5.8704.$$

EXERCISE 105

Write the characteristic and the mantissa of each of the following logarithms:

1. 2.8129
2. $-1 + 0.3483$
3. 5.7419
4. $-3 + 0.5988$
5. 0.9509
6. $0.8854 - 2$
7. $3.4683 - 4$
8. 1.0043
9. $9.9279 - 10$

Find the characteristic and the mantissa of each of the following logarithms:

10. -0.1543
11. -2.5317
12. -3.8297
13. -1.1926
14. -5.0022
15. -0.3925

Give the characteristics of the logarithms of the following numbers:

16. 32
17. 507
18. 9
19. 0.06
20. 800
21. 0.12
22. 1.497
23. 0.000051
24. 20,000,000
25. 3.1416
26. 0.00001
27. 1.002
28. 12345
29. 0.7902
30. 10.01

If the logarithm of 344 is 2.5366, write the logarithms of the following numbers:

31. 3.44
32. 0.0344
33. 3440
34. 0.000344
35. 0.344
36. 344,000
37. 3.44×10^5
38. 3.44×10^{-2}

Place the decimal point in the sequence of digits 3465 so that the logarithm of the number will have the following characteristics:

39. 3
40. -2
41. 0
42. 5
43. -4
44. $9 - 10$
45. 1
46. $7 - 10$
47. $5 - 10$
48. 2
49. 6
50. 4

133. Tables of Logarithms. We have seen that the mantissa of a logarithm depends only on the sequence of figures and not on the position of the decimal point. By methods which are developed in more advanced branches of mathematics, these mantissas can be computed for any given number to any accuracy which is desired. Such computed values are given in tables of logarithms known as four-place tables, five-place tables, etc., according to the number of decimal places computed in the mantissas.

On pp. 300–301 is a four-place table which contains the mantissas of the logarithms of all the integers from 1 to 1000. Although mantissas are decimals less than 1, they are shown in the tables without decimal points for convenience in printing.

Illustration. To find the mantissa of log 56.3, we look in the table on p. 301 for the mantissa of 563. We locate 56 in the column headed "No." Along the horizontal line of numbers opposite 56 in the column headed 3, we find 7505. The required mantissa is thus 0.7505.

NOTE. If a number has less than three figures, add zeros until it has three figures. Thus to find the mantissa of log 47, locate the mantissa of log 470 in the table; for log 5, find the mantissa of log 500.

In summary, *to find the logarithm of a number,*

1. Find its characteristic by the rule given in Article 132.
2. In the tables find the mantissa corresponding to the given sequence of figures.

Example 1. Find log 6750, log 0.47, and log 0.005.

Solution. (1) By the rule for characteristics, the characteristics are 3, -1, -3 respectively. (2) In the table we find the respective mantissas of 675, 470, and 500 to be 0.8293, 0.6721, and 0.6990. Hence, log 6750 = 3.8293, log 0.47 = 9.6721 − 10, log 0.005 = 7.6990 − 10.

EXERCISE 106

Find the logarithms of each of the following numbers:

1. 277	**2.** 94.1	**3.** 0.233	**4.** 6.91
5. 3270	**6.** 409	**7.** 62,900	**8.** 0.0376
9. 429	**10.** 7.01	**11.** 0.00305	**12.** 0.111
13. 95	**14.** 6.3	**15.** 0.042	**16.** 3800
17. 0.00014	**18.** 27,000	**19.** 0.99	**20.** 2.5
21. 0.06	**22.** 5000	**23.** 2020	**24.** 0.69
25. 100	**26.** 0.002	**27.** 32.5	**28.** 314
29. 19,000,000	**30.** 1.44	**31.** 0.4	**32.** 0.0001
33. 3.69	**34.** 440	**35.** 0.27	**36.** 30.9

134. Antilogarithms. *A number corresponding to a given logarithm is called an* **antilogarithm.**

In order to find an antilogarithm we reverse the process used in finding a logarithm. That is,

1. In the tables, find the number corresponding to the given mantissa.
2. Place the decimal point in accordance with the value of the characteristic.

Example 1. Find the number whose logarithm is 3.6385.

Solution. In the tables corresponding to the mantissa 6385 we find the sequence of numbers 435.

Since the characteristic is +3, we must have four figures to the left of the decimal. Thus the required number is 4350.

Example 2. Find the antilogarithm of 9.8463 − 10.

Solution. In the tables corresponding to the mantissa 8463 we find the number 702.

Since the characteristic is 9 − 10 or −1, the required number is 0.702.

EXERCISE 107

Find the antilogarithm of each of the following logarithms:

1. 0.0569	**2.** 2.9047	**3.** 1.7380	**4.** 0.9274
5. 9.2833 − 10	**6.** 8.8882 − 10	**7.** 7.9009 − 10	**8.** 9.0414 − 10
9. 5.8363	**10.** 8.9881 − 10	**11.** 2.3139	**12.** 5.7782 − 10
13. 1.4771	**14.** 9.6503 − 10	**15.** 8.7404 − 10	**16.** 3.6990
17. 0.9996	**18.** 7.5911 − 10	**19.** 1.4082	**20.** 5.1106
21. 1.8976	**22.** 2.9415	**23.** 9.8451 − 10	**24.** 8.2148 − 10
25. 4.7324	**26.** 8.5502 − 10	**27.** 0.9217	**28.** 3.8500
29. 9.6484 − 10	**30.** 6.9542 − 10	**31.** 2.9533	**32.** 1.0043
33. 7.3997 − 10	**34.** 9.6191 − 10	**35.** 1.9961	**36.** 8.2900 − 10

135. Interpolation. When a number contains four figures, the mantissa of its logarithm cannot be found directly in the tables. However, by a process called *interpolation* a very close approximation can be made for the mantissa in question. Thus, for example, the mantissa of 2160 is 3345 and the mantissa of 2170 is 3365.

We shall *assume* that the mantissa of the logarithms of the numbers between 2160 and 2170 are evenly distributed from 3345 to 3365. In tabular form we would have

Number	Mantissa	Number	Mantissa
2160	3345	2166	3357
2161	3347	2167	3359
2162	3349	2168	3361
2163	3351	2169	3363
2164	3353	2170	3365
2165	3355		

Suppose that we wish only to find the mantissa corresponding to the number 2167. Since 2167 is $\frac{7}{10}$ of the way from 2160 to 2170, the corresponding mantissa will be $\frac{7}{10}$ of the way from 3345 to 3365. That is, it will be given by

$$3345 + \tfrac{7}{10}(3365 - 3345) = 3345 + 14 = 3359.$$

This process of interpolating is usually carried through without the decimal points and in this form the difference (3365 − 3345) is called the **tabular difference.**

Example 1. By interpolation, find log 28.33.

Solution. The required logarithm is 1 + some mantissa.

Since 2833 lies between 2830 and 2840, the mantissa of log 28.33 is seen to lie between 4518 and 4533 (see the table). The tabular difference between 4518 and 4533 is 15; hence the required mantissa is found by adding 0.3 of 15 to 4518, which gives 4523.

Thus, the required logarithm is 1.4523.

NOTE. In the above example, although 0.3 of 15 equals 4.5, the result is rounded off and called 5. Never retain any more decimal places in a mantissa than are given in the table you use.

Example 2. By interpolation, find log 0.79142.

Solution. Since the process of interpolation is purely approximate, no greater accuracy will be gained by extending the process to numbers of five figures than by merely rounding the given number off to four figures and applying the usual process.

Hence, instead of assuming that the number 79142 is $\frac{42}{100}$ of the way from 79100 to 79200, we merely round off the sequence of numbers 79142 to 7914 and proceed as usual.

The tabular difference for the mantissas of 7910 and 7920 is 5; hence the mantissa of 7914 is $8982 + 0.4 \times 5 = 8984$. Thus,

$$\log 0.79142 = 9.8984 - 10.$$

Example 3. By interpolation, find the antilogarithm of 1.6950.

Solution. The table shows that the mantissa of 6950 lies between 6946 and 6955. These mantissas correspond to 4950 and 4960 respectively. The tabular difference between 6946 and 6955 is 9, and the difference between 6950 (the given mantissa) and 6946 is 4. Hence, the number corresponding to 6950 is found by adding $\frac{4}{9}$ of 10, or 4, to 4950; this gives the sequence of figures 4954.

Then since the given characteristic is +1, we have the final result

$$\text{antilog } 1.6950 = 49.54.$$

LOGARITHMS OF NUMBERS—Table of Mantissas

No.	0	1	2	3	4	5	6	7	8	9
10	0000	0043	0086	0128	0170	0212	0253	0294	0334	0374
11	0414	0453	0492	0531	0569	0607	0645	0682	0719	0755
12	0792	0828	0864	0899	0934	0969	1004	1038	1072	1106
13	1139	1173	1206	1239	1271	1303	1335	1367	1399	1430
14	1461	1492	1523	1553	1584	1614	1644	1673	1703	1732
15	1761	1790	1818	1847	1875	1903	1931	1959	1987	2014
16	2041	2068	2095	2122	2148	2175	2201	2227	2253	2279
17	2304	2330	2355	2380	2405	2430	2455	2480	2504	2529
18	2553	2577	2601	2625	2648	2672	2695	2718	2742	2765
19	2788	2810	2833	2856	2878	2900	2923	2945	2967	2989
20	3010	3032	3054	3075	3096	3118	3139	3160	3181	3201
21	3222	3243	3263	3284	3304	3324	3345	3365	3385	3404
22	3424	3444	3464	3483	3502	3522	3541	3560	3579	3598
23	3617	3636	3655	3674	3692	3711	3729	3747	3766	3784
24	3802	3820	3838	3856	3874	3892	3909	3927	3945	3962
25	3979	3997	4014	4031	4048	4065	4082	4099	4116	4133
26	4150	4166	4183	4200	4216	4232	4249	4265	4281	4298
27	4314	4330	4346	4362	4378	4393	4409	4425	4440	4456
28	4472	4487	4502	4518	4533	4548	4564	4579	4594	4609
29	4624	4639	4654	4669	4683	4698	4713	4728	4742	4757
30	4771	4786	4800	4814	4829	4843	4857	4871	4886	4900
31	4914	4928	4942	4955	4969	4983	4997	5011	5024	5038
32	5051	5065	5079	5092	5105	5119	5132	5145	5159	5172
33	5185	5198	5211	5224	5237	5250	5263	5276	5289	5302
34	5315	5328	5340	5353	5366	5378	5391	5403	5416	5428
35	5441	5453	5465	5478	5490	5502	5514	5527	5539	5551
36	5563	5575	5587	5599	5611	5623	5635	5647	5658	5670
37	5682	5694	5705	5717	5729	5740	5752	5763	5775	5786
38	5798	5809	5821	5832	5843	5855	5866	5877	5888	5899
39	5911	5922	5933	5944	5955	5966	5977	5988	5999	6010
40	6021	6031	6042	6053	6064	6075	6085	6096	6107	6117
41	6128	6138	6149	6160	6170	6180	6191	6201	6212	6222
42	6232	6243	6253	6263	6274	6284	6294	6304	6314	6325
43	6335	6345	6355	6365	6375	6385	6395	6405	6415	6425
44	6435	6444	6454	6464	6474	6484	6493	6503	6513	6522
45	6532	6542	6551	6561	6571	6580	6590	6599	6609	6618
46	6628	6637	6646	6656	6665	6675	6684	6693	6702	6712
47	6721	6730	6739	6749	6758	6767	6776	6785	6794	6803
48	6812	6821	6830	6839	6848	6857	6866	6875	6884	6893
49	6902	6911	6920	6928	6937	6946	6955	6964	6972	6981
50	6990	6998	7007	7016	7024	7033	7042	7050	7059	7067
51	7076	7084	7093	7101	7110	7118	7126	7135	7143	7152
52	7160	7168	7177	7185	7193	7202	7210	7218	7226	7235
53	7243	7251	7259	7267	7275	7284	7292	7300	7308	7316
54	7324	7332	7340	7348	7356	7364	7372	7380	7388	7396

LOGARITHMS OF NUMBERS — Table of Mantissas

No.	0	1	2	3	4	5	6	7	8	9
55	7404	7412	7419	7427	7435	7443	7451	7459	7466	7474
56	7482	7490	7497	7505	7513	7520	7528	7536	7543	7551
57	7559	7566	7574	7582	7589	7597	7604	7612	7619	7627
58	7634	7642	7649	7657	7664	7672	7679	7686	7694	7701
59	7709	7716	7723	7731	7738	7745	7752	7760	7767	7774
60	7782	7789	7796	7803	7810	7818	7825	7832	7839	7846
61	7853	7860	7868	7875	7882	7889	7896	7903	7910	7917
62	7924	7931	7938	7945	7952	7959	7966	7973	7980	7987
63	7993	8000	8007	8014	8021	8028	8035	8041	8048	8055
64	8062	8069	8075	8082	8089	8096	8102	8109	8116	8122
65	8129	8136	8142	8149	8156	8162	8169	8176	8182	8189
66	8195	8202	8209	8215	8222	8228	8235	8241	8248	8254
67	8261	8267	8274	8280	8287	8293	8299	8306	8312	8319
68	8325	8331	8338	8344	8351	8357	8363	8370	8376	8382
69	8388	8395	8401	8407	8414	8420	8426	8432	8439	8445
70	8451	8457	8463	8470	8476	8482	8488	8494	8500	8506
71	8513	8519	8525	8531	8537	8543	8549	8555	8561	8567
72	8573	8579	8585	8591	8597	8603	8609	8615	8621	8627
73	8633	8639	8645	8651	8657	8663	8669	8675	8681	8686
74	8692	8698	8704	8710	8716	8722	8727	8733	8739	8745
75	8751	8756	8762	8768	8774	8779	8785	8791	8797	8802
76	8808	8814	8820	8825	8831	8837	8842	8848	8854	8859
77	8865	8871	8876	8882	8887	8893	8899	8904	8910	8915
78	8921	8927	8932	8938	8943	8949	8954	8960	8965	8971
79	8976	8982	8987	8993	8998	9004	9009	9015	9020	9025
80	9031	9036	9042	9047	9053	9058	9063	9069	9074	9079
81	9085	9090	9096	9101	9106	9112	9117	9122	9128	9133
82	9138	9143	9149	9154	9159	9165	9170	9175	9180	9186
83	9191	9196	9201	9206	9212	9217	9222	9227	9232	9238
84	9243	9248	9253	9258	9263	9269	9274	9279	9284	9289
85	9294	9299	9304	9309	9315	9320	9325	9330	9335	9340
86	9345	9350	9355	9360	9365	9370	9375	9380	9385	9390
87	9395	9400	9405	9410	9415	9420	9425	9430	9435	9440
88	9445	9450	9455	9460	9465	9469	9474	9479	9484	9489
89	9494	9499	9504	9509	9513	9518	9523	9528	9533	9538
90	9542	9547	9552	9557	9562	9566	9571	9576	9581	9586
91	9590	9595	9600	9605	9609	9614	9619	9624	9628	9633
92	9638	9643	9647	9652	9657	9661	9666	9671	9675	9680
93	9685	9689	9694	9699	9703	9708	9713	9717	9722	9727
94	9731	9736	9741	9745	9750	9754	9759	9763	9768	9773
95	9777	9782	9786	9791	9795	9800	9805	9809	9814	9818
96	9823	9827	9832	9836	9841	9845	9850	9854	9859	9863
97	9868	9872	9877	9881	9886	9890	9894	9899	9903	9908
98	9912	9917	9921	9926	9930	9934	9939	9943	9948	9952
99	9956	9961	9965	9969	9974	9978	9983	9987	9991	9996

EXERCISE 108

Using the process of interpolation, find the logarithms of the following numbers:

1. 2635	2. 4.427	3. 0.9432	4. 405.8
5. 21.21	6. 0.001052	7. 70,650	8. 217.7
9. 12.318	10. 693.23	11. 9.505	12. 0.1306
13. 2.0206	14. 0.07758	15. 929.65	16. 500,290
17. 0.005005	18. 70.94	19. 0.2468	20. 34.694

Using the process of interpolation, find the numbers corresponding to the following logarithms:

21. 3.6740	22. 9.4208 − 10	23. 8.9045 − 10	24. 1.8423
25. 0.2415	26. 9.4659 − 10	27. 4.9985	28. 7.7777 − 10
29. 1.5656	30. 8.9291 − 10	31. 2.8450	32. 5.2190
33. 9.3232 − 10	34. 3.1416	35. 7.8087 − 10	36. 8.8765 − 10
37. 2.3415	38. 9.9000 − 10	39. 1.5050	40. 5.6931 − 10

136. Fundamental Laws of Logarithms. Since by definition logarithms are exponents, the rules of operation which apply to exponents also apply to logarithms. These rules expressed in terms of logarithms have the following form:

Laws of Logarithms

Law I. ***The logarithm of a product equals the sum of the logarithms of its factors. In symbols,***

$$\log MN = \log M + \log N.$$

Law II. ***The logarithm of a quotient equals the difference between the logarithm of the dividend and the logarithm of the divisor. In symbols,***

$$\log \frac{M}{N} = \log M - \log N.$$

Law III. ***The logarithm of a power of a number equals the logarithm of the number multiplied by the exponent of the power. In symbols,***

$$\log M^n = n \log M.$$

Law IV. ***The logarithm of a root of a number equals the logarithm of the number divided by the index of the root. In symbols,***

$$\log \sqrt[r]{M} = \frac{\log M}{r}.$$

Proof of Law I. Let $x = \log M$ and $y = \log N$.
Written in exponential form, $M = 10^x$ and $N = 10^y$.
Multiply, using the law of exponents, $MN = 10^x \cdot 10^y = 10^{x+y}$.
Thus, by definition of logarithms, $\log MN = x + y$.
Hence, by substitution, $\log MN = \log M + \log N$.

Proof of Law II. Let $x = \log M$ and $y = \log N$.
Written in exponential form, $M = 10^x$ and $N = 10^y$.
Divide, using the law of exponents, $\frac{M}{N} = \frac{10^x}{10^y} = 10^{x-y}$.
Thus, by definition of logarithms, $\log \frac{M}{N} = x - y$.
Hence, by substitution, $\log \frac{M}{N} = \log M - \log N$.

Proof of Law III. Let $x = \log M$.
Written in exponential form, $M = 10^x$.
Raise to the nth power, $M^n = (10^x)^n = 10^{nx}$.
Thus, by definition of logarithms, $\log M^n = nx$.
Hence, by substitution, $\log M^n = n \log M$.

Proof of Law IV. By the definition of a fractional exponent, we have $\sqrt[r]{M} = M^{\frac{1}{r}}$. Hence, using Law III, we have

$$\log \sqrt[r]{M} = \log M^{\frac{1}{r}} = \frac{1}{r} \log M = \frac{\log M}{r}.$$

137. Computing a Product by Logarithms. By applying Law I to a product, we can find the numerical value of the product, as is shown in the following examples.

Example 1. Find the product of 38 and 17.
Solution. Let $N = 38 \times 17$; then by Law I,

$$\log N = \log 38 + \log 17.$$

Hence, we have the following computation:

$$\log 38 = 1.5798$$
$$\log 17 = 1.2304$$

Add, $\log N = 2.8102$
Find the antilogarithm, $N = 646$.

NOTE. Since the logarithms of negative numbers are not defined, if a negative sign occurs in a problem of computation, *first find the absolute value of the expression by logarithms and then prefix the proper sign.*

Example 2. Find the product of $32.43 \times (-0.695) \times 0.03025$.

Solution. It is clear that the result will be negative. Hence let N represent the product $32.43 \times (0.695) \times 0.03025$; by Law I,

$$\log N = \log 32.43 + \log 0.695 + \log 0.03025.$$

For the computation

$$\begin{aligned} \log 32.43 &= 1.5109 \\ \log 0.695 &= 9.8420 - 10 \\ \log 0.03025 &= \underline{8.4807 - 10} \end{aligned}$$

Adding, we have $\log N = 19.8336 - 20$

Hence, $N = 0.6817.$

The required product is -0.6817.

NOTE. In problems of computation with logarithms, the student will find it advisable to check all his answers with an approximate estimate of the result. Thus, in Example 2, assuming the product to be in the approximate form $30 \times 0.7 \times 0.03$, we obtain 0.63 by a numerical computation. If an error is made in a computation by logarithms, it is likely to have a great effect on the result. This method of checking guards against such errors.

EXERCISE 109

Find the value of the following products by logarithms:

1. 28×65 **2.** 380×4.5 **3.** 72.7×0.66
4. 0.062×0.147 **5.** 5250×491 **6.** 9.27×428
7. 0.004×6.02 **8.** 49×6.07 **9.** 0.312×0.7916
10. $(-3.76) \times 436.5$ **11.** $0.0457 \times (-0.388)$
12. 325×0.002623 **13.** $42{,}900 \times 528$
14. $(-17.1) \times (0.923)$ **15.** $(-3.24) \times (-42.8)$
16. $147 \times 268 \times 541$ **17.** $0.17 \times 3.73 \times 0.01493$
18. $0.569 \times (-8.92) \times 10.32$ **19.** $62.1 \times 9.92 \times 0.00032$
20. $472 \times 3.19 \times 0.26$ **21.** $(-906) \times 0.035 \times (-2.08)$
22. $0.913 \times 6.87 \times 91.4 \times 0.51$ **23.** $0.465 \times 4810 \times 0.06 \times 0.4547$
24. $629 \times 0.8572 \times 9.11 \times 0.042$ **25.** $6.39 \times 0.3214 \times 99.9 \times 57.68$

138. Computing a Quotient by Logarithms. By applying Law II to a quotient, we can find the numerical value of the quotient, as is shown in the following examples.

Example 1. Divide 1520 by 95.

Solution. Let $N = \frac{1520}{95}$; then, by Law II,

$$\log N = \log 1520 - \log 95.$$

Hence, we have the following computation:

$$\begin{aligned} \log 1520 &= 3.1818 \\ \log 95 &= \underline{1.9777} \end{aligned}$$

Subtract, $\log N = 1.2041$

Find the antilogarithm, $N = 16.$

Example 2. Find the quotient $\dfrac{0.0695}{0.347}$.

Solution. Let N represent the value of the given quotient; then, by Law II,

$$\log N = \log 0.0695 - \log 0.347.$$

For the computation we have the following to subtract:

$$\begin{aligned} \log 0.0695 &= 8.8420 - 10 \\ \log 0.347 &= \underline{9.5403 - 10} \end{aligned}$$

Since the mantissa for this difference *must* be positive, we write the logarithm of 0.0695 in the equivalent form 18.8420 − 20 before subtracting. Thus,

$$\begin{aligned} \log 0.0695 &= 18.8420 - 20 \\ \log 0.347 &= \underline{\ \ 9.5403 - 10} \end{aligned}$$

Subtracting gives $\log N = 9.3017 - 10$

Hence, $N = 0.2003.$

EXERCISE 110

Find the value of the following quotients by logarithms:

1. 444 ÷ 37	**2.** 65.1 ÷ 1.42	**3.** 4.55 ÷ 3.25
4. 6520 ÷ 40.8	**5.** 44.7 ÷ 192	**6.** 9.39 ÷ 42.7
7. 0.372 ÷ 0.158	**8.** 0.53 ÷ 0.069	**9.** 4.65 ÷ 0.594
10. 0.831 ÷ 2.56	**11.** 392 ÷ 0.0645	**12.** 7.21 ÷ 7.82
13. 45.3 ÷ 0.3626	**14.** 0.0071 ÷ 0.08249	**15.** 0.52 ÷ 4925
16. 7.065 ÷ 0.329	**17.** 92.2 ÷ 6950	**18.** 1 ÷ 0.3678
19. 0.241 ÷ 3.153	**20.** 5.297 ÷ 14.82	**21.** 93,000 ÷ 42.77
22. 0.9 ÷ 11.82	**23.** 1 ÷ 93.22	**24.** 0.0005 ÷ 0.0074

139. Computing a Power by Logarithms. By applying Law III, we can find any power of a given number as is shown in the following examples.

Example 1. Find the value of 1.52^6.

Solution. Let $N = 1.52^6$, then by Law III,

$$\log N = 6 \log 1.52.$$

Hence, we have the computation:

$$\log 1.52 = 0.1818$$
$$\underline{\times 6}$$

Multiply, $\log N = 1.0908$

Find the antilogarithm, $N = 12.33.$

Example 2. Find the value of $\left(\dfrac{0.0695}{0.483}\right)^3$.

Solution. Let N represent the required number; then, by Law III,

$$\log N = 3 \log \frac{0.0695}{0.483}.$$

Applying Law II to the logarithm on the right side of this equation gives

$$\log N = 3(\log 0.0695 - \log 0.483).$$

The computation may be arranged as follows:

$$\log 0.0695 = 18.8420 - 20$$
$$\log 0.483 = \underline{9.6839 - 10}$$

Subtract, $\log(\text{quotient}) = 9.1581 - 10$

$$\underline{\times 3}$$

Multiply, $\log N = 27.4743 - 30$

Hence, $N = 0.002981.$

EXERCISE 111

Find the value of each of the following by logarithms:

1. 34^3
2. 1.06^8
3. 21.6^5
4. 0.53^4
5. 6.93^2
6. 0.0537^3
7. 2.09^{10}
8. $(-0.945)^5$
9. 78.1^4
10. 2.43^7
11. 0.00061^3
12. 12.7^5
13. $(3.837 \times 5.27)^2$
14. $(0.1391 \times 68.3)^2$
15. $(64.5 \times 0.0056)^3$
16. $(9.126 \times 0.571)^3$
17. $\left(\dfrac{3195}{55.1}\right)^2$
18. $\left(\dfrac{3.657}{0.93}\right)^2$
19. $\left(\dfrac{0.367}{1.051}\right)^3$
20. $\left(\dfrac{65.4}{271}\right)^3$
21. 2.465×3.19^2
22. $1.25^5 \times 38.27$
23. $0.23^3 \times 55.53^2$
24. $2.576^2 \times 3.24^3$

140. Computing a Root by Logarithms. By applying Law IV, we can find any root of a given number as is shown in the following examples.

Example 1. Find the value of $\sqrt[3]{23.3}$.

Solution. Let $N = \sqrt[3]{23.3}$; then, by Law IV,

$$\log N = \frac{\log 23.3}{3}.$$

Hence, we have the computation:

$$\log 23.3 = 1.3674.$$

Divide by 3, $\log N = 0.4558.$

Find the antilogarithm, $N = 2.856.$

Example 2. Find the value of $\sqrt[4]{5.26 \times 0.049}$.

Solution. Let N represent the required root; then, by Law IV and Law I,

$$\log N = \frac{\log 5.26 + \log 0.049}{4}.$$

The computation may be arranged as follows:

$$\begin{aligned} \log 5.26 &= 0.7210 \\ \log 0.049 &= 8.6902 - 10 \\ \hline \log (\text{product}) &= 9.4112 - 10 \end{aligned}$$

In order to avoid decimals in the "minus" term of this logarithm, we change the characteristic $9 - 10$ to the form $39 - 40$ before dividing by 4. Thus,

$$\log (\text{product}) = 39.4112 - 40.$$

Dividing by 4 gives $\log N = 9.8528 - 10.$

Hence, $N = 0.7125.$

EXERCISE 112

Find the value of each of the following by logarithms:

1. $\sqrt{5.27}$
2. $\sqrt{375}$
3. $\sqrt[3]{92.7}$
4. $\sqrt[3]{50{,}000}$
5. $\sqrt[6]{1.64}$
6. $\sqrt[4]{31.8}$
7. $\sqrt{0.051}$
8. $\sqrt[3]{0.835}$
9. $\sqrt[5]{0.00004}$
10. $\sqrt[3]{-9.07}$
11. $\sqrt[4]{0.097}$
12. $\sqrt[9]{41.6}$
13. $\sqrt[3]{0.31 \times 5.725}$
14. $\sqrt{41.9 \times 3.87}$
15. $\sqrt[4]{0.21 \times 0.04376}$
16. $\sqrt[3]{0.00054 \times 2.349}$
17. $\sqrt{\dfrac{9049}{62.2}}$
18. $\sqrt[4]{\dfrac{5.14}{0.6924}}$
19. $\sqrt[5]{\dfrac{3.15}{41.9}}$
20. $\sqrt[3]{\dfrac{0.0043}{0.5247}}$
21. $\sqrt{\dfrac{3.66 \times 52.14}{13.62}}$

22. $\sqrt[3]{\dfrac{0.34 \times 6.978}{4.5}}$ **23.** $\sqrt[4]{\dfrac{14.9}{23.75 \times 0.024}}$ **24.** $\sqrt{\dfrac{6980}{510 \times 821.5}}$

141. Arrangement of Logarithmic Computations. In performing more complex computations with logarithms, for the sake of speed and accuracy it is desirable to make a systematic arrangement of the details of the computation before looking up any mantissas in the table. When such an outline is completed, all the mantissas needed for the problem may be obtained from the tables at the same time.

Since problems of computation vary considerably, there is no set outline for their arrangement. The following example illustrates one of many acceptable forms.

Example. Evaluate $\dfrac{0.677\sqrt[3]{3680}}{(9.27)^2}$.

Solution. Let N represent the required value. The problem may then be outlined as follows:

$$\begin{aligned}
\log 0.677 \qquad\qquad &= 9. \qquad - 10 \\
\log 3680 = 3. \qquad & \\
\tfrac{1}{3}\log 3680 \qquad\qquad &= \underline{\qquad\qquad} \\
\log(\text{numerator}) &= \\
\log 9.27 = 0. \qquad & \\
2\log 9.27 \qquad\qquad &= \underline{\qquad\qquad} \\
\log N &= \\
N &=
\end{aligned}$$

After the outline has been completed, the mantissas of 0.677, 3680, and 9.27 are obtained from the tables. When these values have been put into the outline, the computations may be completed.

The final form of the computation will appear as follows:

$$\begin{aligned}
\log 0.677 \qquad\qquad &= 9.8306 - 10 \\
\log 3680 = 3.5658 \qquad & \\
\tfrac{1}{3}\log 3680 \qquad\qquad &= \underline{\ 1.1886 \qquad\ } \\
\log(\text{numerator}) &= 11.0192 - 10 \\
\log 9.27 = 0.9671 \qquad & \\
2\log 9.27 \qquad\qquad &= \underline{\ 1.9342 \qquad\ } \\
\log N &= 9.0850 - 10 \\
N &= 0.1216.
\end{aligned}$$

EXERCISE 113

Make a computation outline for problems of the following form:

1. $\dfrac{abc}{def}$
2. $a^2b^3c^4$
3. $\dfrac{a}{b\sqrt{c}}$
4. $\dfrac{ab^2}{\sqrt[3]{c}}$
5. $\sqrt[3]{\dfrac{1}{a^2b}}$
6. $\dfrac{a\sqrt{b}}{c\sqrt[3]{d}}$

Find the value of each of the following by logarithms:

7. $\dfrac{3.62 \times 41.94}{0.64 \times 107}$
8. $(0.627)^{\frac{3}{5}}$
9. $\sqrt{\dfrac{2.8}{0.7247}}$
10. $\dfrac{6.41 \times \sqrt{0.95}}{(3.5)^2}$
11. $\dfrac{14}{539 \times 0.83^2}$
12. $\sqrt[3]{\dfrac{317 \times 4.925}{94.5 \times 28.3}}$
13. $53.2\sqrt{\dfrac{21.62}{0.43}}$
14. $0.567^2 \times \sqrt[3]{1.97}$
15. $\dfrac{\sqrt{8.51 \times 3.69}}{2.07}$
16. $\dfrac{1}{\sqrt[3]{52 \times 48^2}}$
17. $\sqrt[3]{0.216\sqrt{1000}}$
18. $(0.365)^{0.2}$
19. $\sqrt[3]{\dfrac{-0.6218}{45^2}}$
20. $\dfrac{47.8}{5.69}\sqrt{\dfrac{6.3}{4280}}$
21. $\dfrac{4.01^2 \times \sqrt{619}}{8}$
22. $\dfrac{3.26^2 \times 80.95}{47.2 \times \sqrt{130}}$
23. $\dfrac{\sqrt[5]{2.17^3} \times 0.7384}{\sqrt{0.3142}}$
24. $\sqrt{\dfrac{9.29}{387}} \times \sqrt[3]{\dfrac{62.73}{0.4715}}$

142. Exponential Equations. (Optional.) An equation in which the unknown occurs in an exponent is called an *exponential equation.*

In Article 130 we solved certain types of exponential equations by inspection. Thus, for the exponential equation $2^x = 8$, since we knew that $2^3 = 8$, it followed that $2^x = 2^3$, and hence $x = 3$.

The above method of solving exponential equations is sufficient for particular examples. However, the general method of solving such equations is illustrated in the following examples.

Example 1. Solve the equation $2^x = 5$.

Solution. Taking the logarithms of both members, we have

$$\log 2^x = \log 5,$$

or, by Law III, $$x \log 2 = \log 5.$$

Solving, we have $$x = \frac{\log 5}{\log 2} = \frac{0.6990}{0.3010} = 2.322.$$

Note. In evaluating $\frac{\log 5}{\log 2}$, observe particularly that we *actually divide* the value log 5 by the value log 2. Do not make the common error of subtracting these values.

Example 2. Solve the equation $3^{2x-1} = 0.07$.

Solution. Taking the logarithms of both members, we have

$$\log 3^{2x-1} = \log 0.07,$$

or, by Law III, $(2x - 1) \log 3 = \log 0.07$.

Solving gives $2x \log 3 - \log 3 = \log 0.07$,

$$x = \frac{\log 0.07 + \log 3}{2 \log 3}.$$

Since $\log 3 = 0.4771$ and $\log 0.07 = 8.8451 - 10 = -1.1549$,

$$x = \frac{-1.1549 + 0.4771}{2(0.4771)} = \frac{-0.6778}{0.9542} = -0.7103.$$

Note. Observe that, when $a < 1$, the numerical value of log a, is *negative*.

EXERCISE 114

Solve each of the following (a) by inspection without tables and (b) by the general method with tables:

1. $5^x = 25$ **2.** $2^x = 32$ **3.** $3^x = 27$
4. $2^{x+1} = 8$ **5.** $7^{2x} = 49$ **6.** $5^{x+2} = 125$
7. $3^{2x} = \frac{1}{81}$ **8.** $2^{x-1} = \frac{1}{4}$ **9.** $9^x = 27$
10. $2^{x+1} = (\frac{1}{8})^2$ **11.** $100^x = 1000$ **12.** $16^x = \frac{1}{8}$

Solve each of the following exponential equations:

13. $3^x = 25$ **14.** $5^x = 100$ **15.** $10^{2x} = 65$
16. $15^x = 30$ **17.** $7^{3x} = 50$ **18.** $2^{x+3} = 58$
19. $(1.72)^x = 3.64$ **20.** $(5.25)^x = 18.9$ **21.** $(3.41)^x = 10$
22. $(4.19)^x = 2$ **23.** $(1000)^x = 57$ **24.** $(5.2)^x = 0.004$
25. $(0.9)^x = 0.523$ **26.** $(4.25)^{3x} = 0.097$ **27.** $3^{2x+1} = 5^{3x-1}$
28. $(0.2)^{x+1} = (0.5)^x$ **29.** $(8.48)^{x+1} = (6.27)^{x-1}$ **30.** $44^{2x+3} = 66^{3x+2}$

REVIEW OF CHAPTER XIII

A

Write each of the following relations in its equivalent exponential form and solve for the unknown N, b, or x:

1. $\log_3 N = 2$ **2.** $\log_2 \frac{1}{4} = x$ **3.** $\log_b 125 = 3$
4. $\log_4 8 = x$ **5.** $\log_b \frac{1}{3} = \frac{1}{3}$ **6.** $\log_{0.1} N = 3$

7. $\log_{\frac{1}{9}} N = -\frac{1}{2}$
8. $\log_b 2\frac{1}{4} = -2$
9. $\log_{\sqrt{2}} 4\sqrt{2} = x$
10. $\log_b \sqrt[5]{3} = \frac{2}{5}$
11. $\log_{27} \frac{1}{9} = x$
12. $\log_4 N = 2.5$

Evaluate each of the following:

13. $\log_2 2 + \log_2 4 + \log_2 8$
14. $\log_2 2 + \log_4 2 + \log_8 2$
15. $\log_{10} 100 + \log_{10} 0.1 - 2 \log_{10} 1$
16. $\log_{\sqrt{3}} 3 - \log_3 \sqrt{3}$
17. $\log_a a^3 - \log_b b^2 + \log_c \sqrt{c}$
18. $\log_{\frac{2}{3}} \frac{4}{9} + \log_{\frac{1}{2}} 4 - \log_x 1$
19. $\log_4 32 + \log_x x\sqrt{x} - \log_5 625$
20. $\log_{\frac{3}{2}} 3\frac{3}{8} + \log_{\frac{4}{9}} 1\frac{1}{2}$

B

Using the laws of logarithms, express each of the following in terms of $\log a$, $\log b$, and $\log c$:

1. $\log abc$
2. $\log ab^2c^3$
3. $\log \dfrac{a^2b^3}{\sqrt{c}}$
4. $\log \dfrac{a^2}{\sqrt[3]{b^2c}}$
5. $\log a \dfrac{\sqrt{b}}{c^2}$
6. $\log \sqrt{\dfrac{a^3\sqrt{c}}{b}}$
7. Given $\log 2 = 0.301$, by using the laws of logarithms find (*a*) $\log 4$, (*b*) $\log 16$, (*c*) $\log \frac{1}{8}$, (*d*) $\log 5$. Hint: $\log 5 = \log \frac{10}{2}$.
8. Given $\log 3 = 0.477$, by using the laws of logarithms find (*a*) $\log 9$, (*b*) $\log 81$, (*c*) $\log \frac{1}{27}$, (*d*) $\log 3\frac{1}{3}$. Hint: $\log 3\frac{1}{3} = \log \frac{10}{3}$.

Given $\log 2 = 0.301$ and $\log 3 = 0.477$, find the following logarithms:

9. $\log 6$
10. $\log 18$
11. $\log 120$
12. $\log 300$
13. $\log \frac{8}{9}$
14. $\log \sqrt[3]{9}$
15. $\log \sqrt{24}$
16. $\log 2\frac{1}{4}$

Given $\log 17.7 = 1.248$ and $\log 6.73 = 0.828$, find the following logarithms:

17. $\log (17.7)(6.73)^2$
18. $\log \dfrac{\sqrt{17.7}}{(6.73)^3}$
19. $\log \dfrac{100}{17.7\sqrt[3]{6.73}}$
20. $\log \dfrac{(6.73)^2}{\sqrt[5]{(17.7)^3}}$

C

Find the value of each of the following by logarithms:

1. 627×0.925
2. 1.93×65.4
3. 42.35×6.724
4. 0.8362×451.9
5. $\dfrac{492}{6.56}$
6. $\dfrac{4.27}{35.6}$

7. $\dfrac{0.1927}{0.07395}$ 8. $\dfrac{0.7254}{6.912}$ 9. $(3.26)^5$

10. $(0.635)^3$ 11. $(76.15)^4$ 12. $(26180)^2$

13. $\sqrt[3]{295}$ 14. $\sqrt{0.0575}$ 15. $\sqrt[5]{0.1598}$

16. $\sqrt[4]{8566}$ 17. $2.19 \times 45.3 \times 0.635$ 18. $\dfrac{21.3 \times 0.0506}{4.48 \times 1790}$

19. $\sqrt[3]{\dfrac{32.8}{777}}$ 20. $(52.3 \times 0.619)^2$

D

Find the value of each of the following by logarithms:

1. $\dfrac{2.69 \times 3.47^2}{19.3}$ 2. $\dfrac{0.671 \times \sqrt{285}}{(3.21)^2}$

3. $1.67^2 \times 0.39^3 \times 2.62^2$ 4. $\sqrt{\dfrac{61.9 \times 35.2}{17.7}}$

5. $\dfrac{26.1}{13.9} \times \sqrt[3]{\dfrac{75}{294}}$ 6. $\sqrt[5]{\dfrac{27\sqrt{1.96}}{(6.7)^2}}$

7. $54.54 \times \sqrt[3]{\dfrac{100}{3.9 \times 6.006}}$ 8. $\left(\dfrac{19.05}{2.3}\right)^2 \times \sqrt{\dfrac{530}{91.25}}$

9. $\dfrac{(27.9)^2}{0.5197} \times \sqrt{607.63}$ 10. $63.72 \times \sqrt[5]{\dfrac{31.06}{222.2}}$

11. $(3.96)^{2.52}$ 12. $(5.47)^{1.93}$

Solve each of the following exponential equations:

13. $5^x = 35$ 14. $2^x = 50$ 15. $3^{x+1} = 37$

16. $4^{2x} = 36$ 17. $(2.38)^x = 6.95$ 18. $(42.6)^x = 0.827$

19. $(3.29)^{x+1} = (4.37)^{x-1}$ 20. $(65.4)^{x+1} = (13.1)^{x-1}$

Fifty-Fifty

What is the remainder when $x^{51} + 51$ is divided by $x + 1$?

Logarithmic Query

If $a^2 + b^2 = 7ab$, show that $\log\left(\dfrac{a+b}{3}\right) = \dfrac{1}{2}(\log a + \log b)$.

CHAPTER XIV

PROGRESSIONS

143. Progressions. A succession of numbers formed according to some fixed law is called a **progression.** The separate numbers are called terms of the progression, and are named from left to right as the first term, the second term, and so on. Thus, $-2, 3, 8, 13, 18$ is a progression of five terms, the first term being -2, the second term 3, and the last or fifth term 18.

Although a progression in general may be determined in many ways, two types of progressions are of particular importance. These are usually referred to as arithmetic progressions and geometric progressions.

Definition. *An* **arithmetic progression** *is a progression in which the terms increase or decrease by a constant amount, which is called the* **common difference.**

Thus, each of the following is an Arithmetic Progression, or A. P.:

$$2, 5, 8, 11, \cdots,$$
$$9, 4, -1, -6, \cdots.$$

The common difference is found by subtracting any term of the progression from the term which *follows* it. Hence the common differences for the above progressions are readily seen to be 3 and -5 respectively.

Definition. *A* **geometric progression** *is a progression in which the terms increase or decrease by a constant factor, which is called the* **common ratio.**

Thus, each of the following is a Geometric Progression, or G. P.:

$$2, -6, 18, -54, \cdots,$$
$$4, 2, 1, \tfrac{1}{2}, \cdots.$$

The common ratio is found by dividing any term of the progression by the term which *precedes* it. Hence it is evident that -3 and $\frac{1}{2}$ are the respective common ratios for the given progressions.

In order for three quantities to form either an A. P. or a G. P., it follows from the definitions that one of the following conditions must be satisfied:

Necessary condition that A, B, C form an A. P.:

$$B - A = C - B \quad \text{or} \quad A + C = 2B$$

Necessary condition that A, B, C form a G. P.:

$$\frac{B}{A} = \frac{C}{B} \quad \text{or} \quad B^2 = AC$$

NOTE. Three quantities A, B, C cannot form an A. P. and a G. P. at the same time except for the trivial case in which $A = B = C$.

Example 1. Find the value of x for which the three quantities $3x - 2$, $x + 5$, and $2x - 3$ form an A. P. Check your result.

Solution. We have $A = 3x - 2$, $B = x + 5$, $C = 2x - 3$. Hence applying the condition $A - B = B - C$, we have

$$(3x - 2) - (x + 5) = (x + 5) - (2x - 3).$$

Solving, we obtain $x = 5$.

Check. Substituting 5 for x in the given expressions, we obtain 13, 10, and 7, which is clearly an A. P. with the common difference -3.

Example 2. Find the values of x for which the three quantities $x^2 + x - 4$, $x + 1$, and 2 form a G. P. Check your results.

Solution. We have $A = x^2 + x - 4$, $B = x + 1$, $C = 2$. Hence applying the condition $B^2 = AC$, we have

$$(x + 1)^2 = 2(x^2 + x - 4).$$

Solving, we obtain $x = 3$ or -3.

Check. Substituting 3 for x, we obtain 8, 4, 2, which is a G. P. with $\frac{1}{2}$ as the common ratio. Setting $x = -3$, we obtain 2, -2, 2, which is a G. P. with -1 as the common ratio.

EXERCISE 115

Determine which of the following progressions are arithmetic progressions and which are geometric progressions:

1. 5, 8, 11, 14, 17
2. 7, 3, −1, −4, −8
3. $\frac{1}{2}$, −2, 8, −32, 128
4. 16, 24, 36, 54, 81
5. 0, 1, 2, 4, 8
6. −8, −5, −2, 2, 5
7. 27, 9, 3, −3, −9
8. 2, $\frac{7}{2}$, 5, $\frac{13}{2}$, 8
9. 1, $\sqrt{2}$, 2, $2\sqrt{2}$, 4
10. −2.4, −1.2, 0, 1.2, 2.4

Determine the values of x for which the following form an A. P. and check your results:

11. $-5, 3, x$
12. $2, x - 1, -16$
13. $1, 2x + 3, 3x + 2$
14. $x + 5, 7, 2x - 3$
15. $2x + 1, x + 4, 4x + 5$
16. $x + 1, -\frac{1}{8}x, 2 - 2x$
17. $x - \frac{2}{3}, x + \frac{1}{2}, \frac{1}{3}(5x - 1)$
18. $x + 1, 2x + 9, 4x + 27$
19. $4 - 3x, x^2 - 1, 9x - 10$
20. $x^2, 2x + 1, 8 - x$

Determine the values of x for which the following form a G. P. and check your results:

21. $27, -18, x$
22. $3, x, 12$
23. $x - 2, x, x + 4$
24. $4, 2(x + 1), (x + 4)(x - 1)$
25. $x + 1, 2x + 1, 4x - 1$
26. $x^2 + 3x + 1, 3x + 2, 9$
27. $x + 2, x + 5, x + 7$
28. $x - 5, x - 2, x + 3$
29. $1 - x, 3 - x, 5 - 3x$
30. $x - 3, x + 1, 5x - 7$

144. Finding the Last, or *n*th, Term of an A. P. If the first term of an A. P. is denoted by a and the common difference by d, the terms of the progression will be represented by

$$a,\ a + d,\ a + 2d,\ a + 3d,\ a + 4d,\ a + 5d, \text{ etc.}$$

It is evident that in any term the coefficient of d is always 1 less than the number of the term. Hence if l represents the last, or nth, term, we have the formula

$$l = a + (n - 1)d.$$

If any three of the numbers l, a, n, d are known, the remaining one can be determined by means of this formula.

Example 1. Find the 37th term of the progression −8, −5, −2, 1, · · · .

Solution. This is an A. P. in which the first term (a) is −8 and the common difference (d) is 3. Since $n = 37$, on substituting in the formula, we obtain

$$l = -8 + (37 - 1)3 = 100.$$

Example 2. Which term of the progression 1, 4, 7, 10, $\cdots$ is 49?

Solution. In this progression $a = 1$, $d = 3$, and $l = 49$. Substituting these values in the formula, we have

$$49 = 1 + (n - 1)3.$$

Solving, we obtain $n = 17$.

Example 3. If the 3rd and 15th terms of an A. P. are respectively 2 and -34, what are the first five terms of the progression?

Solution. Let a be the first term and d the common difference; then $a + 2d$ and $a + 14d$ represent respectively the 3rd and the 15th terms. Thus, we have

$$a + 14d = -34,$$
$$a + 2d = 2.$$

Solving this system of equations, we have $a = 8$ and $d = -3$. Hence, the first five terms of the progression are 8, 5, 2, -1, -4.

145. Arithmetic Means. In any A. P. all the terms between the first and the last are called *arithmetic means* between those two terms.

Thus, in the progression 2, 5, 8, 11, 14, the numbers 5, 8, and 11 are the *three* arithmetic means between 2 and 14. Likewise, in the progression 2, 6, 10, 14, the numbers 6 and 10 are the *two* arithmetic means between 2 and 14.

Example. Insert four arithmetic means between 2 and 14.

Solution. The two given terms, together with the four means, make an A. P. of 6 terms, that is, $n = 6$. Also it is given that $a = 2$ and $l = 14$. Hence, substituting in $l = a + (n - 1)d$, we have

$$14 = 2 + 5d.$$

Solving, we obtain $d = 2\frac{2}{5}$.

Hence the progression is 2, $4\frac{2}{5}$, $6\frac{4}{5}$, $9\frac{1}{5}$, $11\frac{3}{5}$, 14, and the four means are $4\frac{2}{5}$, $6\frac{4}{5}$, $9\frac{1}{5}$, $11\frac{3}{5}$.

The most important instance of arithmetic means is that of *one* arithmetic mean between two given numbers. This mean is also called the *arithmetic average* of the two numbers. If M is the arithmetic mean between a and b, then by the definition of an A. P.,

$$\text{M} - a = b - M.$$

Solving for M gives
$$M = \frac{a + b}{2}.$$

Thus, the arithmetic mean of the numbers 2 and 14 is $\frac{1}{2}(2 + 14) = 8$.

146. Finding the Sum of n Terms of an A. P. Let S represent the sum of n terms of a general arithmetic progression, that is

$$S = a + (a + d) + (a + 2d) + \cdots + (l - 2d) + (l - d) + l.$$

Writing the terms of this series in reverse order, we have

$$S = l + (l - d) + (l - 2d) + \cdots + (a + 2d) + (a + d) + a.$$

Adding the corresponding terms of these two series, we obtain

$$2S = (a + l) + (a + l) + (a + l) + \cdots + (a + l) + (a + l) + (a + l)$$
$$= n(a + l).$$

Therefore,

$$\text{(1)}\quad \boldsymbol{S = \frac{n}{2}(a + l).}$$

Since

$$\text{(2)}\quad \boldsymbol{l = a + (n - 1)d,}$$

$$\text{(3)}\quad \boldsymbol{S = \frac{n}{2}[2a + (n - 1)d].}$$

Example 1. Find the sum of the first 17 terms of the progression $2\frac{1}{2}$, $3\frac{1}{4}$, 4, $\cdots$.

Solution. In the given progression $a = 2\frac{1}{2}$, $d = \frac{3}{4}$, $n = 17$. Substituting in (3), we have

$$S = \tfrac{17}{2}[2(\tfrac{5}{2}) + 16(\tfrac{3}{4})] = \tfrac{17}{2}[5 + 12] = \tfrac{17}{2} \cdot 17 = 144\tfrac{1}{2}.$$

Example 2. The first term of an A. P. is 4, the last term is 36, and the sum is 500. Find the number of terms and the common difference.

Solution. If n is the number of terms, then from (1),

$$500 = \frac{n}{2}(4 + 36).$$

Solving, we have $\qquad n = 25.$

Now substituting in (2), we obtain

$$36 = 4 + 24d.$$

Solving gives $\qquad d = 1\frac{1}{3}.$

Example 3. How many terms of the progression $-1, 1, 3, 5, \cdots$ must be taken in order to make 360?

Solution. In the given progression, $a = -1$, $d = 2$, $S = 360$. Substituting in (3), we have

$$360 = \frac{n}{2}[2(-1) + (n - 1)2].$$

Simplifying gives $\qquad n^2 - 2n - 360 = 0.$

Solving, we have $\qquad n = 20 \quad \text{or} \quad -18.$

Since $n = -18$ has no meaning, we reject it as a solution. Thus 20 terms of the progression must be taken.

EXERCISE 116

In each of the following progressions, find the term indicated:

1. 2, 5, 8, $\cdots$; 13th

2. 9, 11, 13, $\cdots$; 28th

3. 29, 24, 19, $\cdots$; 18th

4. 71, 59, 47, $\cdots$; 9th

5. 4, $5\frac{1}{2}$, 7, $\cdots$; 21st

6. $-\frac{3}{2}$, $-\frac{1}{2}$, $\frac{1}{2}$, $\cdots$; 40th

7. $6\frac{1}{2}$, $5\frac{3}{4}$, 5, $\cdots$; 64th

8. 2.8, 3.1, 3.4, $\cdots$; 35th

9. 3.3, 2.1, 0.9, $\cdots$; 15th

10. 1.005, 1.219, 1.433, $\cdots$; 33rd

Between each pair of numbers, insert the number of means indicated:

11. 11, 35; 5 means

12. 8, -17; 4 means

13. 5, 6; 3 means

14. $2\frac{1}{2}$, $14\frac{1}{2}$; 7 means

15. 7.2, 9.8; 12 means

16. -1.6, 11; 8 means

17. $2\frac{1}{3}$, $6\frac{1}{2}$; 4 means

18. -22, 56; 25 means

19. Find the arithmetic mean of (*a*) 7 and 21, (*b*) 1.2 and 6.6, (*c*) $2\frac{1}{6}$ and $3\frac{1}{3}$.

20. Find the arithmetic mean of (*a*) -15 and 57, (*b*) 8.3 and 3.2, (*c*) $5\frac{1}{4}$ and $1\frac{1}{2}$.

In each of the following progressions, find the sum of the terms indicated:

21. 5, 9, 13, $\cdots$; 17 terms

22. -3, 2, 7, $\cdots$; 25 terms

23. 31, 22, 13, $\cdots$; 9 terms

24. 7, -1, -9, $\cdots$; 16 terms

25. $2\frac{2}{3}$, 3, $3\frac{1}{3}$, $\cdots$; 30 terms

26. $3\frac{1}{4}$, $2\frac{3}{4}$, $2\frac{1}{4}$, $\cdots$; 50 terms

27. $-1\frac{1}{2}$, $\frac{2}{3}$, $2\frac{5}{6}$, $\cdots$; 27 terms

28. -1.1, 1.9, 4.9, $\cdots$; 38 terms

29. 1.52, 4.31, 7.10, $\cdots$; 12 terms

30. 7.5, 5.7, 3.9, $\cdots$; 16 terms

31. In an A. P. whose common difference is 2, the 16th term is 24. Find the first term.

32. In an A. P. whose common difference is -3, the 12th term is 24. Find the first three terms of the series.

33. The 6th term of an A. P. is 17 and the 15th term is 35. Find a and d.

34. The 10th term of an A. P. is -8 and the 37th term is 73. Find a and d.

35. The 4th term of an A. P. is 0 and the 18th term is 21. Find the 50th term.

36. The 12th term of an A. P. is -7 and the 25th term is 6. Find the 4th term.

37. The 17th term of an A. P. is 7 and the 47th term is 31. Find the 30th term.

38. The 8th term of an A. P. is 43.2 and the 43rd term is 4.7. Find the 100th term.

39. The 6th term of an A. P. is 5 and the 19th term is $11\frac{1}{2}$. Find the 20th term.

40. The 67th term of an A. P. is 4.83 and the 76th term is 5.10. Find the 13th term.

41. If $a = 4$, $d = 3$, $l = 52$, find n and S.

42. If $a = 3$, $n = 8$, $l = 45$, find d and S.

43. If $d = 2$, $n = 7$, $l = 40$, find a and S.

44. If $a = -1$, $n = 12$, $d = 2\frac{1}{2}$, find l and S.

45. If $a = 5$, $l = 19$, $S = 84$, find d and n.

46. If $a = 3$, $n = 9$, $S = 66$, find d and l.

47. If $d = 4$, $n = 8$, $S = 80$, find a and l.

48. If $n = 7$, $l = 20$, $S = 84$, find a and d.

49. If $d = -2$, $l = 11$, $S = 231$, find a and n.

50. If $a = 11$, $d = -3$, $S = -150$, find n and l.

147. Finding the Last, or *n*th, Term of a G. P. If the first term of a G. P. is denoted by a, and the common ratio by r, the terms of the progression will be represented by

$$a,\ ar,\ ar^2,\ ar^3,\ ar^4,\ ar^5,\ \text{etc.}$$

It is evident that in any term the exponent of r is always 1 less than the number of the term. Thus the 10th term is ar^9 and the 18th term is ar^{17}, and so on. Hence, if l represents the last, or nth term, we have the formula

$$l = ar^{n-1}.$$

If any three of the numbers l, a, n, r are known, the other one can be determined by means of this formula.

Example. Find the 8th term of the progression 16, 8, 4, $\cdots$.

Solution. In the given progression $a = 16$, $r = \frac{1}{2}$, $n = 8$; hence, by substitution in the formula, we have

$$l = 16(\tfrac{1}{2})^{8-1} = \tfrac{1}{8}.$$

NOTE. In arithmetical computations involving exponents, cancel as much as is possible before actually computing the result. Thus in the above example, we would proceed as follows

$$l = 16(\tfrac{1}{2})^7 = \frac{16}{2^7} = \frac{2^4}{2^7} = \frac{1}{2^3} = \frac{1}{8}.$$

148. Geometric Means. In any G. P. all the terms between the first term and the last term are called *geometric means* between those two terms.

Thus, in the progression $\frac{1}{16}$, $\frac{1}{2}$, 4, 32, 256, the numbers $\frac{1}{2}$, 4, and 32 are the *three* geometric means between $\frac{1}{16}$ and 256. Likewise, in the progression $\frac{1}{16}$, 1, 16, 256, the numbers 1 and 16 are the *two* geometric means between $\frac{1}{16}$ and 256.

Example. Insert two geometric means between 2 and 54.

Solution. The two given terms, together with the two means, make a G. P. of 4 terms; that is, $n = 4$. Also it is given that $a = 2$ and $l = 54$. Hence, substituting in $l = ar^{n-1}$, we have

$$54 = 2 \cdot r^3.$$

Solving, we have $r = 3.$

The progression is 2, 6, 18, 54, and the two means are 6 and 18.

The most important instance of geometric means is that of *one* geometric mean between two given numbers. This mean is also called the *mean proportional* between two numbers. If M is the geometric mean between a and b, then by the definition of a G. P., we have

$$\frac{M}{a} = \frac{b}{M}.$$

Solving for M gives $M = \pm\sqrt{ab}.$

Thus, the geometric mean of 3 and 12 is $\sqrt{3 \cdot 12} = 6.$

NOTE. Although 3, -6, 12 also form a G. P., the value -6 does not fall between 3 and 12. It is called an *improper mean.* On the other hand, -6 is the *proper geometric mean* between -3 and -12.

149. Finding the Sum of *n* Terms of a G. P. Let S represent the sum of n terms of a general geometric progression; that is,

$$S = a + ar + ar^2 + \cdots + ar^{n-2} + ar^{n-1}.$$

Multiplying both sides throughout by r, we have

$$rS = ar + ar^2 + ar^3 + \cdots + ar^{n-1} + ar^n.$$

Hence, by subtraction,

$$S - rS = a - ar^n,$$

$$(1 - r)S = a(1 - r^n).$$

Therefore, (1) $\boldsymbol{S = a\left(\frac{1 - r^n}{1 - r}\right)}$.

Since (2) $\boldsymbol{l = ar^{n-1}}$,

(3) $\boldsymbol{S = \frac{a - rl}{1 - r}}$.

Example 1. Find the sum of 9, -3, 1, $\cdots$ to 6 terms.

Solution. In the given progression $a = 9$, $r = -\frac{1}{3}$, $n = 6$; hence substituting in (1), we have

$$S = 9 \times \frac{1 - (-\frac{1}{3})^6}{1 - (-\frac{1}{3})} = 9 \times \frac{1 - \frac{1}{729}}{1 + \frac{1}{3}} = 9 \times \frac{\frac{728}{729}}{\frac{4}{3}} = 9 \times \frac{728}{729} \times \frac{3}{4} = \frac{182}{27}.$$

Example 2. If $a = 3$, $l = 96$, $S = 189$, find r and n.

Solution. Substituting in (3) gives

$$189 = \frac{3 - 96r}{1 - r}.$$

Solving, we have $r = 2$.

Now substituting in (2), we have

$$96 = 3 \times 2^{n-1}.$$

Hence, $32 = 2^{n-1}$.

Since $32 = 2^5$, we have $n - 1 = 5$, or $n = 6$.

EXERCISE 117

In each of the following progressions, find the term indicated:

1. 1, 2, 4, $\cdots$; 7th

2. 1, 3, 9, $\cdots$; 6th

3. $\frac{1}{2}$, -1, 2, $\cdots$; 10th

4. $\frac{3}{2}$, -3, 6, $\cdots$; 8th

5. 81, 27, 9, $\cdots$; 12th

6. $\frac{1}{1000}$, $\frac{1}{100}$, $\frac{1}{10}$, $\cdots$; 9th

7. $\frac{1}{64}$, $\frac{1}{32}$, $\frac{1}{16}$, $\cdots$; 14th

8. 81, 54, 36, $\cdots$; 6th

9. 0.008, 0.04, 0.2, $\cdots$; 9th

10. 4.096, 2.048, 1.024, $\cdots$; 12th

11. The first term of a G. P. is 3, and the fourth term is 24. Find the common ratio.

12. The first term of a G. P. is $\frac{1}{2}$, and the third term is 2. Find the seventh term.

13. The 4th and 5th terms of a G. P. are respectively 3 and 6. Write the first three terms of the progression.

14. A G. P. contains four terms, of which the first is 5 and the last is 135. Find the sum of the second and third terms.

15. The first term of a G. P. is 5 and the common ratio is -3. Which term is equal to -135?

16. The first term of a G. P. is 0.0002 and the common ratio is 10. Which term is equal to 2,000,000?

17. The sixth term of a G. P. is 27 and the common ratio is 3. Write the first five terms of the progression.

18. In a G. P. the first term is $\frac{1}{2}$, the second term is -2, and the last term is 32,768. Find the next to last term.

19. Insert two geometric means between 5 and 40.

20. Insert two geometric means between -4 and 108.

21. Insert three geometric means between 162 and 2.

22. Insert three geometric means between $2\frac{1}{2}$ and 40.

23. Find the proper geometric mean between (*a*) 2 and 8, (*b*) $3\frac{1}{3}$ and $7\frac{1}{2}$, (*c*) 2.8 and 6.3.

24. Find the proper geometric mean between (*a*) 20 and 125, (*b*) $\frac{5}{8}$ and $2\frac{1}{2}$, (*c*) 7.5 and 14.7.

In each of the following progressions, find the sum of the terms indicated:

25. 3, 6, 12, $\cdots$; 8 terms

26. $\frac{1}{8}, \frac{1}{4}, \frac{1}{2}, \cdots$; 8 terms

27. $\frac{1}{3}, -1, 3, \cdots$; 6 terms

28. 64, 32, 16, $\cdots$; 7 terms

29. 16, 24, 36, $\cdots$; 6 terms

30. 27, -18, 12, $\cdots$; 6 terms

31. 0.0035, 0.035, 0.35, $\cdots$; 8 terms

32. 100, -10, 1, $\cdots$; 9 terms

Find the sum of each G. P.:

33. $-16, 8, -4, \cdots, -\frac{1}{16}$

34. $4, 6, 9, \cdots, 30\frac{3}{8}$

35. $5\frac{1}{3}, 8, 12, \cdots, 60\frac{3}{4}$

36. $81, -27, 9, \cdots, -\frac{1}{27}$

37. 1.024, 5.12, 25.6, $\cdots$, 16000

38. 600, 60, 6, $\cdots$, 0.0006

39. If $S = 121\frac{1}{3}$, $r = \frac{1}{3}$, $l = \frac{1}{3}$, find a and n.

40. If $S = 513$, $r = -2$, $l = 768$, find a and n.

41. If $a = 4$, $r = 2$, $S = 508$, find l and n.

42. If $a = \frac{4}{3}$, $r = \frac{1}{2}$, $S = 2\frac{5}{8}$, find l and n.

150. Infinite Geometric Series. The sum of any geometric series of n terms may be written:

$$S = a + ar + ar^2 + \cdots + ar^{n-1} = \frac{a - ar^n}{1 - r}.$$

If r is less than 1 in numerical value, the successive terms continually decrease in numerical value. Thus, if $r = \frac{1}{2}$, we have $r^2 = \frac{1}{4}$, $r^3 = \frac{1}{8}$, $r^4 = \frac{1}{16}, \cdots, r^{10} = \frac{1}{1024}$, and so on.

By increasing n, the value of r^n may be made as small as we please. Hence as n increases indefinitely, the limiting value of ar^n is zero, and the formula for the sum of the **infinite geometric series** becomes

$$\boldsymbol{S = \frac{a}{1 - r}}.$$

Example 1. Find the sum of the infinite series $18 - 6 + 2 - \cdots$.

Solution. In this series $a = 18$, $r = -\frac{1}{3}$; hence

$$S = \frac{18}{1 - (-\frac{1}{3})} = 18 \times \frac{3}{4} = 13\tfrac{1}{2}.$$

Example 2. Express the infinite repeating decimal $0.4181818 \cdots$ as a numerical fraction.

Solution. The given decimal may be written in the form

$$0.4 + 0.018 + 0.00018 + 0.0000018 + \cdots$$

where the first term contains the nonrepeating part of the decimal. It is evident that the terms following the first term form an infinite geometric series with $a = 0.018$ and $r = 0.01$. Therefore the value of the given decimal may be reduced to

$$0.4 + \frac{0.018}{1 - 0.01} = \frac{4}{10} + \frac{18}{990} = \frac{2}{5} + \frac{1}{55} = \frac{23}{55}.$$

EXERCISE 118

Find the sum of the following infinite geometric progressions:

1. $12, 6, 3, \cdots$

2. $-9, -6, -4, \cdots$

3. $6, -2, \frac{2}{3}, \cdots$

4. $\frac{1}{2}, \frac{1}{4}, \frac{1}{8}, \cdots$

5. $70, 7, 0.7, \cdots$

6. $\frac{6}{5}, -1, \frac{5}{6}, \cdots$

7. $-4, 1.6, -0.64, \cdots$

8. $0.9, 0.03, 0.001, \cdots$

9. $2, \sqrt{2}, 1, \cdots$

10. $\sqrt{3}, 1, \frac{1}{3}\sqrt{3}, \cdots$

Find the fractions equivalent to the following repeating decimals:

11. $0.777777 \cdots$

12. $0.363636 \cdots$

13. $0.818181 \cdots$

14. $0.252252 \cdots$

15. $0.253333 \cdots$

16. $0.402222 \cdots$

17. $3.569696 \cdots$

18. $1.324545 \cdots$

19. $2.335444 \cdots$

20. $5.408111 \cdots$

151. Problems Involving Progressions. Arithmetic and geometric progressions occur in the solution of many stated problems. The following examples indicate some of the methods used.

Example 1. The sum of eight numbers in arithmetic progression is 100 and the sum of the last three numbers is 60. What are the numbers?

Analysis. Let a, $a + d$, $a + 2d$, $\cdots$, $a + 7d$ represent the eight numbers; then,

$$(a + 5d) + (a + 6d) + (a + 7d) = 60$$

and from the formula $S = \frac{1}{2}n[2a + (n - 1)d]$,

$$100 = \tfrac{8}{2}[2a + (8 - 1)d].$$

Solution. The above equations reduce to

$$3a + 18d = 60$$
$$8a + 28d = 100$$

and have the solution, $a = 2$, $d = 3$.

Answer. The numbers are 2, 5, 8, 11, 14, 17, 20, 23.

Example 2. At the end of every year the value of a certain machine decreases 30% of what its value was at the beginning of the year. If the machine costs \$10,000 when new, find its value after 5 years of use.

Analysis. Let V represent the initial value of the machine. At the end of one year the machine is worth 70% of its initial value, or $(0.70)V$. At the end of the next year, it retains 70% of this new value, or 70% of $(0.70)V = (0.70)^2V$.

Thus, the terms of the geometric progression

$$(0.70)V,\ (0.70)^2V,\ (0.70)^3V,\ (0.70)^4V,\ \cdots$$

represent the value of the machine at the end of each succeeding year.

Solution. When $V = \$10{,}000$, the value at the end of the fifth year is given by

$$(0.70)^5 \times 10{,}000 = (0.16807) \times 10{,}000 = \$1680.70.$$

Example 3. Three numbers whose sum is 24 are in A. P. If they are increased by 3, 4, and 7 respectively, the resulting numbers will be in geometric progression. Find the original numbers.

Analysis. Let $a - d$, a, and $a + d$ represent any three numbers in A. P. It is evident that the sum of these numbers is $3a$. Hence, in general, if three numbers are in A. P., the middle number always equals $\frac{1}{3}$ the sum of the numbers.

Thus, in this problem the three numbers whose sum is 24 may be represented by $8 - d$, 8, and $8 + d$.

Solution. Increasing these numbers by 3, 4, and 7 respectively we have $11 - d$, 12, and $15 + d$. Applying the condition, $B^2 = AC$, that three numbers form a G. P., we have

$$12^2 = (11 - d)(15 + d).$$

Solving, we have $d = 3$ or -7.

Answer. The numbers are 5, 8, and 11 or 15, 8, and 1.

EXERCISE 119

1. In boring a well 300 feet deep the cost is 29 cents for the first foot and an additional cent for each subsequent foot. What is the cost of boring the entire well?
2. A radio is sold by selling tickets marked from 1 to 100. The tickets are sealed and drawn from a box, the price in cents per draw being equal to the number drawn. How much money is received from the sale?
3. In the infinite progression $6\frac{2}{5}$, $4\frac{4}{5}$, $3\frac{3}{5}$, $\cdots$, find the ratio of the sum of the even terms with respect to the sum of the odd terms.
4. At each stroke of an air pump one-half of the air remaining in a container is removed. What percentage of the original quantity of air remains in the container after the fifth stroke?
5. How many integers are there between 300 and 800 that are multiples of 7?
6. A ball rolls down an inclined plane 4.2 feet the first second, and in each succeeding second 8.4 feet more than in the preceding second. How far will it roll in 10 seconds?
7. The population of a city is 50,000 and it increases 5% every year. What will be the population in 10 years? Compute by logarithms.
8. A man borrows \$500 and agrees to pay back at the end of each month one-half of the amount still due. How long is it until he owes less than \$1?
9. The 12th, 85th, and last terms of an A. P. are 25, 244, and 328 respectively. Find the number of terms in the A. P.
10. A certain Christmas club fund collects 5 cents the first week, 10 cents the second week, and so on, increasing by 5 cents each week for 50 weeks. How much is in the fund at the end of that time?
11. A used-car dealer offered to sell a car for \$300 cash or on the following terms: 1 cent the first day, 2 cents the second day, 4 cents the third day, and so on for 15 days, the amount paid each day being twice that paid the previous day. Which is the better proposition for the dealer?

12. Find the A. P. whose first term is 3 and whose first, third, and seventh terms form a geometric progression.

13. How far must a boy run in a potato race if there are 10 potatoes in a straight line at a distance 3 feet apart, the first potato being 10 feet from the basket?

14. Find the sum of 20 terms of the progression whose nth term is $3n + 1$.

15. The arithmetic mean of two numbers is 15 and their geometric mean is 12. What are the numbers?

16. If $(1.06)^{10} = 1.7908$, find the value of

$$(a)\ 1 + (1.06) + (1.06)^2 + \cdots + (1.06)^9,$$
$$(b)\ 1 + (1.06)^{-1} + (1.06)^{-2} + \cdots + (1.06)^{-9}.$$

17. If the 5th term of an A. P. is 10 and the 10th term is 5, find the 15th term.

18. If the sum of 5 terms of an A. P. is 10 and the sum of 10 terms is 5, find the sum of 15 terms.

19. If the present value of an automobile is 60% of its value the preceding year, what is the value at the end of four years of a car that cost $1800?

20. Is the progression $\frac{1}{3}, \frac{1}{4}, \frac{1}{6}, \cdots$ arithmetical or geometrical? What are the next three terms?

21. A carpenter wishes to obtain a pole of sufficient length for 18 rungs of a ladder. If the rungs diminish uniformly from 24 inches at the base to 16 inches at the top, what length will be needed?

22. A and B start together, traveling in the same direction. A goes $2\frac{1}{2}$ miles the first hour, 3 miles the next, $3\frac{1}{2}$ miles the next, and so on. B goes 3 miles the first hour, $3\frac{1}{3}$ miles the next, $3\frac{2}{3}$ miles the next, and so on. In how many hours will they be together again? How far will each have traveled?

23. The sum of three numbers in A. P. is 42. If 2 is added to the second number and 12 is added to the third number, the resulting numbers will form a geometric progression. What are the numbers?

24. A company offers a clerk a job, starting at an annual salary of $2000 with a 10% increase in salary every year for 10 years. How much will he receive in his sixth year of service?

25. A contractor who fails to complete a building in a certain specified time is compelled to forfeit $100 a day for the first 6 days of extra time required, and for each additional day thereafter the forfeit is increased by $10 each day. If he loses $1760, by how many days did he overrun the contract time?

26. An industrial firm offers an engineer a choice of two contracts. He could start at \$4000 and receive an annual increase of \$400 thereafter, or he could start at the same rate and receive a semiannual increase of \$100. Which contract offers more money over a period of 10 years, and how much more?

27. What number must be added to each of the numbers a, b, and c, so that the resulting numbers will form a G. P.?

28. An A. P. and a G. P. each have the same first term 9, and their third terms are also equal. The second term of the G. P. is 2 less than the second term of the A. P. Find the two progressions.

29. A man arranges to pay off a debt of \$3600 by 20 annual payments which form an arithmetic progression. When 15 of the payments are paid, he dies, leaving a third of the debt unpaid. What was the value of the first payment?

30. Find three numbers in arithmetic progression such that their sum is 15 and the product of the first and third is 16 greater than the second.

31. Insert two numbers between 2 and 9 so that the first three numbers form an A. P. and the last three numbers form a G. P.

32. The width, the length, and the area of a rectangle form a G. P. If the perimeter of the rectangle is 112, find its dimensions.

33. Two cyclists 65 miles apart start toward each other. The first travels 5 miles the first [illegible] the next, 6 miles the next, and so on. The second [illegible] the first hour, 6 miles the next, 7 miles the next [illegible] how many hours will they meet?

34. Thr[illegible] form an A. P., the common difference being 5. If the first number is increased by 1, the second by 2, and the third by 4, the resulting numbers will form a geometric progression. Find the original numbers.

35. Find an infinite geometric progression whose sum is 4 and such that each term is three times the sum of all the terms which follow it.

36. The sum of three terms in A. P. beginning with $\frac{2}{3}$ is equal to the sum of three terms in G. P. beginning with $\frac{2}{3}$. If the common difference of the A. P. equals the common ratio of the G. P., what are the two progressions?

37. A man is offered a job starting at \$3600 for the first year, and with the following choice of advancement: (1) a 10% increase per year for 10 years, or (2) \$400 increase per year for 10 years. Which is the better proposition for a service period of (a) 5 years, (b) 10 years?

38. Derive a formula for the value of an insurance policy n years after January 1, 1935, from the following data: (Consider n greater than 5.)
(1) On January 1, 1935, its value is $1000.
(2) At the end of every year after January 1, 1940, its value is increased $5, $10, $15, etc., according to the number of years elapsed.

39. The sum of the terms of an infinite G. P. is 1 and the sum of the squares of the terms is $\frac{1}{3}$. What are the first three terms?

40. The sum of three numbers in A. P. is 3 and the squares of the numbers form a G. P. What are the numbers?

REVIEW OF CHAPTER XIV

A

Find the last term and the sum of the terms of each A. P.:

1. 2, 7, 12, $\cdots$, to 21 terms. **2.** -4, 2, 8, $\cdots$, to 36 terms.
3. 32, 29, 26, $\cdots$, to 15 terms. **4.** 8, 5, 2, $\cdots$, to 17 terms.
5. 2.7, 3.9, 5.1, $\cdots$, to 50 terms. **6.** 0.08, 0.17, 0.26, $\cdots$, to 100 terms.
7. $2\frac{1}{4}$, $4\frac{1}{2}$, $6\frac{3}{4}$, $\cdots$, to 35 terms. **8.** $8\frac{1}{4}$, $7\frac{3}{4}$, $7\frac{1}{4}$, $\cdots$, to 30 terms.
9. $1+\sqrt{2}$, 1, $1-\sqrt{2}$, $\cdots$, to 10 terms.
10. 1, $\sqrt{2}$, $2\sqrt{2}-1$, $\cdots$, to 16 terms.

11. Insert 4 arithmetic means between 3 and 23.

12. Insert 6 arithmetic means between -7 and 35.

13. Find the arithmetic mean of (a) 13 and 29, (b) -0.8 and 3.4, (c) 2 and $5\frac{1}{3}$.

14. Find the arithmetic mean of (a) -6 and 42, (b) 2.[illegible]32, (c) $2\frac{1}{2}$ and $9\frac{3}{4}$.

15. The 5th term of an A. P. is 1 and the 25th term is 41. Find the 15th term.

16. The 16th term of an A. P. is 79 and the 37th term is 16. Find the 45th term.

17. If $a = 5$, $d = 3$, $l = 23$, find n and S.

18. If $a = 23$, $l = 37$, $S = 240$, find d and n.

19. If $d = -3$, $n = 9$, $l = 9$, find a and S.

20. If $d = 4$, $n = 12$, $S = 174$, find a and l.

B

Find the last term and the sum of the terms of each G. P.:

1. 2, 6, 18, $\cdots$, to 6 terms. **2.** 32, 16, 8, $\cdots$, to 8 terms.
3. -5, 10, -20. $\cdots$, to 8 terms. **4.** 8, -12, 18, $\cdots$, to 6 terms.

5. $0.64, 0.96, 1.44, \cdots$, to 7 terms.
6. $3\frac{3}{8}, 2\frac{1}{4}, 1\frac{1}{2}, \cdots$, to 6 terms.
7. $\sqrt{2}, 2, 2\sqrt{2}, \cdots$, to 8 terms.
8. $\sqrt{2} - 1, 1, \sqrt{2} + 1, \cdots$, to 5 terms.

Find the sum of the following infinite progressions:

9. $45, -30, 20, \cdots$
10. $7\frac{1}{9}, 5\frac{1}{3}, 4, \cdots$
11. $2.61, 0.087, 0.0029, \cdots$
12. $2 + 2\sqrt{2}, 2 + \sqrt{2}, 1 + \sqrt{2}, \cdots$
13. Insert two geometric means between 24 and 375.
14. Insert three geometric means between 81 and 256.
15. Find the geometric mean between (*a*) 2 and 50, (*b*) $5\frac{1}{4}$ and $9\frac{1}{3}$, (*c*) 3.5 and 1.26.
16. Find the geometric mean between (*a*) -12 and -147, (*b*) $1\frac{1}{3}$ and $5\frac{1}{3}$, (*c*) 2.45 and 3.2.
17. The 2nd term of a G. P. is 54 and the 5th term is -16. Find the 7th term.
18. The sum of an infinite G. P. is 32. If the first term is 48, give the first five terms of the series.
19. If $S = 9.92$, $n = 5$, $r = 0.5$, find a and l.
20. If $S = 364$, $a = 1$, $l = 243$, find r and n.

C

1. The 7th, 36th, and last terms of an A. P. are 13, 158, and 283 respectively. Find the sum of the progression.
2. Find the sum of all multiples of 47 which lie between 1000 and 2000.
3. A sum of money invested at compound interest doubles itself every 18 years. What will $1000 amount to at the end of 90 years?
4. In an A. P. the 9th term is four times the second term and the 15th term is 46. Find the progression.
5. If an electric refrigerator which costs $250 depreciates 10% in value each year, find the value of the refrigerator after five years of use.
6. If the sum of 6 terms of an A. P. is 57 and the sum of 10 terms is 155, find the sum of 15 terms of the progression.
7. The sum of a G. P. whose common ratio is 3 is 484, and the last term is 324. Find the first term.
8. The sum of the first fifty even numbers is how many percent greater than the sum of the first fifty odd numbers?

9. The numbers from 1 to 100 are drawn by a hundred students for a school party, the fee paid being equal in cents to the number drawn. The fee was returned to the holders of five consecutive numbers chosen by lot and the net receipts were \$47.55. Which were the lucky numbers?

10. Out of 39 students receiving C or better in a course, the number making respectively A, B, C formed an increasing A. P. If one who made A had gotten a B and four of those receiving B had gotten C, the numbers in each group would have formed a G. P. How many students received A?

11. An advertising concern holds a contest offering the following prize money:
(1) First twenty prizes: \$1000, \$950, \$900, and so on.
(2) Next one hundred prizes: \$10 each.
(3) Next one hundred prizes: \$5 each.
Find the total value of the prizes.

12. In a certain auditorium the first row contains 20 seats, the second row 22 seats, the third row 24 seats, and so on for the first ten rows. The eleventh row has 41 seats, the twelfth row 42 seats, and so on for the remainder of the rows. If there are thirty rows altogether, find the seating capacity of the auditorium.

13. A and B start out together, going in the same direction, A traveling at a constant rate of 10 miles per hour. B travels 5 miles the first hour, 6 miles the next, 7 miles the next, and so on. In how many hours will they be together again?

14. In an infinite geometric progression, the sum of the odd terms is twice the sum of the even terms, and the sum of the first two terms is $2\frac{1}{4}$. Find the first four terms of the progression.

15. A birthday proposition was offered to a man. He was to receive either (1) 1 cent for the first month of his age, 2 cents for the second month, 3 cents for the next, and so on, or (2) \$1.40 for the first year of his age, \$2.80 for the second year, \$4.20 for the next year, and so on. The man figured that in either case he would receive the same amount of money. Find his age and the amount he was to receive.

16. Two numbers differ by 32. Their geometric mean is 6 less than their arithmetic mean. Find the numbers.

17. Three numbers in G. P. are such that the sum of the first two equals 5 and the sum of the last two equals 10. What are the numbers?

18. Find the sum of all positive integers less than 100 which are not divisible by either 2 or 3.

19. The set of positive integers is divided into groups as follows: 1, (2,3), (4,5,6), (7,8,9,10), and so on. Find the sum of the integers in the 25th group.

20. If the sides of a right triangle are in A. P., show that the sides must be in the ratio 3 : 4 : 5.

Irregular Workers

A number of men were engaged to do a job which would have taken them 32 hours if they had commenced at the same time. Instead of doing this, however, they began at equal intervals and then continued to work until the job was finished. How many hours did the first man work if he spent seven times as many hours on the job as the last man?

Another G.P.

In a manner similar to that used in summing a geometric progression, derive a formula for the sum $S = a + 2ar + 3ar^2 + \cdots + nar^{n-1}$.

One Up on Pythagoras

If a, b, c, and d form a geometric progression, show that

$$(a - c)^2 + (b - c)^2 + (b - d)^2 = (a - d)^2.$$

Partial Sums

If S_1, S_2, S_3 are the sums of n, $2n$, $3n$ terms respectively of an arithmetic progression, show that $S_3 = 3(S_2 - S_1)$.

O Happy Prime

I'm a little prime as happy as can be,
Ill fortune is the sum of my digits three.
If I'm reversed, I look the same,
Can you tell me, "What's my name?"

Ad Infinitum

If $S_1, S_2, S_3, \cdots, S_p$ are the sums of n terms of arithmetic progressions whose first terms are $1, 2, 3, \cdots, p$, and whose common differences are $1, 3, 5, \cdots, 2p - 1$, respectively, find the value of $S_1 + S_2 + S_3 + \cdots + S_p$.

ANSWERS

CHAPTER I

Exercise 1, Page 3

9. 72 **11.** 9 **13.** 36 **15.** 6 **17.** 36 **19.** 14 **21.** 1
23. 5 **25.** 2 **27.** 3 **29.** 4 **31.** 9 **33.** 1 **35.** 19
37. 19

Exercise 2, Page 6

1. 94,612 **3.** 1306 **5.** 12,474 **7.** (a) 80 mi. (b) 20 mi.
(c) $6\frac{2}{3}$ mi. **9.** 336 sq. in. **11.** \$914 **13.** \$490 **15.** 10.34 in.
17. 9856 cu. in. **19.** 75.46 sq. ft. **21.** 47.916 cu. ft. **23.** 45.54

Exercise 3, Page 13

1. $x = 9$ **3.** $x = 23$ **5.** $x = 9$ **7.** $x = 13$ **9.** $x = 45$
11. $x = 6$ **13.** $x = 11$ **15.** $x = 126$ **17.** $x = 24$
19. $x = 81$ **21.** $x = 2\frac{1}{2}$ **23.** $x = 13\frac{1}{3}$ **25.** $x = 8\frac{1}{2}$
27. $x = 6$ **29.** $x = 1\frac{3}{4}$ **31.** $x = 3\frac{5}{6}$ **33.** $x = 8\frac{1}{4}$
35. $x = 13\frac{3}{4}$ **37.** $x = 3$ **39.** $x = 5\frac{1}{2}$ **41.** $x = 6.3$
43. $x = 5.6$ **45.** $x = 0.14$ **47.** $x = 10.8$ **49.** $x = 14$
51. $x = 2.5$ **53.** $x = 6.14$ **55.** $x = 1.04$ **57.** $x = 4.15$
59. $x = 9.62$

Exercise 4, Page 15

1. $x = 6$ **3.** $x = 28$ **5.** $x = 10$ **7.** $x = 5$ **9.** $x = 7$
11. $x = 32$ **13.** $x = 12$ **15.** $x = 4$ **17.** $x = 3$ **19.** $x = 4$
21. $x = 18$ **23.** $x = 6$ **25.** $x = 4$ **27.** $x = 8$ **29.** $x = 6$
31. $x = 27$ **33.** $x = 24$ **35.** $x = 6$ **37.** $x = 2$ **39.** $x = 4$
41. $x = 24$ **43.** $x = 8$ **45.** $x = 6$ **47.** $x = 4$ **49.** $x = 9$
51. $x = 44$ **53.** $x = 11\frac{1}{2}$ **55.** $x = \frac{1}{2}$ **57.** $x = 12$
59. $x = 5\frac{1}{3}$

Exercise 5, Page 18

1. $2x + 3$ **3.** $a + b$ **5.** $3x + 1$ **7.** $y - 9$ **9.** $2n + 2$
11. $b - a$ **13.** $2x + 3$ **15.** $20/x$ **17.** $0.95x$ **19.** $(a + b)/2$

21. $\frac{1}{2}P - 6$ **23.** $\frac{1}{3}l$ **25.** $\frac{3}{2}D$ **27.** $x - 4$ **29.** $2x + 3$
31. $5d + 32$

Exercise 6, Page 20

1. $C = np$ **3.** $P = abc$ **5.** $A = x + 7$ **7.** $A = s^2$
9. $I = 10r$ **11.** $D = \frac{5}{2}R$ **13.** $V = 25q + 50$ **15.** $V = \frac{1}{10}d + \frac{1}{20}n$
17. $C = 9s + 70t$ **19.** $A = 0.35S$ **21.** $D = 52{,}800t$ **23.** $V = e^3$
25. $A = \pi R^2 - \pi r^2$ **27.** $C = 90 + 7(n - 15)$
29. $C = 5 + \frac{1}{4}(n - 10)$ **31.** $C = 100 + 0.8N$
33. $C = 500 + 4(n - 100)$ **35.** $C = 200 + 3(n - 35)$

Exercise 7, Page 23

1. 13 and 19 **3.** 14″ and 28″ **5.** 38 and 39 **7.** 18
9. 46 and 23 **11.** 65 **13.** 10 yrs. **15.** 18, 18, and 13
17. 15, 16, 17, 18, and 19 **19.** 31 **21.** City $13,500, County $27,000, State $54,000 **23.** 37 years **25.** Boy $7.50, girl $3.75 **27.** 13, 13, and 21 **29.** 72 **31.** 3 **33.** 29 and 43
35. 69 papers **37.** 6 and 12 **39.** Man 30 yrs., wife 28 yrs., son 6 yrs.

Review of Chapter I

A, Page 26

1. 10 **3.** 9 **5.** 2 **7.** 25 **9.** 18 **11.** 8 **13.** 1
15. 14/15 **17.** 18 **19.** 1

B, Page 26

1. 396 in. **3.** $195.75 **5.** 108 miles **7.** 165 cu. in.
9. $A = 2704$

C, Page 27

1. $x = 12$ **3.** $x = 14$ **5.** $x = 5$ **7.** $x = 4$ **9.** $x = 8$
11. $x = 2$ **13.** $x = 16$ **15.** $x = 2$ **17.** $x = 28$ **19.** $x = 40$

D, Page 27

1. $x^2 + x$ **3.** $100y/x$ **5.** $2x - 16$ **7.** $2m + 12$ **9.** $\frac{7}{15}c$
11. $C = 220N$ **13.** $V = \frac{1}{2}h + \frac{1}{4}q$ **15.** $A = 2ab + 2ac + 2bc$
17. $C = 72a + 84b$ **19.** $C = 4n$

E, Page 28

1. 65 and 74 3. 26, 28, and 30 5. 22 7. 17 and 37
9. Man 42, son 9, daughter 7 11. 91 games 13. 7′, 9′, and 10′
15. 13,022 and 15,635 17. $2160 19. 7.38 and 16.22

CHAPTER II

Exercise 8, Page 33

1. $+9$ 3. -16 5. -18 7. -15 9. -55 11. -14
13. $+2$ 15. -71 17. $+8$ 19. -10 21. 0 23. $+17$
25. -33 27. -85 29. 0 31. -11 33. $+40$ 35. -10

Exercise 9, Page 35

1. $-3m$ 3. $4a^2$ 5. 0 7. $-3b^2$ 9. x^3 11. 0
13. $-3x + 5y$ 15. $19a - b$ 17. $4xy - 2xy^2$ 19. $9x + y$
21. a 23. $-4y$ 25. $7x^2 - 2y^2$ 27. $18x^2 - 9x - 9$
29. $12u + 5v - w$ 31. $-2a + 11b$ 33. $-3x - 3z$
35. $-2x^2 + 6xy - 4y^2$ 37. $\frac{1}{4}x^2 + \frac{4}{5}xy - \frac{1}{3}y^2$ 39. $-\frac{3}{5}b$
41. $\frac{11}{6}x + \frac{7}{6}y - \frac{13}{6}z$ 43. $-1.1x + 0.3y + 1.9$
45. $-2.02a^2 + 0.45b^2 + 5.53c^2$ 47. $-3.13x^2 - 0.18xy - 4.52y^2$

Exercise 10, Page 37

1. -40 3. $+76$ 5. 0 7. $+46$ 9. -56 11. $+46$
13. $+39$ 15. -18 17. -24 19. 0 21. $+8$ 23. $+77$
25. $+88$ 27. -58 29. $+10$ 31. -3 33. -10
35. $+12$

Exercise 11, Page 39

1. $x + 5y$ 3. a 5. $16m - 8n$ 7. $2x^2y^2 + 2x^2y$ 9. $8a^2 + a$
11. $2a - b + 4c$ 13. $-x + 10y$ 15. $11x^2 - xy - 5y^2$
17. $x^4 - 2x^3 + x^2 - 2x + 1$ 19. $-3x^3 + x^2 + x + 3$
21. $5x^4 + x^3 - 2$ 23. $\frac{1}{6}a + \frac{5}{6}b$ 25. $-\frac{7}{6}x - \frac{2}{15}$
27. $-0.9x - 8.9y - 1.6z$ 29. $-15.48a^2 + 3.27ab + 0.98b^2$

Miscellaneous Exercises, Page 40

1. $2a + b$ 3. $x^3 + 8x - 4$ 5. $-2x - 3y$
7. $4m^3 + 10m^2 - 7m - 1$ 9. a^3 11. $2a + 4b + c$
13. $-2ac + 3bc$ 15. $-x^3 + x + 10$ 17. $x^2 + 2xy - 6y^2$

Exercise 12, Page 42

1. $5x + 4y$ **3.** $-a - 2b$ **5.** $-2x - 4$ **7.** $-2x + 3y$
9. $2a$ **11.** $2x^2 - 3x + 4$ **13.** $-5x + 4y$ **15.** b **17.** $4p - q$
19. $x^2 + 18x - 11$ **21.** $5a - 5b + 7c$ **23.** $4x^2 - 3x + 7$
25. $8x - 2$ **27.** $6a + 4b$ **29.** $3x - 1$ **31.** $-3x - y$
33. $4a + 4b$

Review of Chapter II

A, Page 43

1. $4a^2$ **3.** $-9b$ **5.** $(x - y)$ **7.** $10xy^2$ **9.** $3bc$
11. $(x + y)$ **13.** $5x^2 - 5x - 2$ **15.** $a - c$ **17.** $7m + 2n$
19. $9x^2 - 5xy + 3y^2$ **21.** $xy^2 + 7x^2y^2$

B, Page 44

1. $a^2 - 3ab$ **3.** $3a - 3c$ **5.** $7m + n + 23$ **7.** $2a - 3b + 6c$
9. $-7x + 4y$ **11.** $-8a^2 + ab$ **13.** y **15.** $-m^2 - m + 4$
17. $4x^2 + x + 3$

C, Page 45

1. $7a + 3b$ **3.** $-3x^2 + 4xy - 6y^2$ **5.** $4c$ **7.** $4x + y$
9. $4x^2 - 6y^2$ **11.** $x = 4$ **13.** $x = 6$ **15.** $x = 3$ **17.** $x = 4$
19. $x = 9$

CHAPTER III

Exercise 13, Page 47

1. -42 **3.** -8 **5.** $+20$ **7.** -49 **9.** -120 **11.** $+144$
13. -30 **15.** $+18$ **17.** $+36$ **19.** -48 **21.** -120
23. -14 **25.** -18 **27.** -1 **29.** -32

Exercise 14, Page 48

1. -6 **3.** -2 **5.** $+9$ **7.** -5 **9.** $+7$ **11.** -29
13. -21 **15.** -3 **17.** $+3$ **19.** -4 **21.** $+3$ **23.** -2
25. -10 **27.** $+4$ **29.** -3

Exercise 15, Page 49

1. -3 **3.** -29 **5.** -5 **7.** 13 **9.** 1 **11.** 9 **13.** -1
15. -35 **17.** -5 **19.** -81 **21.** 4 **23.** -5 **25.** -2
27. -1 **29.** 1 **31.** 4 **33.** 0 **35.** -4 **37.** -7
39. 48

Exercise 16, Page 51

1. -32 **3.** 20 **5.** -27 **7.** -5 **9.** 13 **11.** 7
13. -1 **15.** 36 **17.** 8 **19.** 38 **21.** 18 **23.** -13
25. -48 **27.** -3 **29.** 4 **31.** 19 **33.** -5 **35.** 0
37. $6\frac{1}{2}$ **39.** 0.71 **41.** 5.814 **43.** -21.83 **45.** -0.175
47. $-\frac{1}{2}$ **49.** 9

Exercise 17, Page 54

1. $-8x$ **3.** $14x^2$ **5.** $16x^2y$ **7.** $-3x^2y^3$ **9.** $-24a^2b$
11. $-3p^2q$ **13.** $-15x^2$ **15.** $9x^3$ **17.** $-12x^2y^2$ **19.** $-6a^2b^3c$
21. $-6x^3y^2z^2$ **23.** $63u^2v^2w^2$ **25.** $36x^3$ **27.** $-8a^2b^2$ **29.** $6x^2y^2$
31. $42a^2b^2c^2$ **33.** $72xyz$ **35.** $-108ab^2c^3$ **37.** $-30x^4$
39. $-180x^3y^4$ **41.** $-42a^2b^3$ **43.** $90x^4y^3$ **45.** $-12a^2b^3c^3$
47. $-108x^3y^4z^2$

Exercise 18, Page 55

1. $21x - 14y$ **3.** $-12a - 14$ **5.** $-8x^3 + 14x^2$
7. $10a^2b^2 + 6ab^3$ **9.** $15p - 35q$ **11.** $-12x^2 + 10xy$
13. $10a^2c - 35ac^2$ **15.** $-10ab + 5b^2$ **17.** $21x - 35y - 7z$
19. $-12a^3 + 18a^2b - 6ab^2$ **21.** $54pqr + 36q^2r - 6qr^2$
23. $-x^3y + x^3y^2 + x^3y^3$ **25.** $-6x^3y + 12xy^3 + 2xy^2$
27. $12a^2b^2 - 27ab^4 + 6a^2b^3$ **29.** $-3s^2t + 2st^2 + st$
31. $-30y^3 + 15y^2 - 5y$ **33.** $-10x^3 + 12x^2y + 18xy + 8x$
35. $63p^2q^3 + 42p^2q^2 - 14pq^3 + 7pq^2$ **37.** $21a^3 - 42a^2b + 35a^3b - 14a^2$
39. $-6x^4 + 12x^3 + 4x^2 + 10x$ **41.** $-6x^2 + 9xy - 3x^2y + 21x$
43. $27pq^3 - 54p^2q^2 - 18q^3 - 9q^2$ **45.** $-8a^6 + 4a^5 + 6a^4 - 2a^3$
47. $18x^3yz - 36x^2yz^2 - 9x^3y - 3x^2yz$

Exercise 19, Page 57

1. $6x^2 + x - 15$ **3.** $3m^2 - 10mn - 8n^2$ **5.** $5x^4 - 7x^2y^2 + 2y^4$
7. $3b^2 - 8b - 35$ **9.** $a^2 + ab + ac + bc$ **11.** $7p^2 + 6pq - q^2$
13. $15x^3 - 31x^2 + 9x + 7$ **15.** $3a^2 - 8ab + 5b^2 + 7a - 7b$
17. $35m^3 - 12m^2n - 9mn^2 + 2n^3$ **19.** $15x^3 - 28x^2 + 10x - 1$
21. $m^3 - n^3$ **23.** $3a^6 - 4a^4b^2 - 3a^2b^4 + 4b^6$
25. $10x^4 - 47x^3 + 19x + 2$ **27.** $a^4 + a^2b^2 + b^4$
29. $6m^4 - 33m^3n + 14m^2n^2 + 32mn^3 - 7n^4$
31. $9x^5 - 10x^3 + 15x^2 - 9x - 5$ **33.** $5p^5 - 3p^3 + p^2 + 4p - 7$
35. $18x^5 - 21x^4y + 24x^3y^2 - 10x^2y^3 - 18xy^4 - 5y^5$

Exercise 20, Page 59

1. $x^2 + 2x - 15$ **3.** $x^2 + 10x + 24$ **5.** $x^2 + 5x - 6$
7. $6x^2 - 5x + 1$ **9.** $16x^2 - 9$ **11.** $3x^2 + 22x + 35$
13. $20x^2 - 9x + 1$ **15.** $2x^2 + 5x - 187$ **17.** $12x^2 + 17xy + 6y^2$
19. $15x^2 + 2xy - 45y^2$ **21.** $30x^2 - xy - 42y^2$ **23.** $x^2 + 10x + 25$
25. $x^2 - 6x + 9$ **27.** $9x^2 - 6x + 1$ **29.** $9x^2 + 12x + 4$
31. $25x^2 - 20x + 4$ **33.** $100x^2 - 20x + 1$ **35.** $16x^2 + 24xy + 9y^2$
37. $25x^2 + 40xy + 16y^2$ **39.** $x^2 + y^2 + z^2 + 2xy - 2xz - 2yz$
41. $4x^2 + 9y^2 + 4z^2 - 12xy + 8xz - 12yz$
43. $25x^2 + 9y^2 + z^2 + 30xy - 10xz - 6yz$

Exercise 21, Page 61

1. $-5x$ **3.** $-2a$ **5.** $-5x$ **7.** $-7p$ **9.** $-26ac$ **11.** $4xy$
13. $2 - 3m^2$ **15.** $-x + 1$ **17.** $-y + 2x$ **19.** $-2x^2 + 5$
21. $2 - 3n^2$ **23.** $b - 5a$ **25.** $x^2 - 2x + 4$ **27.** $n - 2m - 3mn$
29. $-c + b - a$ **31.** $-3xyz + 5x^2y - 9yz^2$
33. $7x^2y^2 - 5x^2z^2 + 3y^2z^2$ **35.** $3mn^2 + 5m^2n + 7m$ **37.** $-3x + 2$
39. $2m^2n - 4mn^2 - 3$ **41.** $-4a + 2b + \frac{4}{3}c$ **43.** $5x - 9y - 7$
45. $2m^2 - 4mn + 7n^2$ **47.** $11z + 7x + 13xz$

Exercise 22, Page 64

1. $2x + 3$ **3.** $3m + 4n$ **5.** $4a^2 + 2$ **7.** $5x - 1 - \dfrac{5}{x - 1}$
9. $2a + 5b$ **11.** $2p - q$ **13.** $4x^2 + 3x + 8$ **15.** $5x^2 + 21x + 4$
17. $p^2 + 5pq - 4q^2$ **19.** $2a^2 + a - 1$ **21.** $7a^2 - 2ab + 3b^2$
23. $y^2 - 2y + 3 - \dfrac{4}{y + 1}$ **25.** $x^3 - 4x^2 - 4x + 16$
27. $a^3 + a^2b - 4ab^2 - 4b^3$ **29.** $12x^3 + 8x^2 + 4x + 1$ **31.** $x - 4$
33. $a - b$ **35.** $2y + 3$ **37.** $x^2 - xy + y^2$ **39.** $2x^2 - 3x + 2$

Review of Chapter III

A, Page 65

1. -120 **3.** 32 **5.** 4 **7.** 21 **9.** 40 **11.** 45 **13.** -3
15. 10 **17.** $\frac{5}{6}$ **19.** -35

B, Page 66

1. $12x^2y$ **3.** $-10ax$ **5.** $30p^3q^3$ **7.** $30x^2y^3z$ **9.** $-12u^4v^4w^3$
11. $18a^2 - 15ab$ **13.** $a^5b + a^3b^3 - 5a^2b^3$ **15.** $10a^3 - 15a^2b - 5ab^2$
17. $5a^2b^3 - 3a^3b^3 + 2a^3b^2$ **19.** $-12p^3q + 18p^2q^2 - 3pq^3 + 15pq$
21. $2a^2 + ab - 3b^2$ **23.** $x^3 - y^3$ **25.** $4m^4 - 9m^2 + 30m - 25$
27. $x^3 - 5xy^2 + 2y^3$ **29.** $3m^3 - 5m^2n + 5mn^2 - 2n^3$

C, Page 66

1. $x^2 + 5x + 6$ **3.** $x^2 + 20x + 100$ **5.** $2x^2 - 13x + 15$
7. $4x^2 + 12x + 9$ **9.** $3x^2 + 2x - 5$ **11.** $6x^2 + 5xy + y^2$
13. $25x^2 - 10xy + y^2$ **15.** $5x^2 + 13xy + 6y^2$
17. $9x^2 - 42xy + 49y^2$ **19.** $100x^2 + 40xy - 21y^2$

D, Page 67

1. $-9a$ **3.** $-2m^2n$ **5.** $-3pq^2$ **7.** $7xy$ **9.** $-2m^3n$
11. $-2x^3 + 3x$ **13.** $2 - 5xy$ **15.** $-2p - 3p^2 + 4q$
17. $-5x^2 + 4xy$ **19.** $-3p^2q + 2p^2q^3 + p$ **21.** $x + 8$
23. $2x - 5y$ **25.** $5c - 2$ **27.** $3a^2 - 2ax - x^2$
29. $a^2 + 2a - 2$

CHAPTER IV

Exercise 23, Page 69

1. $x = -10$ **3.** $x = 5$ **5.** $x = -6\frac{1}{3}$ **7.** $x = -4\frac{1}{4}$
9. $x = 7\frac{1}{2}$ **11.** $x = -8$ **13.** $x = -5$ **15.** $x = 2$
17. $x = -1\frac{5}{6}$ **19.** $x = 4$ **21.** $x = 9$ **23.** $x = -9.45$
25. $x = -3.77$ **27.** $x = 5.265$ **29.** $x = -2.11$ **31.** $x = -1.1$
33. $x = -11\frac{2}{3}$ **35.** $x = 401.19$ **37.** $x = 8.55$ **39.** $x = -0.002$

Exercise 24, Page 72

1. $x = 2$ **3.** $x = 5$ **5.** $x = -4$ **7.** $x = -7$ **9.** $x = \frac{1}{2}$
11. $x = 2$ **13.** $x = 2$ **15.** $x = 0$ **17.** $x = -8$ **19.** $x = 5$
21. $x = 5$ **23.** $x = 1$ **25.** $x = \frac{2}{3}$ **27.** $x = 7\frac{2}{3}$ **29.** $x = -2\frac{1}{5}$
31. $x = 5$ **33.** $x = 9\frac{1}{2}$ **35.** $x = -7$ **37.** $x = -2\frac{3}{4}$
39. $x = \frac{1}{2}$ **41.** $x = 2$ **43.** $x = -2$ **45.** $x = 4$
47. $x = -2$ **49.** $x = 1$

Exercise 25, Page 75

1. 32 lb. 68¢, 28 lb. 83¢ **3.** 5 lb. **5.** $8\frac{1}{2}$ gals. **7.** $16\frac{2}{3}$ lb.
9. 18 oz. of 90%, 27 oz. of 60% **11.** 32 lb. 43¢, 28 lb. 58¢
13. $5\frac{2}{3}$ gals. **15.** 20 quarts 2:3 sol., 10 quarts 7:3 sol. **17.** 80 lb.
19. 15 lb.

Exercise 26, Page 79

1. 27 m.p.h., 33 m.p.h. **3.** 3 m.p.h., 4 m.p.h. **5.** 21 m.p.h., 42 m.p.h. **7.** 126 mi. **9.** 9 hrs., 11 hrs. **11.** 6 mi. **13.** 0.4 mi.
15. 360 mi. **17.** 100 mi. **19.** 10 mi.

Exercise 27, Page 82

1. Son 12, father 36 **3.** Mary 7, Edna 5 **5.** 10 years **7.** 8 years old **9.** George 18, father 42 **11.** Boyd 10, Jimmy 8 **13.** Doris 7, sister 4 **15.** 5 and 10 **17.** 4, 8, and 32 **19.** *A* 20, *B* 10, and *C* 4

Exercise 28, Page 85

1. \$2100 at 4%, \$1700 at 5% **3.** \$2300 at 4%, \$1700 at 6% **5.** \$2000 at $2\frac{1}{2}$%, \$2300 at 4% **7.** \$384 **9.** \$7000 at 4%, \$14,000 at 7% **11.** \$2300 at 4%, \$1100 at 5% **13.** $2\frac{1}{2}$% **15.** 4.8% **17.** \$1000 at 3%, \$500 at 4%, and \$1000 at 6% **19.** \$4000 at $3\frac{1}{2}$%, \$3500 at 5%

Exercise 29, Page 88

1. 32 nickels, 23 dimes **3.** 17 quarters, 15 halves **5.** 6 nickels, 10 dimes, 13 quarters **7.** 16 dimes, 8 quarters **9.** 2261 grandstand, 597 bleacher **11.** 38 fives, 62 threes **13.** Silk \$4.96, linen \$1.24 **15.** 147 apricots **17.** \$4.45 **19.** 7600 gen. adm., 15,200 reserved seats, 2200 box seats

Exercise 30, Page 91

1. 32 **3.** 85 **5.** 72 **7.** 67 **9.** 18 **11.** 52 **13.** 35 **15.** 21 **17.** 142 **19.** 588

Review of Chapter IV

A, Page 93

1. $x = 5$ **3.** $x = -4$ **5.** $x = 1$ **7.** $x = 4$ **9.** $x = 2$ **11.** $x = 6$ **13.** $x = 4$ **15.** $x = 3\frac{3}{7}$ **17.** $x = 6$ **19.** $x = 5$ **21.** $x = \frac{3}{2}$ **23.** $x = 1$

B, Page 93

1. 36 m.p.h., 43 m.p.h. **3.** 14 and 5 **5.** 73 **7.** \$1850 at 4%, \$2250 at 5% **9.** 60 cc. of 65%, 40 cc. of 20% **11.** 15, 15, and 12 **13.** 72 **15.** 4 years **17.** 1 mile **19.** 17 by 22 **21.** 87 adults **23.** \$5000

C, Page 95

1. 1 P.M. **3.** 128 **5.** 68 **7.** \$2700 at 4%, \$2300 at 5% **9.** 2 gals. 1st sol., 3 gals. 2nd sol. **11.** 225 sq. ft. **13.** 360 **15.** Seven times older **17.** 3 m.p.h. **19.** 5 inches **21.** 70 pictures **23.** \$2100 at 4%, \$1000 at 6%

CHAPTER V

Exercise 31, Page 99

1. $3(x + 4y)$ **3.** $2x(4 - x)$ **5.** $4a(a - 4)$ **7.** $9ax(9ax - 1)$
9. $x(4x^2 - x + 1)$ **11.** $2ab(a^2 + 2ab + 4b^2)$
13. $x^2y^2z^3(xz + y + x^2yz)$ **15.** $3b(2x^2 + 5x - 6)$
17. $2x(x^3 - 2x^2 + 5x - 4)$ **19.** $3a(3x^2 - 4ax^2 - 2x + 1)$
21. $(a + b)(x - y)$ **23.** $(x - y)(2a + 3b)$ **25.** $(x - y)(x + y)$
27. $(a - b)(7m - n)$ **29.** $(x - y)(a + b)$ **31.** $(3x - y)(2x - y)$
33. $(x - y)(a + b - c)$ **35.** $(a + b)(a - 3b + 5)$
37. $(x^2 + y^2)(ab + ac + bc)$ **39.** $(x - 2)(3x^2 - x + 4)$

Exercise 32, Page 100

1. $(a + b)(a + c)$ **3.** $(a - b)(s - t)$ **5.** $(2x + y)(3x - a)$
7. $(x^2 + a^2)(x^2 + b^2)$ **9.** $(x - y)(1 + x)$ **11.** $(a - b)(5c - 1)$
13. $(a - 2b)(a^2 + 2b^2)$ **15.** $(2x - 3)(x^2 + 1)$
17. $(y + 1)(5y^2 - 1)$ **19.** $(4b - 5)(3b^2 - 4)$
21. $(3 - a)(4 + a^2)$ **23.** $(x + 1)(x^2 + 1)$
25. $(x + y)(a + b - c)$

Exercise 33, Page 102

1. $x^2 + 2x - 35$ **3.** $x^2 + 7x - 44$ **5.** $x^2 - 5xy + 6y^2$
7. $x^4 - 2x^2 - 24$ **9.** $x^2 + \frac{1}{4}x - \frac{1}{8}$ **11.** $(x + 1)(x + 2)$
13. $(b - 2)(b + 1)$ **15.** $(c - 3)(c - 3)$ **17.** $(x - 8)(x + 3)$
19. $(m + 3)(m + 7)$ **21.** $(y - 10)(y + 3)$ **23.** $(a - 6)(a + 3)$
25. $(x - 10)(x + 1)$ **27.** $(y - 5)(y + 4)$ **29.** $(x + 6)(x + 15)$
31. $(a + 26)(a - 10)$ **33.** $(x - 3y)(x - 15y)$
35. $(12 + x)(3 - x)$ **37.** $(6 + b)(14 + b)$ **39.** $(xy - 30)(xy + 2)$
41. $(x^2 - 19)(x^2 + 8)$ **43.** $(19 - m^2)(19 - m^2)$
45. $(x + 1)(x + \frac{1}{2})$ **47.** $(x + 0.5)(x - 0.3)$

Exercise 34, Page 104

1. $4x^2 + 20x + 21$ **3.** $9x^2 - 1$ **5.** $6x^2 - 7x + 2$
7. $12a^2 - ab - b^2$ **9.** $15 + 7x - 2x^2$ **11.** $(3x + 2)(x + 1)$
13. $(3x - 2)(x + 3)$ **15.** $(2x + 1)(x - 1)$ **17.** $(4x - 7)(x + 2)$
19. $(3x - 1)(x - 3)$ **21.** $(3a + 4)(a - 1)$ **23.** $(5a - 2)(a - 2)$
25. $(4y - 5)(y + 1)$ **27.** $(3c + 4)(c + 3)$ **29.** $(2a + 7)(3a - 2)$
31. $(3a + b)(9a - 4b)$ **33.** $(2m - n)(2m - 3n)$
35. $(2xy + 1)(3xy - 4)$ **37.** $(2a^2 - 3)(3a^2 - 2)$
39. $(3xy - 7z)(3xy - z)$ **41.** $(3 - 4x)(2 - x)$
43. $(5 + x)(1 - 2x)$ **45.** $(5 - 2x)(1 + 3x)$
47. $(2m + 3n)(12m - 7n)$ **49.** $(2 + 3x)(7 + x)$

Exercise 35, Page 105

1. $4a^2 - b^2$ **3.** $49m^2 - 16n^2$ **5.** $1 - x^6$ **7.** $\frac{4a^2}{9} - \frac{9b^2}{4}$
9. $4y^2 - 0.25$ **11.** $(x + 4)(x - 4)$ **13.** $(n^2 + 8)(n^2 - 8)$
15. $(3a + 5b)(3a - 5b)$ **17.** $\left(\frac{x}{2} + y\right)\left(\frac{x}{2} - y\right)$
19. $(ab + 1)(ab - 1)$ **21.** $(6p + 7q)(6p - 7q)$
23. $(1 + 10x)(1 - 10x)$ **25.** $(x^3 + 5)(x^3 - 5)$
27. $(ax^2 + 7)(ax^2 - 7)$ **29.** $(a + 8x^3)(a - 8x^3)$
31. $(4x^8 + 5y^{12})(4x^8 - 5y^{12})$ **33.** $(ab^2 + c^2d^3)(ab^2 - c^2d^3)$
35. $(8a^8 + 7b^7)(8a^8 - 7b^7)$ **37.** $(\frac{1}{2}p + \frac{1}{3}q)(\frac{1}{2}p - \frac{1}{3}q)$
39. $(0.1x + 1)(0.1x - 1)$ **41.** 50,000 **43.** 400 **45.** 500,000
47. 416,000 **49.** 192,000

Exercise 36, Page 106

1. $(a + b + c)(a + b - c)$ **3.** $(a - b + c)(a - b - c)$
5. $(x + 2y - z)(x - 2y + z)$ **7.** $(a + 3b)(a - b)$
9. $(3x + 2y - 1)(3x - 2y + 1)$ **11.** $(3m - n + 4)(3m - n - 4)$
13. $y(2x + y)$ **15.** $(a + b + c + d)(a + b - c - d)$
17. $(a - b + c + d)(a - b - c - d)$ **19.** $(4a + 3b)(2a + 7b)$
21. $(9x - 2y)(x - 4y)$ **23.** $(4x - 7)(10x + 3)$
25. $5(a + 1)(a - 1)$ **27.** $(3m - 2n)(m + 4n)$
29. $-8(x - y)(x + 3y)$ **31.** $a(9a - 2x)$ **33.** $-(2a + 11)(2a + 3)$
35. $8(x - 1)$

Exercise 37, Page 108

1. $x^2(x + y)(x - y)$ **3.** $2(x + 3yz)(x - 3yz)$
5. $2x(3x + 2)(x + 5)$ **7.** $6y^2(4x + 3y)(x - 2y)$
9. $(m^2 + 4)(m + 2)(m - 2)$ **11.** $(x + y)(x + y)(x - y)$
13. $(a + b)(a - b)(a + 2b)(a - 2b)$ **15.** $a(a + b)(x + y)(x - y)$
17. $(1 + x - y)(1 - x + y)$ **19.** $4x^2y(7x - 5)(x + 3)$
21. $(a - 19)(a + 13)$ **23.** $(x^4 + 1)(x^2 + 1)(x + 1)(x - 1)$
25. $(x - 2y)(x + 2y + 1)$ **27.** $(5x + 2y)(x + 4y)$
29. $abc(1 + a^3b^2c)(1 - a^3b^2c)$ **31.** $(8 + x)(5 - x)$
33. $(a + bc)(xy - z)$ **35.** $2(6x - 7y)(5x + 2y)$
37. $(a + 2)(a - 2)(a - 2)$ **39.** $2(x + 2y)(x - y)$
41. $(2x + 1)(x - 1)(x - 1)$ **43.** $(ax + b)(cx + d)$
45. $(7x - 3)(3x + 13)$ **47.** $(a + b)(a + b + 1)$
49. $x(3x - 2)(x^2 + 1)$

Exercise 38, Page 109

1. $(a - 1)(a^2 + a + 1)$ **3.** $(3x + y)(9x^2 - 3xy + y^2)$
5. $(y + \frac{1}{4})(y^2 - \frac{1}{4}y + \frac{1}{16})$ **7.** $(xy + 1)(x^2y^2 - xy + 1)$
9. $(ab + c)(a^2b^2 - abc + c^2)$ **11.** $(m^2 - n)(m^4 + m^2n + n^2)$
13. $(a + b - c)(a^2 + 2ab + b^2 + ac + bc + c^2)$
15. $(a - b - c)(a^2 + ab + ac + b^2 + 2bc + c^2)$
17. $(1 - x)(7x^2 + 4x + 1)$ **19.** $(3a + 2)(3a^2 + 3a + 1)$
21. $2a(x - 3a)(x^2 + 3ax + 9a^2)$
23. $(x + 2)(x^2 - 2x + 4)(x - 2)(x^2 + 2x + 4)$
25. $(x + 1)(x + 1)(x^2 - x + 1)$ **27.** $ab(ab + 1)(a^2b^2 - ab + 1)$
29. $(x + 1)(2x - 1)(x^2 - x + 1)(4x^2 + 2x + 1)$
31. $(c^3 + 2)(c - 1)(c^2 + c + 1)$ **33.** $(a + b)(x - 2)(x^2 + 2x + 4)$
35. $(a - 1)^2(a^2 + a + 1)^2$

Exercise 39, Page 111

7. $(x + 1)(x^2 + x - 1)$ **9.** $(x + 2)(x^2 - 2x + 5)$
11. $(x - 3)(x^2 + x + 4)$ **13.** $(x - 1)(x^4 + x^3 + x^2 + x + 2)$
15. $(x + 1)(x + 2)(x + 3)$ **17.** $(x - 1)(x - 2)(x^2 - 2)$
19. $(x + 1)(x - 2)(x^3 + 2)$ **21.** $(x + 1)(x - 2)(x + 3)$
23. $(x + 1)^2(x - 2)$ **25.** $(x + 1)(x^2 - 3x + 3)$
27. $(x + 2)(x^2 + 2x + 3)$ **29.** $(x + y)^3$
31. $(x - 1)^2(x - 2)(x + 4)$ **33.** $(x + 1)(x - 2)(x^2 - x + 2)$
35. $(x - 1)^3(x + 2)$ **37.** $(x - 1)^2(x + 2)(x^2 + x + 3)$
39. $(x + 1)^4(x - 4)$

Review of Chapter V

A, Page 112

1. $2x^2(x - 4)$ **3.** $pq(p^2 - q)$ **5.** $3x(x^2 + x + 1)$
7. $a^4b^3(b^2 + a^2 - ab)$ **9.** $2u^2v^2(2u - u^2v - 4uv + 5v^3)$
11. $(2x - y)(x + y)$ **13.** $(a - 3)(7a - 4)$ **15.** $(m - 3n)(3m - n)$
17. $(a + b)(x + y + z)$ **19.** $(x - 3)(x^2 - 3x + 1)$

B, Page 112

1. $(a + b)(x + y)$ **3.** $(a + 2)(3a^2 - 2)$ **5.** $(2m - 3)(m^2 - 2)$
7. $(x + 1)(x + 1)(x - 1)$ **9.** $(3a - 1)(a + 1)(a - 1)$
11. $(a - b)(a + c)$ **13.** $(a + 2)(b + 3)$ **15.** $(2m + 1)(3n - 1)$
17. $(x + 2y)(2x^2 - 3y^2)$ **19.** $(a - 4y)(a^2 - 2y^2)$

C, Page 113

1. $(x - 7)(x + 6)$ **3.** $(m + 2n)(m + 5n)$ **5.** $(7 + c)(5 - c)$
7. $(xy - 3)(xy - 14)$ **9.** $(x - 16)(x + 12)$ **11.** $(3x + 4)(x - 3)$
13. $(2a + 1)(2a + 3)$ **15.** $(4xy + 1)(xy - 3)$
17. $(5a + 3b)(5a - b)$ **19.** $(5x - 8)(10x + 13)$

D, Page 113

1. $(x + 3y)(x - 3y)$ **3.** $(xy + ab)(xy - ab)$
5. $(10 + 9x^2)(10 - 9x^2)$ **7.** $(\frac{2}{3}x + \frac{3}{4})(\frac{2}{3}x - \frac{3}{4})$
9. $(0.9a + 0.7b)(0.9a - 0.7b)$ **11.** $(x + y + 3)(x + y - 3)$
13. $3(a - b)(a - 3b)$ **15.** $8(x + 2)$ **17.** $(9a + 5b)(a - 9b)$
19. $-(x + 8)(5x + 2)$

E, Page 113

1. $x^2(3a + 7)(3a - 7)$ **3.** $2x(x + 5)(x - 3)$ **5.** $4y^2(y + 1)(y - 1)$
7. $5ab(a - 4b)(a + 3b)$ **9.** $(x + 3)(x + 1)(x - 1)$
11. $x(x + 2)^2(x - 2)^2$ **13.** $x(9x^2y^2 + 1)(3xy + 1)(3xy - 1)$
15. $(a - 1)(a + 2)(a - 2)$ **17.** $4(2a + 1)(a + 4)$
19. $(3x - 2)(x + 1)(2 + 3x)(1 - x)$ **21.** $(x + 1)(x + 8)$
23. $(a + b + c)(a + b - c)$ **25.** $(2x + 1)(2x - 1)(3x + 1)(3x - 1)$
27. $2ax^3(8ax + 9)(4ax - 3)$ **29.** $(x + y + z)(x + y - z)$

F, Page 114

1. $(p - 3)(p^2 + 3p + 9)$ **3.** $(ax + 1)(a^2x^2 - ax + 1)$
5. $(x - 3)(x^2 + 3)$ **7.** $a^2x(x + a)(x^2 - ax + a^2)$
9. $2x^2y(3y - 1)(9y^2 + 3y + 1)$ **11.** $(x - 3)(x^2 + x + 3)$
13. $(x - 1)(x^3 - x^2 - x - 1)$ **15.** $(x - 1)^2(x^3 + 3x^2 + 5x + 7)$
17. $x(x - 5)(x^2 + x + 5)$ **19.** $(x + 1)^2(x - 2)^2$

CHAPTER VI

Exercise 40, Page 116

1. $\frac{4}{7}$ **3.** $5a/3$ **5.** $4x/7z$ **7.** $2c^2x/3a^2$ **9.** $3x^2/5a^2$
11. $(x - 1)/(x + 1)$ **13.** $7/a$ **15.** $1/(ax - 1)$ **17.** $2b/(a - 1)$
19. $(x - y)/(x + y)$ **21.** b/c **23.** $(x - 1)/(x + 1)$
25. $1/(2c + 3y)$ **27.** $(a + b)/(a - 2b)$ **29.** $(x - 3)/(x - 1)$
31. $(x - 1)/(x + 1)$ **33.** $(3x + 5)/(x - 2)$ **35.** $a(x - 4)/(x + 5)$
37. $xy/(xy + 1)$ **39.** $(6m + 5n)/(7m - 4n)$ **41.** $(a + b)/(a - b)$
43. $(x + 2)/(x - 5)$ **45.** $(a + 1)/(a - 1)$

Exercise 41, Page 118

1. $\frac{1}{6}$ **3.** $\frac{2}{3}$ **5.** $9x/5a$ **7.** $4a^2/15d^2$ **9.** $25p^2/36m$
11. $9ax/4c^2z^2$ **13.** $\frac{4}{15}$ **15.** $(c-d)/2$ **17.** $(2a-b)/(a-b)$
19. $3x^2/5a$ **21.** $(x-8)(x-5)/(x+7)^2$ **23.** $(x-3)/6x$
25. $\frac{1}{2}$ **27.** 1 **29.** $(x+6)/(x-2)$ **31.** a
33. $6a^2(c+x)/c(a+x)$ **35.** $2a/5$ **37.** $(x+2)/(x-2)$
39. $1/3y^2$

Exercise 42, Page 120

1. $\frac{2}{3}$ **3.** $21b^2/20c^2$ **5.** $4xz^2/3y$ **7.** $7m^2/8q$ **9.** $4ab^2/3cmp^2$
11. ad/bc **13.** $1/(x+4)(x+2)$ **15.** $\frac{9}{10}$ **17.** $\frac{5}{6}$
19. $(x-3)/(x+7)$ **21.** $(x+2)(x-3)/(x-2)(x+3)$
23. $(2x+3)/3$ **25.** $2a(a+x)/x(a-x)$ **27.** $(2x-1)/(2x+1)$
29. 3 **31.** $3/2x$ **33.** $4x-7$ **35.** $x+y$

Exercise 43, Page 123

1. $(3x+y+2z)/6$ **3.** $ab/8$ **5.** $x/20$ **7.** $(yz+xz+xy)/xyz$
9. $-1/30m$ **11.** $(a^2+b^2+c^2)/abc$ **13.** $(3a+8)/4$
15. $(5m-6)/6$ **17.** $-(4a+b)/ab$ **19.** $(3x^2-2x+1)/x^3$
21. $(19x-22)/12$ **23.** $(19x-152)/225$ **25.** $(x+45)/70$
27. $(12x^2-24x-9)/8x^2$ **29.** 0 **31.** $(18x^2-15x-2)/30x^3$
33. $(3c-b)/bc$ **35.** $(m+n)/mn$

Exercise 44, Page 125

1. $9/2(a+3)$ **3.** $-3/14(x-4)$ **5.** $(3+2x)/x(a-x)$
7. $19/6(x+1)$ **9.** $19/6(2a-b)$ **11.** $(18x-13)/(2x-3)(3x-1)$
13. $38/(7a-2)(5a+4)$ **15.** $(x^2+y^2)/(x-y)(x+y)$
17. $(4x-3)/2(x-1)(2x-1)$ **19.** $15x/2(x-1)(x+2)$
21. $2(x^2-2)/(x-1)(x+1)$ **23.** $6/(a-2)(a-5)$
25. $y(7y-12)/(y-2)(3y-2)$ **27.** $5x^2/(x-2y)(x+2y)$
29. $(7x^2-4x+1)/2(x+1)(x-1)$
31. $(3x^2-12x+11)/(x-1)(x-2)(x-3)$
33. $-b(13a+17b)/(a+b)(a-b)(a+2b)$
35. $2x/(x+1)(x+2)(x+3)$

Exercise 45, Page 128

1. $ax/(x+a)(x-a)$ **3.** $2x/(x-y)(x+y)^2$ **5.** $a/(a-b)$
7. $1/(y-1)$ **9.** 1 **11.** $2/(x-1)(x-3)$
13. $1/(2-3a)(1+2a)(1-a)$ **15.** $4x/(x-1)^2(x+1)^2$
17. $x/(x-y)(x+y)(x+2y)$ **19.** $2/(x-1)(x-2)$
21. $2/(x-3)$ **23.** $1/(x+1)(x+2)$ **25.** $25/(2x-5)(3x+5)$
27. $2/(x-1)(x+1)$ **29.** $8x/(x-1)(x+1)^2$
31. $12/(x-1)(x-3)$

Exercise 46, Page 131

1. $19/20(1-x)$ **3.** $(7x+2)/6(x+2)(x-2)$
5. $1/(1+c)(1-c)$ **7.** 0 **9.** $-2/(a+5)$
11. $4/(3+x)(1-x)$ **13.** $(5x+8)/(x+1)(x+2)(2x-1)$
15. a/b **17.** $-2/(x+1)(x-1)$ **19.** 0

Exercise 47, Page 132

1. $(x+1)/x$ **3.** $ab/(a-b)$ **5.** $1/(1-y)$
7. $2(x-3)/(x+1)$ **9.** $(x^2-3x+1)/(x-2)$
11. $-2b^2/(a-b)(a+b)$ **13.** $(2x-1)(2x-3)$ **15.** $x-y$
17. $(x-2)/(x-5)$ **19.** x

Exercise 48, Page 134

1. $1/y$ **3.** a/b^2 **5.** $(y+x)/(y-x)$ **7.** y **9.** a
11. a/b **13.** $(x-1)/x(x+3)$ **15.** $(x+2)/(x+3)$
17. $(a+b)/a$ **19.** $(1+x)/2$ **21.** y **23.** 2

Exercise 49, Page 136

1. $x = 72$ **3.** $x = 9$ **5.** $x = -1$ **7.** $x = 3$ **9.** $x = 1$
11. $x = 3$ **13.** $x = 5$ **15.** $x = 6$ **17.** $x = 2$ **19.** $x = 9$
21. $x = 0$ **23.** $x = -\frac{2}{3}$ **25.** $x = 2$ **27.** $x = \frac{13}{3}$ **29.** $x = 4$
31. $x = -3$

Exercise 50, Page 139

1. $x = 3a$ **3.** $x = a+b$ **5.** $x = 3b$ **7.** $x = a$ **9.** $x = 4a$
11. $x = -3a/2$ **13.** $x = -14b$ **15.** $x = 2b/3$ **17.** $x = -5a/3$
19. $x = -3a$ **21.** $x = 4$ **23.** $x = a+b$ **25.** $x = b+a$
27. $x = 1/c$ **29.** $x = m+n$ **31.** $x = a+2b$ **33.** $x = a+b$
35. $x = (a-b)/2$ **37.** $x = 1$ **39.** $x = (a+b)/2$

Exercise 51, Page 140

1. $t = I/pr$ **3.** $B = 90 - A$ **5.** $h = 3V/B$ **7.** $l = (p-2w)/2$
9. $t = (A-p)/pr$ **11.** $C = (5F-160)/9$ **13.** $B = (2A-hb)/h$
15. $h = 4V/\pi d^2$ **17.** $h = Fl/W$ **19.** $M = (6V-hB-hb)/4h$
21. $c = b^2/4a$ **23.** $s = 360A/\pi r^2$ **25.** $m = Fr^2/kM$
27. $g = 4sa^3bM/ml^3$ **29.** $r = (nE-IR)/In$ **31.** $p = fq/(q-f)$
33. $r = (S-a)/(S-l)$ **35.** $p = (PV-Pv-K)/(V-v)$

Exercise 52, Page 143

1. 56 min. **3.** 6 min. **5.** A 10 hrs., B 15 hrs. **7.** A 42 min., B 56 min. **9.** 1 hour **11.** 42 min. **13.** 36 min. **15.** $2\frac{1}{2}$ hours **17.** 36 min. **19.** 15 days

Exercise 53, Page 145

1. 60 **3.** 150 **5.** 38, 39, and 40 **7.** 15′ by 21′ **9.** 10 and 15 **11.** 72 **13.** $\frac{7}{10}$ **15.** 8 years old **17.** 29 **19.** 13 **21.** 20 hours **23.** $\frac{10}{17}$ **25.** 8′ by 14′ **27.** $a(bc - 1)/(a + b)$ **29.** 25 m.p.h. **31.** $n(a - b)/b$ gals. **33.** $n/t(t - 1)$ m.p.h. **35.** Son \$19, Daughter \$13 **37.** 32 white, 52 black **39.** $\frac{6}{7}$

Review of Chapter VI

A, Page 148

1. $4a^2/5b$ **3.** $-\frac{4}{3}axy^2$ **5.** $5/2a$ **7.** $4a/3b$ **9.** $a(x - y)/b$ **11.** $(a + 3)/(a + 2)$ **13.** $(x - 2)/(2x - 1)$ **15.** $(3x - 2)/(2x - 5)$ **17.** $ax/(ax + 1)$ **19.** $(2x - 3)/(3x + 2)$

B, Page 149

1. $4ax/15b^2y$ **3.** $4a/15b$ **5.** $2(a + b)/c(a - b)$ **7.** 1 **9.** $(x - 2y)/(3x - 2y)$ **11.** $6ay/bx$ **13.** $3x/y$ **15.** $m + n$ **17.** $(x + 2y)/(x - 5y)$ **19.** 6

C, Page 150

1. $(x - 4)/4$ **3.** $(10 - 6x + 7x^2)/10x^3$ **5.** $(4a - 6)/3$ **7.** $(8x^2 + 6x - 5)/8x^3$ **9.** $2/a$ **11.** $1/(2a - 3)$ **13.** $(x + 5y)/(x + y)(x + 2y)$ **15.** $5/(x - 3)(x - 4)$ **17.** $4/3(x^2 - 1)$ **19.** $1/(x - 1)(x - 2)$

D, Page 150

1. $(a + 3)/(a - 2)$ **3.** $-2y$ **5.** $x - 4$ **7.** $x + 1$ **9.** $a + b$ **11.** $a^2 - 9$ **13.** $\frac{2}{3}$ **15.** $(x - y)/(x + y)$ **17.** $1/(1 - x)$ **19.** $(x^2 + 2xy - y^2)/(x^2 + y^2)$

E, Page 151

1. $x = 2$ **3.** $x = -1$ **5.** $x = \frac{1}{2}$ **7.** $x = 2$ **9.** $x = \frac{1}{3}$ **11.** $x = -10$ **13.** $x = -4$ **15.** $x = -3$ **17.** $x = 12$ **19.** $x = 4a/5$ **21.** $x = a$ **23.** $x = c/d$ **25.** $x = (c - d)/2$ **27.** $n = (l - a + d)/d$ **29.** $r = eR/(E - e)$

F, Page 152

1. 26, 28, and 30 **3.** 13 games **5.** 224 students **7.** 24 minutes
9. 1 hour and 12 minutes **11.** 32 days **13.** $Cd/(C + n)$
15. $bm/(b + m)$ **17.** \$6600 **19.** $(sq + r)/(q + 1)$, $(s - r)/(q + 1)$

CHAPTER VII

Exercise 54, Page 157

1. $x = 3, y = 2$ **3.** $x = 2, y = 4$ **5.** $x = -1, y = -1$
7. $x = 2, y = 0$ **9.** $x = 4, y = 5$ **11.** $x = 1, y = -2$
13. $x = 4, y = 4$ **15.** $x = 3, y = 4$ **17.** $x = -3, y = 2$
19. $x = \frac{1}{2}, y = \frac{1}{3}$ **21.** $x = \frac{2}{5}, y = \frac{3}{5}$ **23.** $x = 2, y = 1$
25. $x = 4, y = -3$ **27.** $x = 5, y = 6$ **29.** $x = -1, y = 3$

Exercise 55, Page 159

1. $x = 4, y = 1$ **3.** $x = 2, y = 1$ **5.** $x = -1, y = -3$
7. $x = 4, y = 2$ **9.** $x = -2, y = 5$ **11.** $x = 3, y = -2$
13. $x = -1, y = 3$ **15.** $x = -1, y = -2$ **17.** $x = 3, y = 1$
19. $x = 6, y = 3$ **21.** $x = \frac{1}{2}, y = \frac{3}{2}$ **23.** $x = \frac{3}{5}, y = -\frac{2}{5}$
25. $x = 6, y = 5$ **27.** $x = 6, y = -7$ **29.** $x = 38, y = -34$

Exercise 56, Page 160

1. $x = 6, y = 9$ **3.** $x = 24, y = 20$ **5.** $x = 8, y = -12$
7. $x = 11, y = 6$ **9.** $x = 8, y = 6$ **11.** $x = 5, y = 2$
13. $x = -1, y = 5$ **15.** $x = 3, y = 3$ **17.** $x = 1, y = -3$
19. $x = 2, y = 3$ **21.** $x = 1, y = -1$ **23.** $x = 3, y = 4$
25. $x = 9, y = 1$

Exercise 57, Page 162

1. $x = 2, y = 3$ **3.** $x = -1, y = 1$ **5.** $x = \frac{1}{4}, y = \frac{1}{2}$
7. $x = \frac{1}{5}, y = \frac{1}{7}$ **9.** $x = \frac{1}{5}, y = -1$ **11.** $x = \frac{1}{2}, y = \frac{1}{3}$
13. $x = \frac{1}{14}, y = \frac{1}{15}$ **15.** $x = 2, y = \frac{1}{2}$ **17.** $x = 2, y = -2$
19. $x = 4, y = 5$

Exercise 58, Page 164

1. $x = c + d, y = c - 2d$ **3.** $x = 2m + n, y = 2n - m$
5. $x = p, y = -p$ **7.** $x = 2a + b, y = a + 2b$
9. $x = c - 2d, y = 3c + d$ **11.** $x = 3/a, y = 1/b$
13. $x = 5/c, y = -3/d$ **15.** $x = 6a, y = 8b$ **17.** $x = 2c, y = -3d$
19. $x = 5c/a, y = -c/b$ **21.** $x = a + b, y = 2a - b$
23. $x = 2a - b, y = a - 2b$ **25.** $x = a, y = 0$ **27.** $x = a^2, y = ab$
29. $x = a + b, y = a$

Exercise 59, Page 166

1. $x = 4, y = 2, z = 1$ **3.** $x = 1, y = 2, z = 1$
5. $x = -1, y = -2, z = 2$ **7.** $x = 1, y = 2, z = 3$
9. $x = 2, y = -1, z = -\frac{1}{3}$ **11.** $x = 2, y = 3, z = 4$
13. $x = 2, y = 4, z = 8$ **15.** $x = 2, y = 4, z = 6$
17. $x = \frac{1}{3}, y = -\frac{1}{2}, z = 1$ **19.** $x = 3a, y = -a, z = 2a$

Exercise 60, Page 169

1. Rowing 4 m.p.h., Current 1 m.p.h. **3.** Plane 120 m.p.h., Wind 30 m.p.h. **5.** Launch $17\frac{1}{2}$ m.p.h., Tide $2\frac{1}{2}$ m.p.h. **7.** $\frac{1}{2}$ m.p.h. **9.** 20 m.p.h. **11.** 60 m.p.h. **13.** 4 m.p.h. **15.** 1 m.p.h. and 3 miles **17.** 400 miles **19.** $3\frac{1}{3}$ m.p.h.

Exercise 61, Page 173

1. 53 and 38 **3.** $\frac{17}{30}$ **5.** Horse $55, Cow $145 **7.** *A* 3 miles, *B* $2\frac{2}{3}$ miles **9.** $9500 **11.** 4′ by 8′ **13.** 9 and 2 **15.** 27 dimes, 18 quarters **17.** 36 **19.** Dan 23, Marvin 17 **21.** Man $7.30, Boy $5.10 **23.** 5 m.p.h. and $2\frac{1}{2}$ m.p.h. **25.** Grade A 70¢, Grade B 52¢ **27.** *A* $170, *B* $65, *C* $85 **29.** 3 m.p.h. **31.** $3000 at 4% **33.** *A* 20 days, *B* 25 days **35.** 3′ by 10′ and 6′ by 7′ **37.** *A* 3 m.p.h., *B* 5 m.p.h. **39.** 407 **41.** 10′ and 12′

Review of Chapter VII

A, Page 177

1. $x = 2, y = -3$ **3.** $x = -2, y = 1$ **5.** $x = 4, y = 3$
7. $x = 20, y = 30$ **9.** $x = \frac{1}{2}, y = \frac{1}{3}$ **11.** $x = -\frac{1}{2}, y = \frac{3}{4}$
13. $x = \frac{4}{3}, y = \frac{6}{5}$ **15.** $x = 1.4, y = -0.9$

B, Page 177

1. $x = 4, y = 6$ **3.** $x = -\frac{1}{2}, y = -\frac{1}{5}$ **5.** $x = 2, y = -3$
7. $x = \frac{3}{4}, y = \frac{2}{5}$ **9.** $x = -\frac{5}{3}, y = \frac{5}{26}$

C, Page 178

1. $x = 2a + 3b, y = 3a - 2b$ **3.** $x = 4/m, y = 5/n$
5. $x = a(a + b), y = b(a + b)$ **7.** $x = ab + 3, y = 2ab - 1$
9. $x = a - b, y = 2a + b$

D, Page 178

1. $x = 4, y = 5$ **3.** $x = -6, y = 10$ **5.** $x = -2, y = -1$
7. $x = 3, y = 5$ **9.** $x = -5, y = 7$ **11.** $x = 7, y = 9$

E, Page 179

1. $x = 3, y = -1, z = -2$ **3.** $x = 4, y = 2, z = 3$
5. $x = 3, y = 1, z = -1$ **7.** $x = n, y = n - 1, z = n - 2$

F, Page 179

1. 17 and 12 **3.** Alg. $2.30, Geom. $2.55 **5.** Walking 4 m.p.h., running 8 m.p.h. **7.** $x = 12$, $y = 4$ **9.** First sol. 40%, second sol. 30% **11.** 45 **13.** Tea 70¢, Coffee 60¢ **15.** $\frac{13}{17}$
17. 7′ by 13′ **19.** 15 hours and 20 hours

CHAPTER VIII

Exercise 62, Page 184

1. x^7 **3.** a^6 **5.** y^{23} **7.** a^5x^7 **9.** x^6y^4 **11.** x^3 **13.** 1
15. a/b **17.** a **19.** x^2/z^2 **21.** x^{15} **23.** $-b^{16}$ **25.** $-m^6$
27. x^6y^6 **29.** $a^4b^9c^{16}$ **31.** $8x^9$ **33.** $-a^8b^{12}$ **35.** $8a^3b^6$
37. $a^8b^{12}c^4$ **39.** $16a^4x^8y^{12}$ **41.** $\frac{27}{64}$ **43.** a^4/x^8 **45.** a^{15}/b^{25}
47. $25a^6/36b^4$ **49.** $a^{10}b^{20}c^{10}/x^{20}y^{30}z^{10}$ **51.** a^5x^{13} **53.** y^5/x^{10}
55. $x^{12}y^8$ **57.** $72a^3b^2x^{12}$ **59.** a **61.** y^{10}/x^3 **63.** $1/n^7$
65. x^4y/ab^4 **67.** $a^4b^3x^4y^3$ **69.** $6c$

Exercise 63, Page 186

1. ± 4 **3.** ± 8 **5.** $\pm a$ **7.** $\pm x^2y$ **9.** $\pm 5a^2$ **11.** 2
13. -4 **15.** a **17.** $4xy$ **19.** $-a^2b$ **21.** 11 **23.** 3
25. 2 **27.** -2 **29.** $\frac{1}{4}$ **31.** b^2 **33.** 4 **35.** $3a$ **37.** 10
39. $11a^2$ **41.** $\frac{1}{6}$

Exercise 64, Page 188

1. $\sqrt{x}, \sqrt[3]{x}, \sqrt[3]{y^2}, \sqrt[4]{y^3}, \sqrt[7]{a}, \sqrt[5]{a^3}, \sqrt{b}, \sqrt[5]{b^2}$
3. $3\sqrt{x}, \sqrt{3x}, 8\sqrt[3]{y}, 5\sqrt[5]{y^3}, 2\sqrt[3]{a^4}, 2\sqrt[4]{a^3}, \sqrt{a^2 + b^2}, \sqrt[3]{a^3 - b^3}$
5. $a^{\frac{1}{2}}, a^{\frac{1}{3}}, b^{\frac{2}{3}}, b^{\frac{3}{2}}, x^{\frac{2}{5}}, x^{\frac{5}{2}}, y^{\frac{3}{4}}, y^{\frac{5}{3}}$
7. $2a^{\frac{1}{2}}, (2a)^{\frac{1}{2}}, 7b^{\frac{2}{3}}, 3b^{\frac{3}{5}}, 2x^{\frac{3}{4}}, 5x^{\frac{5}{2}}, 2(x - y)^{\frac{1}{2}}, 4(x^2 - y^2)^{\frac{1}{2}}$ **9.** 5 **11.** 2
13. 8 **15.** 12 **17.** -1 **19.** 27 **21.** $\frac{1}{2}$ **23.** $\frac{4}{9}$ **25.** 0.5
27. 0.3 **29.** $a^{\frac{3}{4}}$ **31.** $b^{\frac{3}{2}}$ **33.** $2^{\frac{5}{6}}$ **35.** n **37.** $a^{\frac{1}{2}}$ **39.** $y^{\frac{5}{12}}$
41. $n^{\frac{7}{12}}$ **43.** $x^{\frac{1}{6}}$ **45.** a^4 **47.** $x^{\frac{1}{3}}$ **49.** 2 **51.** x^3y^2
53. $x^{\frac{5}{3}}/y^{\frac{5}{2}}$ **55.** $b^{\frac{9}{2}}/c^4$ **57.** $x^{\frac{5}{2}}$ **59.** $n^{\frac{10}{3}}$ **61.** $a^{\frac{1}{2}}b$

Exercise 65, Page 190

1. $\frac{1}{4}$ 3. $\frac{1}{8}$ 5. 1 7. 5 9. $\frac{1}{16}$ 11. $\frac{1}{2}$ 13. $\frac{8}{9}$
15. $\frac{3}{10000}$ 17. 2 19. $-\frac{1}{4}$ 21. x 23. m^{-4} 25. x^4
27. x^6 29. y^3 31. ab^{-4} 33. $1/a^4$ 35. $1/5x^2$ 37. x^2/a^3
39. $2y/3x^4$ 41. $(y-x)/xy$ 43. $1/(x-y)$ 45. $2x/y$
47. $(a^2+b^2)/ab$ 49. $(n^2-m^2)/m^2n^2$ 51. $xy(x-y)$ 53. a^3
55. b^{2-n} 57. x^{n^2-n}/y^{3n-3} 59. $a^{2n}b^{n^2}$ 61. a^{k+2} 63. p^rq^{-r}

Exercise 66, Page 193

1. 4 3. 5 5. -2 7. $2a$ 9. $-3b^2$ 11. $5\sqrt{2}$
13. $11\sqrt{2}$ 15. $9\sqrt{2}$ 17. $20\sqrt{2}$ 19. $3\sqrt[3]{2}$ 21. $-5\sqrt[3]{2}$
23. $2\sqrt[4]{3}$ 25. $3ab\sqrt{5a}$ 27. $3a\sqrt{7}$ 29. $3ab^3\sqrt[3]{3a^2}$ 31. $7\sqrt{3}$
33. $15\sqrt{7}$ 35. $5\sqrt{42}$ 37. $2\sqrt[3]{7}$ 39. $20a^3$

Exercise 67, Page 195

1. 2 3. 7 5. 2 7. -2 9. $2a$ 11. x^2 13. $\frac{1}{5}\sqrt{15}$
15. $\frac{1}{6}\sqrt{15}$ 17. $\frac{1}{2}\sqrt{10}$ 19. $\frac{3}{5}\sqrt{15}$ 21. $\frac{1}{2}\sqrt[3]{6}$ 23. $\frac{1}{2b^2}\sqrt{6ab}$
25. $\sqrt[3]{xy^2}/y$ 27. $\sqrt{2ab}$ 29. $\frac{1}{10}\sqrt{70}$ 31. $\frac{2}{5}\sqrt{35}$ 33. $\frac{1}{25}\sqrt{15}$
35. $\frac{1}{6}\sqrt{6}$ 37. $\frac{1}{2}\sqrt[3]{15}$ 39. $\sqrt[3]{4}$ 41. $\sqrt{5b}$ 43. $\frac{1}{2}x\sqrt{2y}$

Exercise 68, Page 196

1. $7\sqrt{2}$ 3. $\sqrt{5}$ 5. $3\sqrt{3}$ 7. $11\sqrt{2}$ 9. $3\sqrt{2}$ 11. $4\sqrt{5}$
13. $18\sqrt{7}$ 15. $12\sqrt{2}$ 17. $-4\sqrt{5}$ 19. $3\sqrt[3]{3}$ 21. $3\sqrt[3]{5}$
23. $13\sqrt[3]{6}$ 25. $2\sqrt[3]{3}$ 27. $2\sqrt{7}+6\sqrt{5}$ 29. $67\sqrt{3}-42\sqrt{5}$
31. $\sqrt{3}+2\sqrt{15}$ 33. 0 35. $8-3\sqrt[3]{6}$

Exercise 69, Page 197

1. $3\sqrt{2}+6$ 3. $2+\sqrt{6}-2\sqrt{3}$ 5. $6\sqrt{3}-6\sqrt{2}-6$
7. $40-48\sqrt{3}-16\sqrt{5}$ 9. $108-40\sqrt{6}-24\sqrt{5}$ 11. 1
13. $12-11\sqrt{21}$ 15. $35+12\sqrt{6}$ 17. $14\sqrt{6}-4$
19. $756-216\sqrt{6}$ 21. $1-2\sqrt[3]{6}+\sqrt[3]{36}$ 23. $132\sqrt{3}-162\sqrt{2}$
25. $x\sqrt{a}-x\sqrt{b}+x\sqrt{c}$ 27. $a+b-2\sqrt{ab}$
29. $1-2\sqrt{n(1-n)}$ 31. $3a-8+a\sqrt{a-3}$
33. $9+5\sqrt{2}$ 35. 0 37. $17+\sqrt{2}-\sqrt{3}-4\sqrt{6}$

Exercise 70, Page 199

1. $\sqrt{2}-3\sqrt{5}+4\sqrt{7}$ 3. $4-3\sqrt{6}+\sqrt{10}$
5. $\frac{5}{6}\sqrt{6}+\frac{10}{3}\sqrt{3}-\frac{15}{2}\sqrt{2}$ 7. $\frac{1}{4}\sqrt{30}+\frac{1}{3}\sqrt{78}-\frac{1}{6}\sqrt{114}$

9. $\sqrt{3}+1$ **11.** $3-\sqrt{6}$ **13.** $\sqrt{5}-2\sqrt{2}$ **15.** $9+4\sqrt{5}$
17. $\sqrt{6}+2$ **19.** $(\sqrt{a}-\sqrt{b})/(a-b)$ **21.** $\sqrt{15}$
23. $6+2\sqrt{2}$ **25.** $\sqrt{2}$

Exercise 71, Page 201

1. $\sqrt{2}$ **3.** $\sqrt{2}$ **5.** $\sqrt[3]{9}$ **7.** $\sqrt{2}$ **9.** $\sqrt{2}$ **11.** $\sqrt[5]{11}$
13. $\sqrt{3xy}$ **15.** $\sqrt{2mn}$ **17.** $\sqrt[6]{27}, \sqrt[6]{25}$ **19.** $\sqrt[20]{32}, \sqrt[20]{81}$
21. $\sqrt[6]{49}, \sqrt[6]{42}$ **23.** $\sqrt[8]{16}, \sqrt[8]{16}, \sqrt[8]{8}$ **25.** $\sqrt[12]{729}, \sqrt[12]{625}, \sqrt[12]{343}$
27. $\sqrt[12]{x^9}, \sqrt[12]{9x^{10}}$ **29.** $\sqrt[6]{1125}$ **31.** $\sqrt[4]{539}$ **33.** $\sqrt[6]{40}$
35. $a\sqrt[10]{a^3b^7}$ **37.** $mn\sqrt[12]{40m^2}$ **39.** $ab\sqrt[30]{a^9b^{13}}$

Exercise 72, Page 202

1. $2i$ **3.** $8i$ **5.** $2\sqrt{3}i$ **7.** $21i$ **9.** $15\sqrt{2}i$ **11.** $2ai$
13. $2x^2\sqrt{5}i$ **15.** $2a^2\sqrt{2ai}$ **17.** 1 **19.** i **21.** -1
23. -1 **25.** 1 **27.** 1 **29.** $-i$

Exercise 73, Page 204

1. $x = 13$ **3.** No root **5.** No root **7.** $x = 2$ **9.** $x = -29$
11. $x = 0$ **13.** No root **15.** $x = 5$ **17.** $x = 4$ **19.** $x = \frac{5}{2}$
21. No root **23.** $x = 50$

Exercise 74, Page 206

1. 2.98 **3.** 70.8 **5.** 0.939 **7.** 109 **9.** 0.497 **11.** 0.022
13. 4.13 **15.** 4.89 **17.** 0.89 **19.** 0.23 **21.** 4.47
23. 4.08

Exercise 75, Page 208

1. 1.4142 **3.** 2.2361 **5.** 3.1623 **7.** 26.30589 **9.** 23.57605
11. 8.06226 **13.** 3.00150 **15.** 6.10066 **17.** 2.449489
19. 11.704699

Exercise 76, Page 210

1. 4.359 **3.** 8.124 **5.** 9.327 **7.** 96 **9.** 44 **11.** 15.10
13. 13.23 **15.** 24.60 **17.** 1.659 **19.** 5.888 **21.** 1.342
23. 1.803 **25.** 0.7798 **27.** 6.854 **29.** 4.560

Exercise 77, Page 211

1. $c = 10$ **3.** $a = 24$ **5.** $a = 27$ **7.** 27.9 **9.** 111.0 ft.
11. 105.3 ft.

Review of Chapter VIII

A, Page 212

1. $11\sqrt{2}$ **3.** $2a\sqrt{3a}$ **5.** $7(x - y)\sqrt{2}$ **7.** $(a + b)\sqrt{a - b}$
9. $4ab\sqrt[3]{a^2c^2}$ **11.** $8\sqrt{3}$ **13.** $\frac{4}{7}\sqrt{3}$ **15.** $\dfrac{2a}{3x}\sqrt[3]{9ax}$ **17.** $6x\sqrt{2}$
19. $\sqrt[3]{2}$ **21.** $(a - b)\sqrt{2}$ **23.** $11\sqrt{5} - 39\sqrt{2}$
25. $\frac{5}{6}\sqrt[3]{36} + 2\sqrt[3]{2}$ **27.** $(a - b)\sqrt{a - b}$ **29.** $(p + q)^2\sqrt[3]{p}$
31. $\sqrt{ab}/b$ **33.** $2\sqrt{x - 2}$

B, Page 212

1. 9 **3.** 120 **5.** 3780 **7.** $12\sqrt{5} - 36 + 24\sqrt{7}$
9. $17 + 4\sqrt{10}$ **11.** $\sqrt[3]{4} - 2\sqrt[3]{2}$ **13.** $2\sqrt{6}$ **15.** $3\sqrt[3]{4}$
17. $4\sqrt{7} - \sqrt{3} - 4$ **19.** $2(\sqrt{7} + 2)$ **21.** $3 + \sqrt{6}$
23. $2\sqrt{x} + 3\sqrt{y}$

C, Page 213

1. 1 **3.** $2\frac{1}{2}$ **5.** 4 **7.** 3 **9.** 1 **11.** $\frac{1}{2}$ **13.** a
15. $35a^{\frac{4}{3}}x^{\frac{1}{6}}$ **17.** m^4n^5 **19.** $a^{\frac{5}{2}}b^4$ **21.** $8x^4y$ **23.** $1/xy$
25. $x/(x - 1)$ **27.** z/y **29.** $x^{n+2}y$

D, Page 214

1. $3\sqrt[6]{24}$ **3.** $3\sqrt[12]{243}$ **5.** $\frac{1}{2}\sqrt[3]{2}$ **7.** $2 - \sqrt[3]{4}$ **9.** $ab\sqrt[20]{a^7b^6}$
11. $\sqrt[6]{3mn}$

E, Page 214

1. $x = -3$ **3.** $x = -4$ **5.** $x = 1$ **7.** $x = \frac{4}{9}$ **9.** $x = 9$
11. $x = 20$

F, Page 214

1. 53 **3.** 1.26 in. **5.** 19.6 ft. **7.** 340 ft. **9.** 85 miles

CHAPTER IX

Exercise 79, Page 222

3. $AB = 5$, $CD = 5$ **5.** (*b*) Parallel **7.** (2, 7) **9.** 13
11. (*a*) Isosceles, (*b*) $6\frac{1}{2}$, $6\frac{1}{2}$, and 5 **13.** (*a*) Yes **15.** $17\sqrt{2}$, $17\sqrt{2}$,
34. It is.

Exercise 81, Page 227

1. $(-2, -1)$ **3.** $(-1, 2)$ **5.** $(1, 0)$ **7.** $(-1, 2)$ **9.** Parallel
11. $(6, -2)$ **13.** $(2, 3)$ **15.** $(4, -6)$ **17.** $(5, -4)$
19. $(2.2, 2.2)$ **21.** $(0.5, -1.5)$ **23.** $(-0.4, 0.1)$

Exercise 82, Page 229

1. $x = 0, V(0, 0)$ **3.** $x = 0, V(0, 4)$ **5.** $x = 2, V(2, 9)$
7. $x = 1, V(1, -1)$ **9.** $x = -2, V(-2, 0)$

Exercise 83, Page 230

1. $x = 1, -5$ **3.** $x = 1, 7$ **5.** $x = 2, -5$ **7.** $x = -3, \frac{1}{2}$
9. $x = -\frac{3}{2}, -\frac{3}{2}$ **11.** $x = -3, \frac{1}{3}$ **13.** $x = 2.7, -0.7$
15. $x = -1.6, -4.4$ **17.** $x = 1.8, -0.3$

Review of Chapter IX

B, Page 234

1. $A(-2, 1), B(-1, 5), C(1, 2), D(3, -1), E(0, -2)$ **3.** $y = 1\frac{2}{3}$
5. $(-3, -3)$ **7.** $y = 2$ **9.** (*b*) Trapezoid, (*c*) 2, (*d*) 7
11. $(3, 2)$ **13.** Parallel **15.** $(-2, 1)$ **17.** $(0.1, 0.6)$
19. Area $= \frac{9}{4}$

C, Page 235

1. $x = 0, V(0, -2)$ **3.** $x = -1, V(-1, 16)$ **5.** $x = -1, V(-1, 1)$
7. $x = 2, -6$ **9.** $x = 2, -\frac{3}{2}$ **11.** $x = 0.4, -1.4$
13. $(-1, 1), (2, 4)$ **15.** $(0, 1), (-1, 0)$

CHAPTER X

Exercise 85, Page 237

1. $x = \pm 8$ **3.** $x = \pm 5$ **5.** $x = \pm\frac{5}{3}$ **7.** $x = \pm 3$
9. $x = \pm 3$ **11.** $x = \pm a$ **13.** $x = \pm 5c$ **15.** $x = \pm 2a$
17. $x = \pm 2\sqrt{2}$ **19.** $x = \pm 2\sqrt{3}/3$ **21.** $x = \pm\sqrt{2}$
23. $x = \pm\sqrt{2}/4$ **25.** $r = \pm\sqrt{A/\pi}$ **27.** $r = \pm\sqrt{3V/\pi h}$
29. $a = \pm bc/\sqrt{b^2 - c^2}$

Exercise 86, Page 239

1. $x = 9, -1$ **3.** $x = 4, 1$ **5.** $x = -4, -5$ **7.** $x = 3, 6$
9. $x = 3, -2$ **11.** $x = -3, \frac{1}{2}$ **13.** $x = 5, \frac{1}{2}$ **15.** $x = -2, -\frac{2}{5}$

17. $x = -1, \frac{2}{3}$ **19.** $x = \frac{2}{3}, -\frac{7}{2}$ **21.** $x = -1, \frac{5}{3}$ **23.** $x = \frac{5}{2}, -\frac{7}{3}$
25. $x = \frac{3}{2}, -\frac{2}{3}$ **27.** $x = \frac{1}{3}, \frac{5}{2}$ **29.** $x = -5, \frac{1}{5}$ **31.** $x = -\frac{3}{2}, -\frac{5}{3}$
33. $x = 40, -30$ **35.** $x = 18, 30$ **37.** $x = \frac{10}{3}, -\frac{15}{2}$
39. $x = 15, -\frac{25}{2}$

Exercise 87, Page 241

1. $x = 7, -13$ **3.** $x = 2, -1$ **5.** $x = 2, -3$ **7.** $x = 7, -9$
9. $x = 5, -4$ **11.** $x = -1, -2$ **13.** $x = 4, -1$
15. $x = 24, -1$ **17.** $x = 13, -2$ **19.** $x = 2, -\frac{1}{3}$ **21.** $x = 5$
23. $x = -2$ **25.** $x = -2$ **27.** $x = 3, 7$ **29.** $x = 1$

Exercise 88, Page 242

1. $x = 7a, -2a$ **3.** $x = -6n, -7n$ **5.** $x = 2b/a, 9b/a$
7. $x = 3/p, -1/2p$ **9.** $x = 6a, -3a$ **11.** $x = -m, -9m$
13. $x = 1/a, -10/3a$ **15.** $x = 3a, -5a$ **17.** $x = a, a$
19. $x = 7c, 2c/3$ **21.** $x = 3y, -5y$ **23.** $x = -b/a, -c/b$
25. $x = 7a$ **27.** $x = b$

Exercise 89, Page 244

1. 6 and 12 **3.** Either 1″ and 4″, or 5″ and 8″ **5.** 75
7. Frank 8, Harold 12 **9.** 1 m.p.h. **11.** 7 in. by 12 in. **13.** 36
15. 10, 11, and 12; or −10, −11, and −12 **17.** 36 days and 45 days
19. 20 m.p.h. **21.** 5 ft. and 12 ft. **23.** 23 **25.** Man 24, son 4
27. 3 m.p.h., No. **29.** 5 ft. **31.** 81 **33.** 14 miles
35. 15 days **37.** 36 **39.** $16; Yes **41.** Either 4 hrs. and 5 hrs., or 5 hrs. and 6 hrs. Both answers are reasonable. **43.** $\frac{4}{9}$
45. 18 m.p.h. **47.** 5″ by 9″ by 2″ **49.** 14¢

Exercise 90, Page 250

1. 4 **3.** 25 **5.** $\frac{1}{4}$ **7.** $\frac{1}{36}$ **9.** a^2 **11.** $4m^2/49$
13. $x = 2, -4$ **15.** $x = -2, -6$ **17.** $x = 3, -2$
19. $x = -3, -5$ **21.** $x = 1, 4$ **23.** $x = 3, -\frac{1}{2}$ **25.** $x = 2, -\frac{2}{3}$
27. $x = -\frac{1}{2}, -\frac{2}{3}$ **29.** $x = \frac{3}{4}, -\frac{4}{3}$ **31.** $x = \frac{9}{2}, -\frac{3}{4}$
33. $x = -\frac{2}{3}, -\frac{2}{3}$ **35.** $x = 5a, -2a$ **37.** $x = -c/2, -4c/3$
39. $x = b/2, 3b/2$

Exercise 91, Page 252

1. $x = 1 \pm \sqrt{6}$ **3.** $x = -4 \pm 2\sqrt{5}$ **5.** $x = \dfrac{1 \pm \sqrt{5}}{2}$
7. $x = \dfrac{2 \pm \sqrt{2}}{2}$ **9.** $x = \dfrac{2 \pm 2\sqrt{3}}{3}$ **11.** $x = \dfrac{4 \pm \sqrt{11}}{5}$

13. $x = \frac{-5 \pm \sqrt{7}}{3}$ **15.** $x = \frac{1}{2}, \frac{4}{3}$ **17.** $x = \frac{-1 \pm \sqrt{6}}{2}$

19. $x = \frac{7 \pm \sqrt{13}}{12}$ **21.** $x = 5.32, -1.32$ **23.** $x = 1.14, -2.64$

25. $x = -1.21, -0.41$ **27.** $x = 1.82, -0.49$ **29.** $x = 0.61, 0.14$

Exercise 92, Page 254

1. $x = 3, -1$ **3.** $x = -5, \frac{1}{2}$ **5.** $x = -2, -\frac{1}{3}$ **7.** $x = \frac{5}{3}, -\frac{3}{2}$

9. $x = \frac{3}{2}, -\frac{7}{5}$ **11.** $x = -2, \frac{9}{8}$ **13.** $x = \frac{1 \pm \sqrt{31}}{5}$

15. $x = \frac{-3 \pm 2\sqrt{6}}{2}$ **17.** $x = \frac{2 \pm \sqrt{7}}{6}$ **19.** $x = \frac{1 \pm \sqrt{101}}{5}$

21. $x = \frac{-3 \pm 3\sqrt{5}}{8}$ **23.** $x = \frac{5 \pm \sqrt{15}}{6}$ **25.** $x = -2, \frac{3}{2}$

27. $x = 0.54, -5.54$ **29.** $x = 2, -1$ **31.** $x = -1, -1$

33. $x = 0.62, -1.62$ **35.** $x = 1, -\frac{7}{3}$

Exercise 93, Page 255

1. Yes **3.** No **5.** Yes **7.** Yes **9.** $x = 3 \pm i$

11. $x = 1 \pm 4i$ **13.** $x = -\frac{1}{2} \pm \frac{\sqrt{3}}{2}i$ **15.** $x = \frac{1}{3} \pm \frac{1}{3}i$

17. $x = -\frac{1}{2} \pm i$ **19.** $x = \frac{5}{4} \pm \frac{\sqrt{7}}{4}i$ **23.** $-\frac{5}{3}, -3$ **25.** $\frac{3}{4}, a/4$

27. $3/2a, 2$ **29.** $-2, (c + 1)/c$

Review of Chapter X

A, Page 256

1. $x = 7, -12$ **3.** $m = \frac{3}{2}, -\frac{3}{2}$ **5.** $b = 7, \frac{3}{2}$ **7.** $x = 6, -\frac{7}{3}$

9. $a = 5, -10$ **11.** $y = 4, -\frac{1}{4}$ **13.** $x = \frac{4}{3}, -\frac{9}{4}$

15. $c = -3, -\frac{1}{3}$ **17.** $r = \frac{5}{6}, -\frac{3}{8}$ **19.** $y = \frac{5}{8}, -\frac{9}{20}$

B, Page 256

1. $x = -5, \frac{1}{2}$ **3.** $a = -2 \pm 2\sqrt{2}$ **5.** $m = \frac{4 \pm \sqrt{10}}{3}$

7. $y = \frac{-3 \pm 3\sqrt{17}}{8}$ **9.** $b = \frac{-3 \pm \sqrt{59}}{5}$ **11.** $a = 8, 16$

13. $y = 0.14, -2.47$ **15.** $p = \frac{3}{2}, -\frac{1}{6}$ **17.** $b = \frac{1}{4}, -\frac{3}{4}$

19. $y = \frac{5}{4}, -\frac{8}{5}$

C, Page 257

1. $x = \dfrac{3 \pm 3\sqrt{3}}{2}$ 3. $m = \dfrac{1 \pm \sqrt{5}}{6}$ 5. $y = \frac{4}{3}, -\frac{2}{3}$
7. $c = \dfrac{4 \pm \sqrt{6}}{4}$ 9. $x = \dfrac{-5 \pm \sqrt{109}}{14}$ 11. $a = 0.26, 5.07$
13. $z = 3.12, -0.12$ 15. $p = 0.05, -2.05$ 17. $c = 1.31, -0.48$
19. $y = 0.17, -0.95$

D, Page 257

1. $x = \frac{5}{3}, -\frac{3}{2}$ 3. $a = \dfrac{15 \pm \sqrt{395}}{10}$ 5. $x = \pm 3\sqrt{5}$ 7. $x = 2, \frac{1}{4}$
9. $x = 5, \frac{1}{26}$ 11. $x = -1, -\frac{56}{15}$ 13. $x = 2, -\frac{3}{2}$ 15. $x = 3, -2$
17. $x = 12$ 19. $x = 2$

E, Page 258

1. $x = 4ab, 5ab$ 3. $x = -2c/3, \frac{1}{2}c$ 5. $x = d, 5d$
7. $x = 0, (a + b)/2$ 9. $x = a, b - a$ 11. $x = a(a + b), b(a + b)$
13. $x = 1, (a - 4)/6$ 15. $x = 2b/a, 3b/a$

F, Page 258

1. 6, 8, and 10 3. Either \$40 or \$60 5. 32 m.p.h. 7. 2 m.p.h.
9. 8 and 9 11. 125 yds. 13. 30 m.p.h. 15. Tony 5, sister 3
17. 3 m.p.h. 19. 10 boys

CHAPTER XI

Exercise 95, Page 266

1. $(3, 4), (0, -2)$ 3. $(3, 2), (-4, -\frac{3}{2})$ 5. $(\pm 2, \pm 1)$
7. No solutions 9. $(-3, 0)$ 11. $(\pm 3, 3)$ 13. No solutions
15. $(2, 2)$ 17. $(-2.1, 3.4), (3.9, -0.6)$ 19. $(2.8, 1.4), (1.1, 3.7), (-2.8, -1.4), (-1.1, -3.7)$ 21. $(1.4, 0.7), (-1.4, -0.7)$
23. $(-2.9, 0.4), (-2.3, -1.3), (-0.6, -2.8), (1.6, 3.7)$

Exercise 96, Page 268

1. $x = 3, y = 2; x = -2, y = -3$ 3. $x = 5, y = 7$
5. $x = 5, y = 2; x = -4, y = -\frac{5}{2}$ 7. $x = 0, y = \frac{9}{4}; x = 1, y = 2$
9. $x = -4, y = -2; x = \frac{16}{5}, y = \frac{8}{5}$ 11. $x = 1, y = 0; x = 3, y = 6$
13. $x = 1, y = -1; x = 5, y = 7$ 15. $x = 3, y = 3; x = \frac{12}{7}, y = \frac{6}{7}$
17. $x = 1, y = 1; x = 49, y = -35$ 19. $x = \frac{1}{2}, y = \frac{1}{2}$

21. $x = 5, y = 4; x = -4, y = -5$ **23.** $x = 2, y = 2; x = \frac{4}{3}, y = 0$
25. $x = 2a, y = a; x = -a, y = -2a$
27. $x = (a + b)/2, y = (a - b)/2; x = (a - b)/2, y = (a + b)/2$
29. $x = a^2, y = a; x = \frac{1}{2}a^2, y = \frac{1}{2}a$

Exercise 97, Page 270

1. $x = \pm 2, y = \pm 3$ **3.** $x = 3, -3, 4, -4; y = 4, -4, 3, -3$
5. $x = \pm 3, y = \pm 1$ **7.** $x = 3, -3; y = 4, -4$
9. $x = \pm 2, y = \pm 1$ **11.** $x = 1, \frac{1}{2}, -\frac{1}{2}; y = 1, 2, -2$
13. $x = \pm 5, y = \pm 6$ **15.** $x = \pm\sqrt{2}, y = \pm\sqrt{3}$
17. $x = 1, -1, 2, -2; y = 2, -2, 1, -1$ **19.** $x = \pm 2, y = \pm 3$
21. $x = \pm a\sqrt{2}, y = \pm a$ **23.** $x = \pm(m + n), y = \pm(m - n)$

Exercise 98, Page 272

1. $x = 4, -4, 0, -5; y = 3, -3, -5, 0$ **3.** $x = 1, -1, 2, -2;$ $y = -2, 2, 3, -3$ **5.** $x = 3, -3, i\sqrt{3}, -i\sqrt{3}; y = 3, -3, -i\sqrt{3}, i\sqrt{3}$
7. $x = 3, -3, 5\sqrt{2}/2, -5\sqrt{2}/2; y = 2, -2, \sqrt{2}/2, -\sqrt{2}/2$
9. $x = 0, 1, -1; y = \pm 2, 1, -1$ **11.** $x = 2, -2, 2i\sqrt{2}, -2i\sqrt{2};$ $y = 1, -1, -2i\sqrt{2}, 2i\sqrt{2}$ **13.** $x = 4, -4, 3\sqrt{3}, -3\sqrt{3}; y = 5, -5,$ $\sqrt{3}, -\sqrt{3}$ **15.** $x = 2, -2; y = 1, -1$ **17.** $x = 2, -2; y = -1, 1$ **19.** $x = 0, 1, 3; y = 0, -1, 1$ **21.** $x = 1, 2; y = 2, 1$ **23.** $x = \frac{3}{2},$ $-2, -6; y = \frac{5}{2}, -1, 5$

Review of Chapter XI

A, Page 273

3. Hyperbola **5.** $x = 4, y = 1$ **7.** $x = \pm 2, y = 0$
9. $x = \pm 2, y = \pm 2$ **11.** $x = -2, 3.5; y = 0, \pm 2.3$

B, Page 274

1. $x = -1, 13; y = 2, -5$ **3.** $x = 4, -4, i, -i; y = \frac{1}{2}, -\frac{1}{2}, -2i, 2i$
5. $x = \pm 2, y = 2$ **7.** $x = 4, \frac{4}{3}; y = 1, -3$ **9.** $x = 8.03, -2.15;$ $y = 3.41, -0.66$ **11.** $x = \pm 1.41, \pm 1.15; y = \mp 1.41, \mp 1.73$
13. $x = 1.20, y = \pm 1.84; x = -2, y = \pm i\sqrt{3}$
15. $x = 3a, -23a/7; y = a, -15a/7$

C, Page 274

1. $x = \frac{5}{4}, -\frac{5}{4}, \frac{5}{2}, -\frac{5}{2}; y = \frac{5}{4}, -\frac{5}{4}, -\frac{5}{2}, \frac{5}{2}$ **3.** $x = 2, 2, 5, \frac{5}{4}; y = 1,$ $4, \frac{5}{2}, \frac{5}{2}$ **5.** $x = 1, -1, \sqrt{2}, -\sqrt{2}; y = 2, -2, -\sqrt{2}/2, \sqrt{2}/2$

7. $x = 1, -1, 4, -4; y = -3, 3, -2, 2$ **9.** $x = 1, -1, 3\sqrt{2}/2, -3\sqrt{2}/2$; $y = -1, 1, -\sqrt{2}, \sqrt{2}$ **11.** $x = 2a, -2a, 2a\sqrt{2}, -2a\sqrt{2}$; $y = 3a, -3a, 2a\sqrt{2}, -2a\sqrt{2}$ **13.** $x = \pm 1, \pm 2; y = \pm 2, \pm 1$
15. $x = \pm 3, y = \pm 2$

D, Page 275

1. $7\frac{1}{2}$ by 10 **3.** 15 ft. and 8 ft. **5.** 5 and 15 **7.** 4 A.M.
9. *A* 6 boxes per hr., *B* 8 boxes per hr.

CHAPTER XII

Exercise 99, Page 278

1. 1:2 **3.** 5:8 **5.** 3:2 **7.** 1:5 **9.** 5:7 **11.** 1:2
13. 4:33 **15.** 1:4 **17.** 4:9 **19.** 52 and 91 **21.** 51 and 85
23. 9″, 15″, and 21″ **25.** 15 and 20 **27.** 33 and 55
29. 18 ft. by 30 ft.

Exercise 100, Page 280

1. $y = 40$ **3.** $a = 65$ **5.** $y = 98$ **7.** $y = 81$ **9.** $y = 26$
11. $x = 28$ **13.** $5\frac{3}{4}$ lbs. **15.** 405 lbs. **17.** $4\frac{1}{6}$ pts.

Exercise 101, Page 282

1. $y = 12$ **3.** $m = 63$ **5.** $y = 4$ **7.** $p = 128$ **9.** $y = 10$
11. 6 days **13.** 64 lbs. **15.** $z = 10$ **17.** 25 amps.

Exercise 102, Page 284

1. 10 **3.** 21 **5.** 18 **7.** bc **9.** b^3/a **11.** 1
21. $x = 2, -\frac{1}{3}$ **23.** $x = 5, y = 1$

Exercise 103, Page 286

1. $a^6 + 6a^5x + 15a^4x^2 + 20a^3x^3 + 15a^2x^4 + 6ax^5 + x^6$
3. $x^4 - 8x^3y + 24x^2y^2 - 32xy^3 + 16y^4$
5. $a^8x^8 - 8a^7bx^7 + 28a^6b^2x^6 - 56a^5b^3x^5 + 70a^4b^4x^4 - 56a^3b^5x^3 + 28a^2b^6x^2 - 8ab^7x + b^8$ **7.** $x^{14} + 7x^{13} + 21x^{12} + 35x^{11} + 35x^{10} + 21x^9 + 7x^8 + x^7$ **9.** $\frac{1}{32}x^5 + \frac{5}{8}x^4 + 5x^3 + 20x^2 + 40x + 32$
11. $x^{10} - 30x^9y + 405x^8y^2 - 3240x^7y^3 + \cdots$
13. $512x^{18} - 1152x^{16} + 1152x^{14} - 672x^{12} + \cdots$
15. $m^{12} - 36m^{11} + 594m^{10} - 5940m^9 + \cdots$
17. $256x^8 + 1024x^6 + 1792x^4 + 1792x^2 + \cdots$

19. $\frac{1}{4096}x^{12} - \frac{1}{512}x^{11}y + \frac{11}{1536}x^{10}y^2 - \frac{55}{3456}x^9y^3 + \cdots$
21. $110{,}565x^{11}$ **23.** $70a^{13}b^3$ **25.** $-489{,}888m^4$ **27.** 59,049
29. 1.030301 **31.** $7 + 5\sqrt{2}$ **33.** 1.26247696 **35.** $49 + 20\sqrt{6}$

Review of Chapter XII

A, Page 287

1. 3:5 **3.** 1:2 **5.** 2:3 **7.** 8:1 **9.** 148 and 185
11. 57, 76, and 152 **13.** 9″, 1′3″, 1′9″, and 2′3″ **15.** 1:3:6:9
17. 38 **19.** 35 ft. by 49 ft.

B, Page 288

1. $y = 38$ **3.** $m = 0.001$ **5.** $x = 41$ **7.** $x = 3\frac{1}{2}$ **9.** $p = 72$
11. $z = 6$ **13.** $14.00 **15.** 2.8 ft. **17.** 24¢ **19.** 0.95 ohms

C, Page 289

1. $81x^4 + 216x^3 + 216x^2 + 96x + 16$
3. $128m^7 - 448m^6 + 672m^5 - 560m^4 + 280m^3 - 84m^2 + 14m - 1$
5. $\frac{1}{81}x^4 - \frac{4}{9}x^3 + 6x^2 - 36x + 81$
7. $x^{13} + 65x^{12}y + 1950x^{11}y^2 + 35{,}750x^{10}y^3 + \cdots$
9. $\frac{1}{1024}x^{10} - \frac{5}{128}x^8 + \frac{45}{64}x^6 - \frac{15}{2}x^4 + \cdots$
11. $x^{18} - 27x^{15} + 324x^{12} - 2268x^9 + \cdots$ **13.** $2{,}050{,}048a^{18}$
15. $\frac{665}{9}x^4$ **17.** 1.2166529024 **19.** $17 - 12\sqrt{2}$

CHAPTER XIII

Exercise 104, Page 292

1. $\log_4 16 = 2$ **3.** $\log_7 49 = 2$ **5.** $\log_6 \frac{1}{36} = -2$ **7.** $\log_8 \frac{1}{4} = -\frac{2}{3}$
9. $2^4 = 16$ **11.** $5^3 = 125$ **13.** $8^{\frac{2}{3}} = 4$ **15.** $10^{-3} = 0.001$
17. 0, 1, −1, 2, −2, 3 **19.** 1, 3, 9, 27, $\frac{1}{3}$, $\frac{1}{9}$, $\frac{1}{27}$ **21.** 2 **23.** −2
25. 1 **27.** $\frac{2}{3}$ **29.** $x = -3$ **31.** $b = 3$ **33.** $b = 27$
35. $N = \frac{1}{32}$ **37.** $2\frac{1}{2}$ **39.** 1

Exercise 105, Page 296

1. 2, .8129 **3.** 5, .7419 **5.** 0, .9509 **7.** −1, .4683
9. −1, .9279 **11.** −3, .4683 **13.** −2, .8074 **15.** −1, .6075
17. 2 **19.** −2 **21.** −1 **23.** −5 **25.** 0 **27.** 0
29. −1 **31.** 0.5366 **33.** 3.5366 **35.** −1 + .5366
37. 5.5366 **39.** 3465 **41.** 3.465 **43.** 0.0003465 **45.** 34.65
47. 0.00003465 **49.** 3,465,000

Exercise 106, Page 297

1. 2.4425 **3.** 9.3674 − 10 **5.** 3.5145 **7.** 4.7987 **9.** 2.6325
11. 7.4843 − 10 **13.** 1.9777 **15.** 8.6232 − 10

17. 6.1461 − 10 **19.** 9.9956 − 10 **21.** 8.7782 − 10
23. 3.3054 **25.** 2.0000 **27.** 1.5119 **29.** 7.2788
31. 9.6021 − 10 **33.** 0.5670 **35.** 9.4314 − 10

Exercise 107, Page 298

1. 1.14 **3.** 54.7 **5.** 0.192 **7.** 0.00796 **9.** 686,000
11. 206 **13.** 30 **15.** 0.055 **17.** 9.99 **19.** 25.6 **21.** 79
23. 0.7 **25.** 54,000 **27.** 8.35 **29.** 0.445 **31.** 898
33. 0.00251 **35.** 99.1

Exercise 108, Page 302

1. 3.4208 **3.** 9.9746 − 10 **5.** 1.3265 **7.** 4.8491 **9.** 1.0906
11. 0.9780 **13.** 0.3056 **15.** 2.9683 **17.** 7.6994 − 10
19. 9.3923 − 10 **21.** 4721 **23.** 0.08026 **25.** 1.744
27. 99,650 **29.** 36.78 **31.** 699.8 **33.** 0.2105 **35.** 0.006437
37. 219.6 **39.** 31.99

Exercise 109, Page 304

1. 1820 **3.** 47.98 **5.** 2,578,000 **7.** 0.02408 **9.** 0.2470
11. −0.01773 **13.** 22,650,000 **15.** 138.6 **17.** 0.009468
19. 0.1971 **21.** 65.97 **23.** 61.03 **25.** 11,840

Exercise 110, Page 305

1. 12.00 **3.** 1.400 **5.** 0.2328 **7.** 2.354 **9.** 7.828
11. 6077 **13.** 124.9 **15.** 0.0001056 **17.** 0.01326
19. 0.07643 **21.** 2175 **23.** 0.01073

Exercise 111, Page 306

1. 39,310 **3.** 4,704,000 **5.** 48.02 **7.** 1589 **9.** 37,230,000
11. 0.0000000002269 **13.** 408.9 **15.** 0.04714 **17.** 3362
19. 0.04259 **21.** 25.08 **23.** 37.51

Exercise 112, Page 307

1. 2.296 **3.** 4.526 **5.** 1.086 **7.** 0.2258 **9.** 0.1319
11. 0.5581 **13.** 1.211 **15.** 0.3096 **17.** 12.06 **19.** 0.5960
21. 3.744 **23.** 2.261

Exercise 113, Page 309

7. 2.217 **9.** 1.966 **11.** 0.03770 **13.** 377.3 **15.** 2.707
17. 1.897 **19.** −0.06747 **21.** 50.00 **23.** 2.097

Exercise 114, Page 310

1. $x = 2$ **3.** $x = 3$ **5.** $x = 1$ **7.** $x = -2$ **9.** $x = 1.5$
11. $x = 1.5$ **13.** $x = 2.930$ **15.** $x = 0.9065$ **17.** $x = 0.6701$
19. $x = 2.383$ **21.** $x = 1.877$ **23.** $x = 0.5853$ **25.** $x = 6.146$
27. $x = 1.029$ **29.** $x = -13.16$

Review of Chapter XIII

A, Page 310

1. $N = 9$ **3.** $b = 5$ **5.** $b = \frac{1}{27}$ **7.** $N = 3$ **9.** $x = 5$
11. $x = -\frac{2}{3}$ **13.** 6 **15.** 1 **17.** $1\frac{1}{2}$ **19.** 0

B, Page 311

1. $\log a + \log b + \log c$ **3.** $2 \log a + 3 \log b - \frac{1}{2} \log c$
5. $\log a + \frac{1}{2} \log b - 2 \log c$ **7.** (*a*) 0.602, (*b*) 1.204, (*c*) 9.097 − 10, (*d*) 0.699 **9.** 0.778 **11.** 2.079 **13.** 9.949 − 10 **15.** 0.690
17. 2.904 **19.** 0.476

C, Page 311

1. 580.0 **3.** 284.8 **5.** 75.00 **7.** 2.606 **9.** 368.2
11. 33,640,000 **13.** 6.657 **15.** 0.6930 **17.** 63.00
19. 0.3482

D, Page 312

1. 1.678 **3.** 1.136 **5.** 1.191 **7.** 88.48 **9.** 36,920
11. 32.08 **13.** $x = 2.209$ **15.** $x = 2.287$ **17.** $x = 2.236$
19. $x = 9.389$

CHAPTER XIV

Exercise 115, Page 315

1. A. P. **3.** G. P. **5.** Neither **7.** Neither **9.** G. P.
11. $x = 11$ **13.** $x = -3$ **15.** $x = \frac{1}{2}$ **17.** $x = 3$
19. $x = 1, 2$ **21.** $x = 12$ **23.** $x = 4$ **25.** $x = -2$
27. $x = -11$ **29.** $x = 2, -1$

Exercise 116, Page 318

1. 38 **3.** −56 **5.** 34 **7.** $-40\frac{3}{4}$ **9.** −13.5
11. 15, 19, 23, 27, 31 **13.** $5\frac{1}{4}, 5\frac{1}{2}, 5\frac{3}{4}$ **15.** 7.4, 7.6, ⋯, 9.6
17. $3\frac{1}{6}, 4, 4\frac{5}{6}, 5\frac{2}{3}$ **19.** (*a*) 14, (*b*) 3.9, (*c*) $2\frac{3}{4}$ **21.** 629 **23.** −45

25. 225 **27.** 720 **29.** 202.38 **31.** -6 **33.** $a = 7$, $d = 2$
35. 69 **37.** $17\frac{2}{5}$ **39.** 12 **41.** $n = 17$, $S = 476$
43. $a = 28$, $S = 238$ **45.** $n = 7$, $d = 2\frac{1}{3}$ **47.** $a = -4$, $l = 24$
49. $n = 11$, $a = 31$

Exercise 117, Page 321

1. 64 **3.** -256 **5.** $\frac{1}{2187}$ **7.** 128 **9.** 3125 **11.** $r = 2$
13. $\frac{3}{8}, \frac{3}{4}, \frac{3}{2}$ **15.** $n = 4$ **17.** $\frac{1}{9}, \frac{1}{3}, 1, 3, 9$ **19.** 10, 20
21. $\pm 54, 18, \pm 6$ **23.** (*a*) 4, (*b*) 5, (*c*) 4.2 **25.** 765 **27.** $-60\frac{2}{3}$
29. $332\frac{1}{2}$ **31.** 38888.8885 **33.** $-10\frac{11}{16}$ **35.** $171\frac{7}{12}$
37. 19999.744 **39.** $a = 81$, $n = 6$ **41.** $l = 256$, $n = 7$

Exercise 118, Page 323

1. 24 **3.** $4\frac{1}{2}$ **5.** $77\frac{7}{9}$ **7.** $-2\frac{6}{7}$ **9.** $4 + 2\sqrt{2}$ **11.** $\frac{7}{9}$
13. $\frac{9}{11}$ **15.** $\frac{19}{75}$ **17.** $\frac{589}{165}$ **19.** $\frac{21019}{9000}$

Exercise 119, Page 325

1. \$535.50 **3.** $\frac{3}{4}$ **5.** 72 **7.** 81,470 **9.** 113 terms
11. Terms by \$27.67 **13.** 470 ft. **15.** 6 and 24 **17.** 0
19. \$233.28 **21.** 30 ft. **23.** 8, 14, 20 or 32, 14, -4 **25.** 14 days
27. $(b^2 - ac)/(a + c - 2b)$ **29.** \$104 **31.** 4 and 6 **33.** 5 hours
35. $3, \frac{3}{4}, \frac{3}{16}, \cdots$ **37.** (*a*) (2) by \$21.64, (*b*) (1) by \$3374.73
39. $\frac{1}{2}, \frac{1}{4}, \frac{1}{8}, \cdots$

Review of Chapter XIV

A, Page 328

1. 102, 1092 **3.** -10, 165 **5.** 61.5, 1605 **7.** $78\frac{3}{4}$, $1417\frac{1}{2}$
9. $1 - 8\sqrt{2}$, $10 - 35\sqrt{2}$ **11.** 7, 11, 15, 19
13. (*a*) 21, (*b*) 1.3, (*c*) $3\frac{2}{3}$ **15.** 21 **17.** $n = 7$, $S = 98$
19. $a = 33$, $S = 189$

B, Page 328

1. 486, 728 **3.** 640, 425 **5.** 7.29, 20.59 **7.** 16, $30 + 15\sqrt{2}$
9. 27 **11.** 2.7 **13.** 60,150 **15.** (*a*) 10, (*b*) 7, (*c*) 2.1
17. $-7\frac{1}{9}$ **19.** $a = 5.12$, $l = 0.32$

C, Page 329

1. 8113 **3.** \$32,000 **5.** \$147.62 **7.** 4 **9.** 57, 58, 59, 60, 61
11. \$12,000 **13.** 11 hours **15.** 32 years, \$739.20
17. $1\frac{2}{3}, 3\frac{1}{3}, 6\frac{2}{3}$ **19.** 7825

INDEX